Gels for Removal and Adsorption

Gels for Removal and Adsorption

Guest Editors

Daxin Liang
Ting Dong
Yudong Li
Caichao Wan

Basel • Beijing • Wuhan • Barcelona • Belgrade • Novi Sad • Cluj • Manchester

Guest Editors

Daxin Liang
College of Materials Science
and Engineering
Northeast Forestry University
Harbin
China

Ting Dong
Key Laboratory of Bio-Fibers
and Eco-Textiles
Qingdao University
Qingdao
China

Yudong Li
College of Materials Science
and Engineering
Northeast Forestry University
Harbin
China

Caichao Wan
College of Materials Science
and Engineering
Central South University of
Forestry and Technology
Changsha
China

Editorial Office
MDPI AG
Grosspeteranlage 5
4052 Basel, Switzerland

This is a reprint of the Special Issue, published open access by the journal *Gels* (ISSN 2310-2861), freely accessible at: www.mdpi.com/journal/gels/special_issues/adsorption.

For citation purposes, cite each article independently as indicated on the article page online and using the guide below:

Lastname, A.A.; Lastname, B.B. Article Title. *Journal Name* **Year**, *Volume Number*, Page Range.

ISBN 978-3-7258-2840-1 (Hbk)
ISBN 978-3-7258-2839-5 (PDF)
https://doi.org/10.3390/books978-3-7258-2839-5

© 2024 by the authors. Articles in this book are Open Access and distributed under the Creative Commons Attribution (CC BY) license. The book as a whole is distributed by MDPI under the terms and conditions of the Creative Commons Attribution-NonCommercial-NoDerivs (CC BY-NC-ND) license (https://creativecommons.org/licenses/by-nc-nd/4.0/).

Contents

About the Editors vii

Preface ix

Ye Zhang and Cheng-An Tao
Metal–Organic Framework Gels for Adsorption and Catalytic Detoxification of Chemical Warfare Agents: A Review
Reprinted from: *Gels* 2023, 9, 815, https://doi.org/10.3390/gels9100815 1

Pavel Yudaev, Irina Butorova, Gennady Stepanov and Evgeniy Chistyakov
Extraction of Palladium(II) with a Magnetic Sorbent Based on Polyvinyl Alcohol Gel, Metallic Iron, and an Environmentally Friendly Polydentate Phosphazene-Containing Extractant
Reprinted from: *Gels* 2022, 8, 492, https://doi.org/10.3390/gels8080492 22

Elodie Guilminot
The Use of Hydrogels in the Treatment of Metal Cultural Heritage Objects
Reprinted from: *Gels* 2023, 9, 191, https://doi.org/10.3390/gels9030191 36

Xuelun Zhang, Feng Li, Xiyu Zhao, Jiwen Cao, Shuai Liu and You Zhang et al.
Bamboo Nanocellulose/Montmorillonite Nanosheets/Polyethyleneimine Gel Adsorbent for Methylene Blue and Cu(II) Removal from Aqueous Solutions
Reprinted from: *Gels* 2023, 9, 40, https://doi.org/10.3390/gels9010040 52

Omar Mouhtady, Emil Obeid, Mahmoud Abu-samha, Khaled Younes and Nimer Murshid
Evaluation of the Adsorption Efficiency of Graphene Oxide Hydrogels in Wastewater Dye Removal: Application of Principal Component Analysis
Reprinted from: *Gels* 2022, 8, 447, https://doi.org/10.3390/gels8070447 70

Nimer Murshid, Omar Mouhtady, Mahmoud Abu-samha, Emil Obeid, Yahya Kharboutly and Hamdi Chaouk et al.
Metal Oxide Hydrogel Composites for Remediation of Dye-Contaminated Wastewater: Principal Component Analysis
Reprinted from: *Gels* 2022, 8, 702, https://doi.org/10.3390/gels8110702 78

Yanjin Tang, Yuhan Lai, Ruiqin Gao, Yuxuan Chen, Kexin Xiong and Juan Ye et al.
Functional Aerogels Composed of Regenerated Cellulose and Tungsten Oxide for UV Detection and Seawater Desalination
Reprinted from: *Gels* 2022, 9, 10, https://doi.org/10.3390/gels9010010 88

Liying Xu, Wenxuan Wang, Yu Liu and Daxin Liang
Nanocellulose-Linked MXene/Polyaniline Aerogel Films for Flexible Supercapacitors
Reprinted from: *Gels* 2022, 8, 798, https://doi.org/10.3390/gels8120798 97

Ivanka Dakova, Penka Vasileva and Irina Karadjova
Cr(III) Ion-Imprinted Hydrogel Membrane for Chromium Speciation Analysis in Water Samples
Reprinted from: *Gels* 2022, 8, 757, https://doi.org/10.3390/gels8110757 107

Yumei Lv, Fei He, Wei Dai, Yulong Ma, Taolue Liu and Yifei Liu et al.
Design of Economical and Achievable Aluminum Carbon Composite Aerogel for Efficient Thermal Protection of Aerospace
Reprinted from: *Gels* 2022, 8, 509, https://doi.org/10.3390/gels8080509 125

Farid Hajareh Haghighi, Roya Binaymotlagh, Laura Chronopoulou, Sara Cerra, Andrea Giacomo Marrani and Francesco Amato et al.
Self-Assembling Peptide-Based Magnetogels for the Removal of Heavy Metals from Water
Reprinted from: *Gels* **2023**, *9*, 621, https://doi.org/10.3390/gels9080621 137

Feng Liu, Xin Di, Xiaohan Sun, Xin Wang, Tinghan Yang and Meng Wang et al.
Superhydrophobic/Superoleophilic PDMS/SiO$_2$ Aerogel Fabric Gathering Device for Self-Driven Collection of Floating Viscous Oil
Reprinted from: *Gels* **2023**, *9*, 405, https://doi.org/10.3390/gels9050405 161

Wenhui Wang, Jia-Horng Lin, Jiali Guo, Rui Sun, Guangting Han and Fudi Peng et al.
Biomass Chitosan-Based Tubular/Sheet Superhydrophobic Aerogels Enable Efficient Oil/Water Separation
Reprinted from: *Gels* **2023**, *9*, 346, https://doi.org/10.3390/gels9040346 174

Afaf N. Abdel Rahman, Basma Ahmed Elkhadrawy, Abdallah Tageldein Mansour, Heba M. Abdel-Ghany, Engy Mohamed Mohamed Yassin and Asmaa Elsayyad et al.
Alleviating Effect of a Magnetite (Fe$_3$O$_4$) Nanogel against Waterborne-Lead-Induced Physiological Disturbances, Histopathological Changes, and Lead Bioaccumulation in African Catfish
Reprinted from: *Gels* **2023**, *9*, 641, https://doi.org/10.3390/gels9080641 186

Sasirot Khamkure, Prócoro Gamero-Melo, Sofía Esperanza Garrido-Hoyos, Audberto Reyes-Rosas, Daniella-Esperanza Pacheco-Catalán and Arely Monserrat López-Martínez
The Development of Fe$_3$O$_4$-Monolithic Resorcinol-Formaldehyde Carbon Xerogels Using Ultrasonic-Assisted Synthesis for Arsenic Removal of Drinking Water
Reprinted from: *Gels* **2023**, *9*, 618, https://doi.org/10.3390/gels9080618 203

Jorge Alberto Cortes Ortega, Jacobo Hernández-Montelongo, Rosaura Hernández-Montelongo and Abraham Gabriel Alvarado Mendoza
Effective Removal of Cu^{2+} Ions from Aqueous Media Using Poly(acrylamide-co-itaconic acid) Hydrogels in a Semi-Continuous Process
Reprinted from: *Gels* **2023**, *9*, 702, https://doi.org/10.3390/gels9090702 232

About the Editors

Daxin Liang

Daxin Liang, an associate professor and doctoral supervisor at Northeast Forestry University, is one of the first recipients of the Heilongjiang Province Outstanding Young Fund. His primary research focuses on the development of advanced functional materials based on biomass. He has been involved in numerous scientific research projects, including the National Natural Science Foundation, the National Key R&D Program, and the Heilongjiang Province Outstanding Youth Fund, as well as educational research projects like those from the Heilongjiang Higher Education Society.

He serves as a committee member of the International Association for Carbon Capture and the editorial board member of Carbon Capture Science and Technology. He has published over 50 peer-reviewed papers in journals including *Applied Catalysis B*, *Nano Letters*, *Advanced Materials*, *Advanced Energy Materials*, *Composites Part B*, *ACS Applied Materials Interfaces*, and *Nanoscale*. He has granted over ten invention patents.

Ting Dong

Ting Dong, an associate professor at Qingdao University, obtained her Ph.D. in Textile Engineering from Donghua University in 2018. She undertook a joint doctoral training program in polymer science and engineering at the University of Massachusetts Amherst from 2015 to 2017. In 2018, she was selected as a Level IV Distinguished Professor at Qingdao University and currently serves at the School of Textile and Apparel, specializing in textile materials and product design. Over the past five years, she has published 10 *SCI* academic papers in journals such as *Marine Pollution Bulletin*, *Industrial Crops and Products*, *Journal of Hazardous Materials*, and *ACS Applied Materials & Interfaces*, among others. Six of these publications are categorized as Zone 1 articles, with four having an impact factor above 7.

Yudong Li

Yudong Li, an associate professor at Northeast Forestry University, has presided over several funds, including the National Natural Science Foundation of China, the Heilongjiang Province Outstanding Youth Fund, and the National Postdoctoral Fund. He has published more than 40 papers and primarily engages in research on the functionalization of biomass, material design, computational simulation, and other related fields. He leads and participates in multiple industrial research projects, such as the development of biomass asphalt and regenerating agents, recycling of waste residue from biomass boilers, disinfectant production, air purification, and electroplating repair of devices. Additionally, he is involved in projects for aerospace and equipment departments.

Caichao Wan

Caichao Wan, professor and doctoral supervisor at Central South University of Forestry and Technology, is a National Talents Special Support Plan for Young Top-notch Talents recipient, fellow of the International VEBLEO Society, beneficiary of China Association for Science and Technology's Youth Lifting Project, National Forestry and Grassland Innovation Young Top-notch Talent, Hunan Province Huxiang Youth Talent, and an expert in the "Science and Innovation China" platform of the China Association for Science and Technology. His main research interests include biomass-based functional materials, wood functional improvement, and intelligent bionic materials. He has published over 60 papers as the first author/corresponding author in *SCI* journals such as *Adv. Energy Mater.* (IF=29.698), *Adv. Funct. Mater.* (IF=19.924), *Chem. Eng. J.* (IF=16.744), and *J. Energy Chem.* (IF=13.599), with eight of his achievements recommended for cover story reports.

He has presided over the publication of two monographs by the Science Press, co-authored four domestic monographs and textbooks, and contributed to two international publishers' monographs, including Elsevier and Wiley. As the first inventor, he has been granted one international patent and four national invention patents, with four more patent applications pending.

Preface

Gels for removal and adsorption have gained significant attention in recent years due to their unique properties and wide range of applications. These materials are three-dimensional networks consisting of polymers that can adsorb and retain large amounts of liquid or solutes, making them ideal for various environmental and industrial processes. Removal and adsorption gels are primarily classified into two categories, i.e., hydrogels and organogels. Hydrogels are water-swollen networks that can absorb and retain water and aqueous solutions, while organogels are composed of organic solvents and find applications in non-aqueous systems. The crosslinked structure of these gels allows them to adsorb and remove target substances such as heavy metals, organic pollutants, and dyes from aqueous or organic phases. One of the key mechanisms behind the removal and adsorption process is the interaction between the gel's functional groups and the target substances. For instance, gels containing amino, carboxyl, or hydroxyl groups can effectively adsorb heavy metals through coordination or ion exchange. Additionally, gels with a high surface area and porosity can enhance the adsorption capacity and selectivity for specific pollutants.

The goal of this Special Issue is to provide a summary of recent important progress in this field, ranging from basic aspects to applications, thereby highlighting both the adsorption and removal to the gel field. The targeted substances include, but are not limited to, gaseous, liquid, and solid states. The gaseous matters can be poisonous gases (e.g., SO_x, NO_x, and CO), greenhouse gases (e.g., CO_2, freon, and CH_4), etc. The liquid matters include wastewaters containing oils, heavy metal ions, radioactive materials, dyes, and antibiotics. The targeted solids are represented by a variety of particulates (e.g., haze, dust, and microorganisms). This Special Issue also covers gels for the absorption and controlled release of drugs in postoperative repair dressing and wound healing. For these applications, the gels have to be constructed responsively or intelligently geared towards certain environmental stimulation, such as pH, light, magnetic field, electric field, and temperature for controlled drug release. In this Special Issue, titled "Gels for Removal and Adsorption", researchers from around the world have presented research highlighting various materials, modifications, structures, and applications of selected gels for removal and adsorption.

As a transition to more sustainable materials exploitation is evident, the contributions to research, as in the present Special Issue, do indicate that gels for removal and adsorption of all kinds of targeted substances offer a versatile approach that can be expected to aid also in this respect.

Daxin Liang, Ting Dong, Yudong Li, and Caichao Wan
Guest Editors

Review

Metal–Organic Framework Gels for Adsorption and Catalytic Detoxification of Chemical Warfare Agents: A Review

Ye Zhang and Cheng-An Tao *

College of Science, National University of Defense Technology, Changsha 410073, China; zhangye8905@foxmail.com
* Correspondence: taochengan@nudt.edu.cn

Abstract: Chemical warfare agents (CWAs) have brought great threats to human life and social stability, and it is critical to investigate protective materials. MOF (metal–organic framework) gels are a class with an extended MOF architecture that are mainly formed using metal–ligand coordination as an effective force to drive gelation, and these gels combine the unique characteristics of MOFs and organic gel materials. They have the advantages of a hierarchically porous structure, a large specific surface area, machinable block structures and rich metal active sites, which inherently meet the requirements for adsorption and catalytic detoxification of CWAs. A series of advances have been made in the adsorption and catalytic detoxification of MOF gels as chemical warfare agents; however, overall, they are still in their infancy. This review briefly introduces the latest advances in MOF gels, including pure MOF gels and MOF composite gels, and discusses the application of MOF gels in the adsorption and catalytic detoxification of CWAs. Meanwhile, the influence of microstructures (pore structures, metal active site, etc.) on the detoxification performance of protective materials is also discussed, which is of great significance in the exploration of high-efficiency protective materials. Finally, the review looks ahead to next priorities. Hopefully, this review can inspire more and more researchers to enrich the performance of MOF gels for applications in chemical protection and other purification and detoxification processes.

Keywords: metal–organic framework; gel; chemical warfare agents; catalytic detoxification

Citation: Zhang, Y.; Tao, C.-A. Metal–Organic Framework Gels for Adsorption and Catalytic Detoxification of Chemical Warfare Agents: A Review. *Gels* **2023**, *9*, 815. https://doi.org/10.3390/gels9100815

Academic Editor: Avinash J. Patil

Received: 16 September 2023
Revised: 6 October 2023
Accepted: 9 October 2023
Published: 13 October 2023

Copyright: © 2023 by the authors. Licensee MDPI, Basel, Switzerland. This article is an open access article distributed under the terms and conditions of the Creative Commons Attribution (CC BY) license (https:// creativecommons.org/licenses/by/ 4.0/).

1. Introduction

Chemical warfare agents (CWAs) are toxic chemicals employed in warfare or related military operations to harm, kill, or paralyze adversaries. Nerve agents and vesicant agents are the most lethal types of chemical warfare agents [1]. Nerve agents are derived from alkyl phosphonate esters, which can cause neurological disorders, damage the nervous regulatory system and respiration processes, and lead to suffocation within minutes. Common nerve agents include tabun (GA), sarin (GB), soman (GD), and VX (left, Figure 1) [2,3]. Vesicant agents can cause severe skin erosion and damage to the respiratory and digestive tracts and have systemic toxic effects, potentially leading to death. Mustard gas (HD) is a commonly used vesicant agent [4–6]. Although chemical weapons are regulated by the Chemical Weapons Convention, the potential for their use by extremist countries or organizations remains. Therefore, the development of effective protective materials against chemical warfare agents remains crucial [7–11]. Due to the highly toxic nature of nerve and erosive agents, their simulants with lower toxicities (right, Figure 1) are often used in research to reduce the risk of accidental poisoning.

Currently, activated carbon is the primary material used for protection against chemical warfare agents. It functions by adsorbing toxic substances, and, in some cases, it can be impregnated with additional substances to enhance its catalytic degradation capabilities, converting CWAs into non-toxic compounds [12]. However, activated carbon materials suffer from several limitations, including a low adsorption capacity, a limited number of active sites, susceptibility to inactivation or destruction of catalytic sites, slow reaction kinetics, and poor structural flexibility.

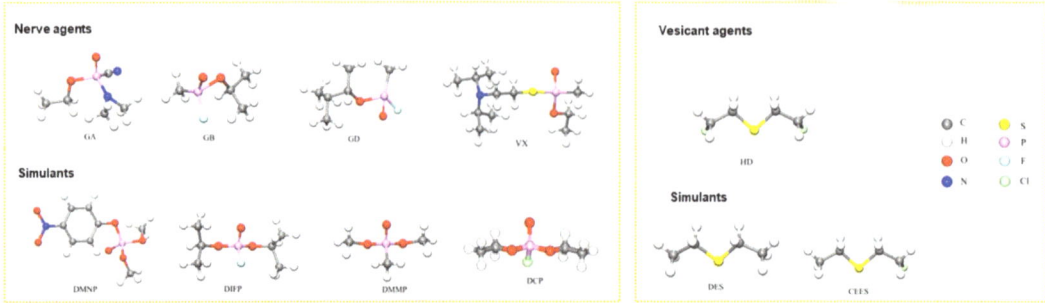

Figure 1. Structure of typical CWAs and simulants [11]. Copyright © 2023, Elsevier.

In recent years, there has been a focus on developing fast, simple, safe, and effective detoxification methods for chemical warfare agents. Scientists have explored various materials with good catalytic performance to achieve better detoxification results [13–17]. These materials typically possess specific structural characteristics, such as larger pore sizes, higher specific surface areas, and flexible structures that provide more active sites. Continued research has led to the discovery of catalytic materials that exhibit excellent performance in the degradation of CWAs, including metal oxides, polyoxometalates, and metal clusters [18–21]. For instance, Wang et al. designed a composite conjugated microporous polymer based on Fe^{2+} for detoxification purposes [22]. Hu et al. developed recoverable amphiphilic polyoxoniobates that catalyze oxidative and hydrolytic decontamination of chemical warfare agents [23]. Zang et al. prepared porphyrinic silver cluster assembled materials for simultaneous capture and photocatalysis of mustard gas simulants [24].

Among the numerous detoxification materials, metal–organic frameworks (MOFs) formed by metal ions or clusters and multidentate ligands have received widespread attention (see Figure 2) [11,25–30]. The metal ions or clusters are mainly derived from transition metal and lanthanide salts, and the multidentate ligands include bridging carboxylic acids, imidazole, porphyrin, etc. Various methods are used to prepare MOFs, and hydrothermal and solvothermal approaches at low temperatures (<250 °C) are the most commonly used at the early stage. Other common methods, such as microwave synthesis and ultrasonic synthesis, have been developed at present for large-scale synthesis, rapid reaction, and reduction of crystallites size. MOFs have an inherently large specific surface area and abundant pore structure, which give them excellent adsorption or solid-phase extraction properties [31–35]. Moreover, metal nodes in MOFs serve as Lewis acid catalytic active centers, which promote the hydrolysis of chemical warfare agents [36]. The photoactive linkers may serve as photocatalysts [37,38]. MOF materials are usually in powdered crystalline states and have poor processability. Agglomerated particles may lead to a decrease in active sites, limiting their practical applications. These problems pose a challenge for practical applications.

MOF (metal–organic framework) gels are a class with an extended MOF architecture that are mainly formed using the metal–ligand coordination as effective force to drive gelation, and these gels combine the unique characteristics of MOFs and organic gel materials [39]. They have the advantages of a hierarchically porous structure, large specific surface area, machinable block structures and rich metal active sites. MOF gels also easily form bulk materials and can be shaped as needed [40]. These materials not only overcome the limitations of MOF powders in practical applications but also contribute to reduce the diffusion barrier between the matrix and active sites, accelerate the mass transfer rate, and enhance the adsorption and catalytic performance [40,41]. MOF gels can be used as adsorbents for removal of hazardous heavy metal ions and organics in water and to capture harmful gases and eliminate particulate matters, such as PM2.5 and PM10. The catalytic applications of MOF gels include electrocatalysis for fuel cells, heterogeneous catalysis for organic chemistry, and photocatalysis for removal of pollutants. The unique structures and composition of MOF gels also inherently meet the requirements for adsorption and catalytic detoxification of CWAs. Metal organic composite aerogels have the advantages of aerogels, showing low density, a high specific surface area and a multistage pore structure, which is conducive to the transport of toxic molecules and degradation of products in the aerogels. It is also possible to retain metal oxygen cluster nodes through a reasonable design and disperse the metal nodes through appropriate organic molecules to ensure catalytic activity, and many related works have been reported. Recently, there have been a few reviews on MOF gels [42–44]; however, research on the application of MOF gels for the adsorption and degradation of CWAs lacks a systematic introduction and review.

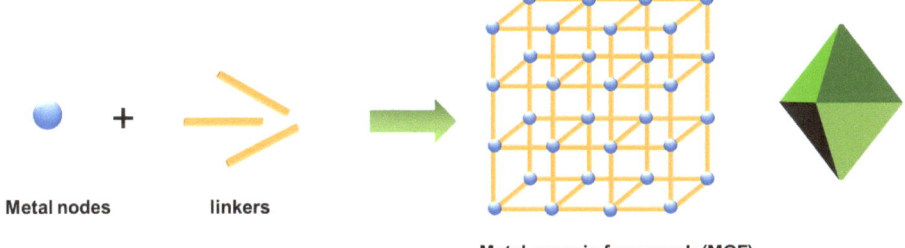

Figure 2. The formation of MOFs.

In this review, pure MOF gels containing different metal ions and the formation process are introduced first. Then, the recent progress of MOF gel composites is summarized. The application of these MOF gel composite materials to protect against nerve agent and vesicant agent CWAs is also discussed. It also looks forward to the next research focuses on the use of MOF gels for CWA protection. The results of this paper provide new ideas for the research and development of novel efficient protective materials.

2. Pure Metal–Organic Framework Gel

Pure MOF gels refer to gel materials consisting of a single MOF material, including xerogel and aerogel, that served as the backbone structure. Pure MOF gels are usually synthesized by directly mixing the metal precursor and organic linker, and the formation process is simple and controllable (see Figure 3) [40,43]. When the coordination polymer separates from the solvent and prevents the solvent from flowing, the gels are obtained. The porous aerogels formed after post-processing have rigid spongy network that consist of nanometer-sized MOF particles. At present, metals with different valence states have been successfully used to prepare MOF gels, typical of which are tetravalent Zr(IV), trivalent Fe(III), Al(III), Cr(III), bivalent Zn(II) and Cu(II), monovalent Ag(I) [45–51]. MOF gels containing a variety of metals have also been prepared [52,53].

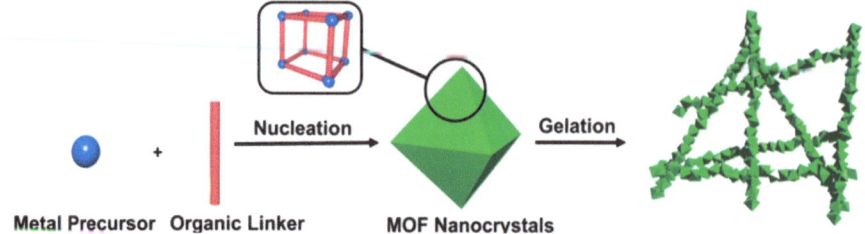

Figure 3. Schematic representation of pristine MOG formation. Reproduced from Ref. [43] with permission from the Royal Society of Chemistry.

2.1. Metal(IV)-MOF Gels

Many early reports describe gel formation during the synthesis of UiO-66, a typical zirconium(IV)-carboxylate MOF [54–56]. For example, Liu et al. reported Zr-MOF gel synthesized from an ethanol–DMF mixture containing aminoterephthalic acid and $ZrCl_4$ [57]. Then, Bueken et al. first reported hierarchically porous, monolithic Zr-MOF xero- and aerogels consisting of several prototypical Zr^{4+}-based MOF nanoparticles, including UiO-66-X (X = H, NH_2, NO_2, $(OH)_2$), UiO-67, MOF-801, MOF-808, and NU-1000 [58]. Among them, the UiO-66 xerogel has a BET surface area of 1459 $m^2 \cdot g^{-1}$, and the total pore volume was 2.09 $cm^3 \cdot g^{-1}$, higher than that of bulk UiO-66 powder.

Moreover, as shown in Figure 4, UiO-66-NH_2 aerogel has been designed as an efficient adsorbent for the trace adsorption of arsenic in water in the full pH range (pH 1–14) [59]. These aerogel have advantages in terms of processability and preventing back pressure during the continuous flow process compared with pristine UiO-66.

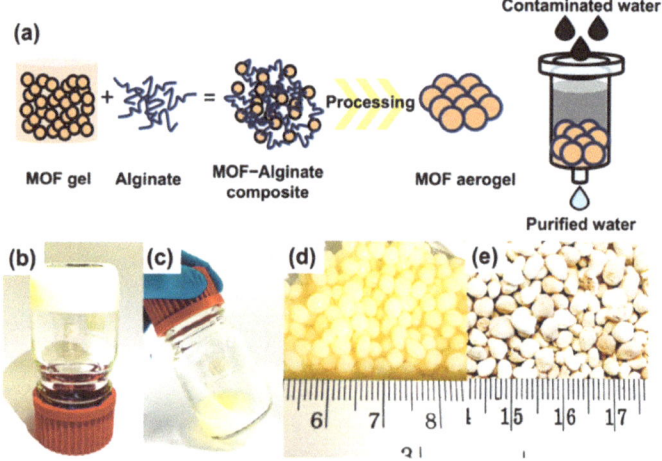

Figure 4. (a) Design concept of MOF aerogel for use in decontamination of arsenic species in water. Optical images of UiO-66-NH_2 in the formation of (b) a nonflowing gel, (c) fluid gel, (d) hydrogel, and (e) aerogel. Reprinted with permission from [59]. Copyright 2022 American Chemical Society.

2.2. Metal(III)-MOF Gels

MOF gels for trivalent metals are the most widely studied, of which iron(III)-based MOF gels were the first to be synthesized. Martin R and coworkers firstly reported metal–organic framework aerogels that were synthesized by mixing $Fe(NO_3)_3$ and trimesic acid in 2009 [60]. The resultant gels have an elemental formula similar to that of MIL-100(Fe)($Fe_3O(BTC)_2F \cdot 2H_2O$) and possess high internal micro- and macroporosity. Their specific surface area and total pore volume can reach as high as 1618 $m^2 \cdot g^{-1}$ and 5.62 $cm^3 \cdot g^{-1}$, respectively. To date, most of the functional pure Fe-MOF gels are still formed by Fe^{3+} and carboxylic acids, especially 1,3,5-benzenetricarboxylic acid (BTC). For example, Hu et al. developed Fe^{3+}–(BTC) metal–organic hybrid gel for online enrichment of trace analytes in a capillary [61]. Zheng et al. synthesized monolithic MIL-100(Fe) with 1,3,5-benzenetricarboxylic acid for energy-efficient removal and recovery of aromatic volatile organic compounds [46].

Su and coworkers reported a series of porous Fe-MOF aerogels produced from Fe^{3+} and bridging carboxylic acids [62] and revealed a simple formation mechanism. The porous aerogels were prepared using three steps: (1) primary nanoparticles were formed via Fe-carboxylate coordination; (2) primary nanoparticles condense together to form networks with an open, continuous and porous structure; and (3) the porous aerogels are produced after a subcritical $CO_2(l)$ extraction process. The highly porous aerogels can be prepared when rigid bridging carboxylates were used, such as 1,4-benzenedicarboxylate, which possesses a higher of BET surface area of 1454 $m^2 \cdot g^{-1}$. A sensitive detection method of dopamine (DA) was proposed given that DA greatly inhibits the Fe-MOX-catalyzed luminol CL (see Figure 5), representing the first example of the use of MOF gels as catalysts for a sensing platform in the CL field [63].

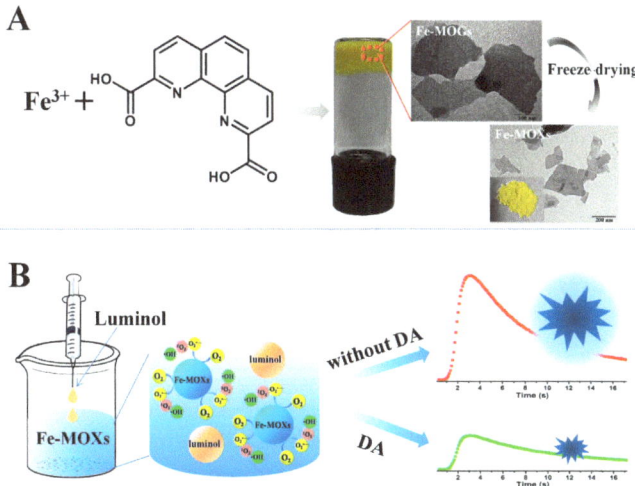

Figure 5. Schematics To Show the Preparation of Fe–MOGs (**A**) and the CL Detection of DA with Fe–MOXs (**B**). Reprinted with permission from [63]. Copyright 2017 American Chemical Society.

Al(III)-based MOF gels have also gained attention. Su et al. prepared gel electrolytes that have a sponge-like porous matrix of metal–organic gel assembled by coordination of Al^{3+} and 1,3,5-benzenetricarboxylate (H_3BTC) for use in highly efficient quasi-solid-state dye-sensitized solar cells (DSSCs) for the first time [64]. Then, a variety of ultralight hierarchically micro/mesoporous Al-MOF aerogels were also first successfully synthesized by Su et al. [47]. As shown in Figure 6, these aerogels are formed through the stepwise assembly of light metal Al(III) with bridging carboxylic acids. In the early stage, the metal ions and ligands assemble into an MOF cluster, which can polymerize or aggregate to trigger nucleation, and the nucleation of new particles is retarded as the concentrations of ligands and metal ions decrease. Then, the consistent epitaxial growth or oriented attachment induced by surface intension will lead to the crystallization of bulky MOFs when the conditions favor the crystal growth of the precursors. However, if the coordination equilibria are perturbed by other competing interactions, non-crystallographic branching may occur, leading to mismatched growth or cross-linking, which provide the opportunity for gelation. The final Al-MOF aerogels were obtained after the careful removal of solvents via sub/supercritical CO_2 extraction.

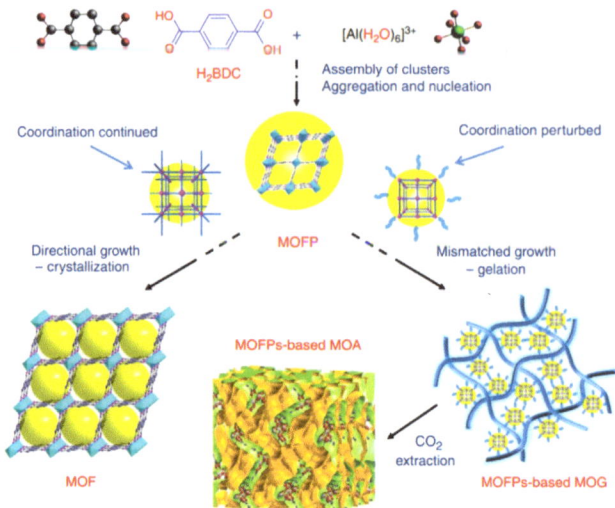

Figure 6. Schematic representation of the formation of MIL-53(Al) MOF versus MOF aerogel [47]. Copyright © 2013, The Authors.

In addition to the above reported MOF gels containing Al(III) and Fe(III), Cr(III) ions can also be used for preparing MOF gels. Su et al. reported on MOF aerogels based on Cr^{3+} and bridging carboxylic acids [62]. Heating induces the formation of these Cr(III)-carboxylate gels, and all of the Cr^{3+}-containing gels could only be formed at temperatures above 80 °C. The texture and porosity of the aerogels are affected by the reactant concentration and organic ligands. At high reactant concentrations (Cr:BDC = 2:3, 0.2 mol·L^{-1}), the Cr-BDC aerogel has a high BET surface area of 737 $m^2·g^{-1}$.

2.3. Metal(II)-MOF Gels

Lee and coworkers developed a luminescent Zn-MOF hydrogel that achieved high sensitivity detection of TNT [65]. Tian and coworkers reported monolithic HKUST-1(Cu-MOFs) with higher volumetric BET areas (1193 m^2/g), pore volumes (0.52 $cm^3·g^{-1}$), and adsorption capacities compared to traditional powdered counterparts [50]. It also has a high bulk density of 1.06 g cm^{-3} and exhibits enhanced methane uptake of 259 cm^3 (STP) cm^{-3} at 65 bar.

2.4. Metal(I)-MOF Gels

Su et al. also reported luminescent coordination polymer gels based on rigid terpyridyl phosphine and Ag(I), and the terpyridine groups could generate interesting photochemical and electronic properties. The gel emits blue luminescence that exhibits an emission intensity comparable with that of the ligand in dilute solution [51].

Cheng et al. synthesized a silver(I) coordination polymeric gelator through the combination of Ag(I) and 2, 7-bis(1-imidazole) fluorene. This coordination polymeric gel exhibited thixotropic behaviors and stimuli responsive to S^{2-}, I^- and displayed antibacterial properties [66].

2.5. Multi-Metal-MOF Gels

A series of bimetallic Co/Fe-MOF xerogels that have sufficient adsorption sites for CO_2 molecules have been prepared, and the metal center of Co acts as a major active site for photocatalysis [67]. This novel bimetallic xerogel exhibits enhanced adsorption and utilization of light energy and improved separation and transfer of carriers. Therefore, the conversion of CO_2 to CO is rapidly promoted, and the Co/Fe xerogel exhibits a high CO yield (67 μmol g^{-1} h^{-1}) when the mole ratio of Co: Fe was set to 1:3, far higher than that of the single Fe center MOF xerogel.

3. Metal–Organic Framework Composite Gel

The formation of pure MOF gels is affected by reaction conditions, such as reactant concentration ratios, temperature, and their structures and applications are limited. The metal–organic framework composite gels formed by growing or aggregating MOF particles into interconnected 3D networks, such as cellulose, graphene, silicon aerogels, etc., exhibit various architectures and are useful in a variety of applications. The MOF composite gels are mainly based on Zr-MOFs, Fe-MOFs, Cu-MOFs, Co-MOFs, and others.

3.1. Zr-MOF Composite Gel

Zr-MOF is one of the most stable MOFs, and many researchers are committed to fabricating Zr-MOF composite gels with other skeleton materials. Currently, Zr-MOF composite gels containing UiO-66 have been most widely studied [45,68–71]. The UiO-66 nanoparticles can still retain their crystallinity and function when integrated within various substrates, such as cellulose nanocrystal (CNC) aerogels, and the obtained flexible and porous composite gels show good processability [72]. The oxygen-containing groups on UiO-66 (Zr-OH) are physically crosslinked with the hydroxyl groups in cellulose by hydrogen bonding. As shown in Figure 7, UiO-66/NC was obtained using nanocellulose as the structural skeleton. This composite gel has a specific surface area of 826 m^2 g^{-1} and can stand on the bristle of grass without observable deformation [73].

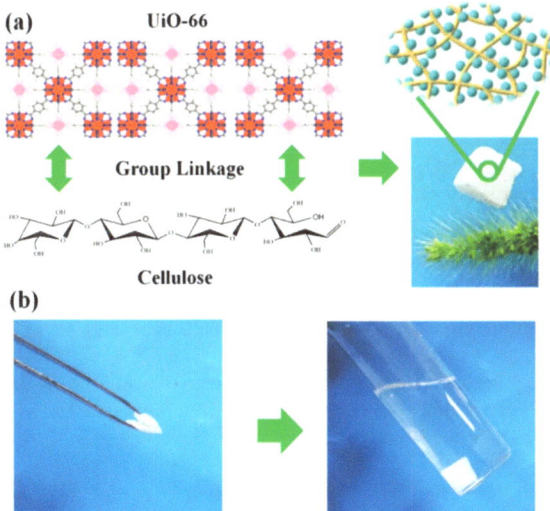

Figure 7. The fabrication process of the MOF/NC aerogel and photographs of the lightweight MOF/NC (**a**). Photographs show the deformed MOF/NC would recover its original shape when put back to solution (**b**) [73]. Copyright © 2019, Elsevier.

In addition, there are also several studies on other Zr MOF composite gels, such as NU-1000. The gel is formed by grafting NU-1000 into agarose (AG) possessing micropores, mesopores, and macropores, and the average pore size is 2.57 nm, which is close to that of NU-1000. This hybrid aerogel has potential applications for adsorbing in water treatment due to the hierarchical pore structure [70].

3.2. Fe-MOF Composite Gel

At present, the Fe-MOF gel is one of the most studied pure MOF gels, and it also attracts much attention for the fabrication of composite gels. Researcher have successfully constructed Fe-MOF composite gels with many suitable porous supports, such as cellulose, graphene, aerogels, etc. As shown in Figure 8, a monolithic iron metal−organic gel/bacterial cellulose (denoted as Fe-MOG/BC) composite has been prepared by the crosslinking of nanoscale Fe-BTC MOG particles with BC nanofibers to form 3D porous networks [74]. These Fe-MOG/BC aerogel possesses many unique structural characteristics, such as a three-dimensional (3D) hierarchically porous microstructure, abundant active sites, and ultralight, water-fast, and mechanically robust features. Therefore, they exhibit a superb saturated sorption capacity (495 mg g^{-1}) for arsenate, higher than that of Fe-MOF/BC.

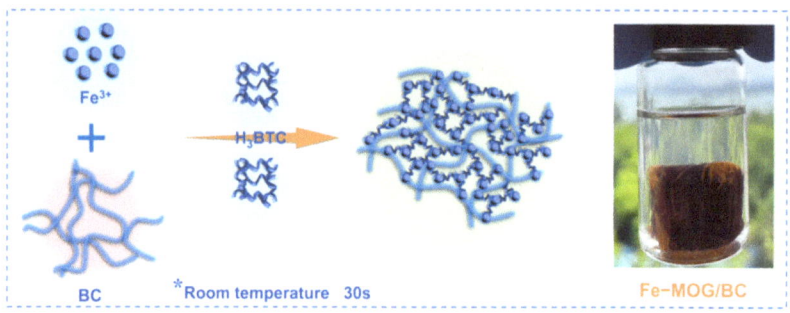

Figure 8. Schematic Illustration of the Fabrication Process of the Fe-MOG/BC Aerogel. Reprinted with permission from [74]. Copyright 2021 American Chemical Society. *: the formation conditions of gel.

The MOF/GA composites can be prepared using growth-oriented MIL-88-Fe synergized with graphene aerogels (GAs), and the oriented composite can be used for high-performance supercapacitors with a specific capacitance as high as 353 F g^{-1} at a scan rate of 20 A g^{-1} [75]. By immobilizing Fe-MOFs on nanofibrous aerogel membranes (NFAMs), a novel Fe-BTC@polyacrylonitrile (PAN)NFAM catalyst was constructed with a 3D interconnected hierarchical porous structure that could be used as a catalytic membrane in a filtration device for the treatment of organic wastewaters [76]. Specially, the combination of Fe-MOFs with a photocatalyst, such as g-C_3N_4, can enhance the visible-light adsorption regions, increase the specific surface areas and prolong the lifetime of the charge carriers. Therefore, the porous g-C_3N_4/NH_2-MIL-53(Fe) aerogel showed excellent recyclability and a higher photocatalytic performance than pure g-C_3N_4 nanosheets [77].

3.3. Cu-MOF Composite Gel

HKUST-1, which is also called Cu-BTC and consists of copper oxide clusters linked by benzene-1, 3, 5-tricarboxylate ligands, is a common material for Cu-MOF composite gels [78,79]. HKUST-1/graphene aerogels, HKUST-1 modified ultrastability cellulose/chitosan composite aerogels, and HKUST-1 silica aerogel composites have been fabricated successfully [78–84]. For example, a core–shell hybrid aerogel sphere material containing Cu-MOF was fabricated using a combined assembly strategy of coordination bonding and ionic cross-linking [85]. The Cu^{2+} ions cross-linked carboxylated cellulose nanocrystals (CNCA) and carboxymethyl chitosan (CMCS) hydrogel spheres to serve as templates for the in situ growth of the MOF-199 crystal using 1,3,5-benzenetricarboxylic acid as ligand (Figure 9). The resultant aerogel spheres showed an excellent adsorption capacity towards methylene blue (MB) with values as high as 1112.2 mg·g^{-1}.

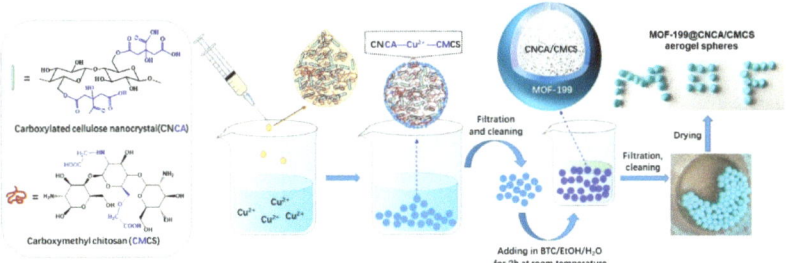

Figure 9. Schematic illustration of the synthetic process of MOF-199@CNCA/CMCS aerogel spheres [85]. Copyright © 2022, Elsevier.

Meanwhile, various methods are used for the synthesis of Cu-MOF composite gel. As shown in Figure 10, a one-droplet synthesis strategy was developed to synthesize functional polysaccharide/MOF(HKUST-1) aerogels [86]. In this one-droplet reaction, the metal ions initiate the cross-linking of polysaccharide molecules and coordinate with organic ligands to form MOFs simultaneously. The resulting composite aerogel has a hierarchical porous structure and exhibits a high adsorption capacity for CO_2.

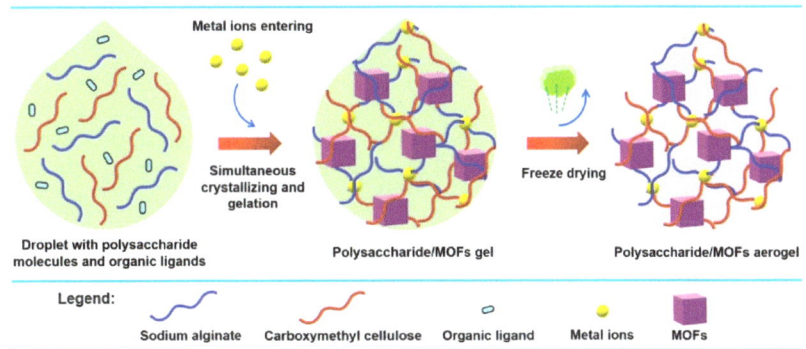

Figure 10. Illustration of the One-Droplet Synthesis of Polysaccharide/MOF Aerogels. Reprinted with permission from [86]. Copyright 2023 American Chemical Society.

3.4. Co-MOF Composite Gel

Most of the Co-MOF composite gels consist of a zeolitic imidazolate framework-67 (ZIF-67) that is formed by 2-methylimidazole [87–90], and their structures are diverse. As shown in Figure 11, the highly hydrophobic ZIF-67@PLA honeycomb aerogel was formed by physically combining ZIF-67 nanoparticles with a PLA solution and a water-assisted heat-induced phase. These ZIF-67@PLA honeycomb aerogels have a multilayer porous structure, a considerably reduced pore size, and an increased honeycomb pore volume and exhibit better oil wettability than pure PLA aerogels [89].

Wood aerogels made from naturally lightweight, high-porosity, thin-walled balsa wood have a lamellar structure and provide sufficient attachment sites for ZIF-67. ZIF-67@WA (wood aerogel) has been prepared successfully through in situ anchoring of ZIF-67 on the wood aerogel, and it exhibits excellent adsorption performance for tetracycline and Cu(II) ions, respectively [91,92]. In addition, Co^{2+} coordinates with the oxygen-containing functional groups of MXene to form a hydrogel and then acts as a nucleation site for the in situ growth of ZIF-67 particles [93]. Porous 3D rGO/ZIF-67 aerogel was prepared via the assembly of ZIF-67 polyhedrons on the 3D rGO framework, which has a specific surface area up to 491 $m^2 \cdot g^{-1}$ and displays excellent adsorption for organic dyes [94].

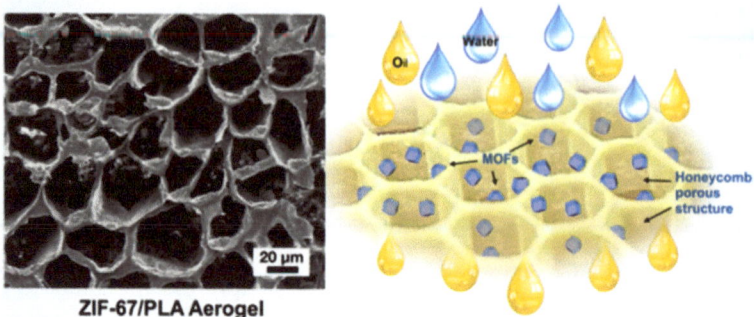

Figure 11. Images of the ZIF-67@PLA honeycomb aerogel structure and the oil–water separation [89]. Copyright © 2022, Elsevier.

3.5. Other MOF Composite Gels

In addition to the above-mentioned MOF composite aerogels, other reported MOF composite aerogels are mainly based on ZIF-8(Zn-MOF) and MIL 101(Cr-MOF). For example, nanocellulose can also serve as template for developing shapeable fibrous ZIF-8 aerogels, which exhibit higher adsorption capacity and rapid adsorption kinetics for different organic dyes [72].

As shown in Figure 12, the graphene aerogel (GA)-supported MIL-101 (Cr-MOF) particles exhibited a three-dimensional (3D) architecture with an interconnected macroporous framework of graphene sheets and uniform dispersion of MOF particles, which could be used as adsorbents for the solid-phase extraction (SPE) of non-steroidal anti-inflammatory drugs (NSAIDs) [95].

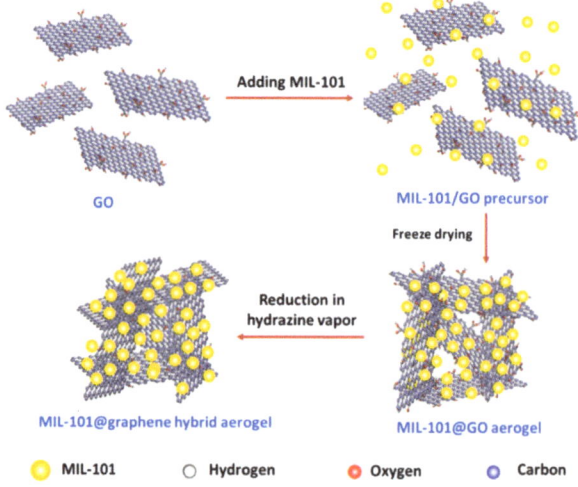

Figure 12. Synthesis of MIL-101@graphene hybrid aerogels. Reproduced from Ref. [95] with permission from the Royal Society of Chemistry.

Specifically, a superhydrophobic aerogel was constructed by fine-tuning the hydrophobicity of MOF (MOF, Eu-bdo-COOH, H_4bdo = 2,5-bis(3,5-dicarboxylphenyl)-1,3,4-oxadiazole) microspheres, and this aerogel exhibits fast and efficient absorption of various oily substances from water [96].

4. Adsorption of CWAs

4.1. MOFs for Adsorption of CWAs

The adsorptive removal of CWAs is an important method of personal protection, and effective adsorbents, such as activated carbons, metal oxides, etc., have been widely explored. A variety of studies have indicated that MOFs are promising materials for the capture of CWAs owing to their high porosity and adjustable reactivity. The selective adsorption of organic phosphonates in MOF-5/IRMOF-1 was investigated first, and the binding energy of DMMP in IRMOF1 was ~19 kcal/mol. The sorption capacity of the CWA simulant DMMP (dimethylmethyl phosphonate) can reach as high as 0.95 g g^{-1} [97]. Both zeolitic imidazolate frameworks ZIF-8 and ZIF-67 have large pores connected through small apertures, and the inner pores exhibit strong hydrophobicity. Therefore, they exhibit excellent performance for rapid adsorption and removal of hydrophobic CEES molecules (Figure 13). The maximum adsorption capacities of ZIF-8 and ZIF-67 for CEES were 456.61 mg g^{-1} and 463.30 mg g^{-1}, respectively, and 100% of HD from from the water/ethanol solution (9:1, v/v) was removed within 1 min in further experiments [98]. Some research suggests that the partial charge of the metal atom induces a higher affinity of CWAs toward the MOF surface [99].

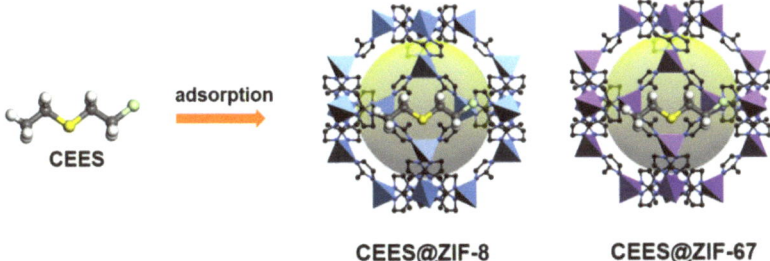

Figure 13. Schematic illustration of CEES adsorption by ZIF-8 and ZIF-67 [98]. Copyright © 2019, Elsevier.

Recently, zirconium-based MOFs have been extensively studied for the adsorption of chemical warfare agents (CWAs) and their simulants [100]. For example, NU-1000 and UiO-67 have been successfully used for capturing chemical warfare agent simulants 2-CEES and DMMP from aqueous media [101]. NU-1000 showed adsorptive capacities of 4.197 and 1.70 mmol g^{-1} for 2-CEES and DMMP, respectively, higher than the results of UiO-67, which can also adsorb 2-CEES and DMMP with capacities of 4.000 and 0.90 mmol g^{-1}, respectively. Zr-MOFs with different surface area/pore volumes, secondary building unit (SBU) connectivity, pore functionalization, and open metal sites for the adsorption of sarin gas and CEES have been examined, and the findings showed that UiO-66, defective UiO-66, and MOF-808 have the highest reactivities toward GB due to the presence of more active sites per unit volume [102].

4.2. MOF Gels for Adsorption of CWAs

Currently, MOF gels containing MOF structures also demonstrate outstanding adsorption properties for CWAs or simulants. We and collaborators prepared granular UiO-66-NH$_2$ xerogels that showed an excellent adsorption capacity of 802 mg/g for CEES vapor in static adsorption and desorption tests, higher than that of many active inorganic nanomaterials [103]. The ability to retain adsorbed CWA on the surface/in the porous structure is a very important feature of protection materials. Static desorption tests monitored their weight change after exposure to 2-CEES vapors for 1 day, and air desorption tests were conducted at 2 d and 7 d. The results demonstrated that these Zr-MOF xerogels have low desorption capacity with only 28 wt%. Moreover, the superelastic hierarchical aerogels composed of MOF-808 and SiO$_2$ nanofibers exhibited hierarchical cellular architectures with interconnected channels. Simultaneously, the additional ceramic constituents in the interconnected channels can generate van der Waals barriers, which are beneficial for nerve agent adsorption in open MOF sites. Therefore, this MOF gel showed efficient adsorption performance against CWAs with a breakthrough extent of 400 L g^{-1} [104].

5. Catalytic Detoxification of CWA

According to recent reports, MOF gels have shown excellent performance in the field of CWA detoxification due to their large specific surface area, hierarchical porous structures and processability. The meso- and macropores facilitate the transport of toxic molecules and degradation products within the gel monoliths. Some substrates introduced for constructing MOF composite gels can adsorb CWAs and exhibit water storage abilities, which promote the degradation process.

Among the numerous chemical agents, nerve agents and vesicant agents are the focus of current study, and Zr-MOFs gels are the most commonly reported materials. The properties of pure zirconium-based MOF gels and zirconium-based MOF composite gels for the degradation of these two CWAs and other simulants are summarized, as shown in Table 1, and details of the analysis are presented below.

Table 1. Various MOF gels Used as Protection Media for the Detoxification of CWAs.

Materials	Amount or Size	Agent Volume	Half-Life	Environment	Mechanism	Refs.
monolithic UiO-66 xerogel	25 mg	HD, 2.5 µL	24.8 min	Liquid-Phase	Catalytic hydrolysis	[105]
monolithic UiO-66 xerogel; monolithic UiO-66-NH$_2$ xerogel	20 mg	VX, 0.4 µL	≤1.5 min	Liquid-Phase	Catalytic hydrolysis	[105]
monolithic UiO-66-NH$_2$ xerogel	25 mg 25 mg	HD, 2.5 µL 2-CEES, 2.5 µL	14.4 min 8.2 min	Liquid-Phase	Catalytic hydrolysis	[105]
granular UiO-66-NH$_2$ xerogel	10 mg	2-CEES, 1 µL	7.6 min	Liquid-Phase	Catalytic hydrolysis	[103]
UiO-66-NH$_2$-loaded cellulose sponge	8.1 mg	DMNP, 4 µL	9 min	Liquid-Phase	Catalytic hydrolysis	[106]
UiO-66/Nanocellulose Aerogel	8 mg	MPO, 2.5 µmol	0.7 min	Liquid-Phase	Catalytic hydrolysis	[107]
MOF-808/BPEIH hydrogel	2.2 mg MOF-808 loading, 6 mol%	DMNP, 4 µL	<1 min	Liquid-Phase	Catalytic hydrolysis	[108]
MOF-808/BPEIH/fiber	1 × 1 cm, containing 1.5 µmol MOF-808	DMNP, 4 µL GD, 3 µL DEMP, 4.2 µL	1 min <10 min <1 min	Liquid-Phase	Catalytic hydrolysis	[108]
MOF-808/SiO$_2$ aerogels	200 mg	DMMP, 4 µL	5.29 min	Liquid-Phase	Catalytic hydrolysis	[104]
UiO-66-NH$_2$@ANF aerogels	20 mg	CEES, 5 µL	8.15 min	Liquid-Phase	Catalytic hydrolysis	[109]
UiO-66-NH$_2$@agarose hydrogels	—	DCP, vapors	—	atmospheric	Catalytic hydrolysis	[110]
UiO-66-AM@PDMAEA@LiCl@PNIPAM aerogel	60 mg of UiO-66-AM loading, 12 mol%	DMNP, 12.5 µmol	1.9 h	atmospheric	Catalytic hydrolysis	[111]
fibrous MOF-808 nanozyme aerogel	1 mm × 1 mm × 1 mm, containing 1.5 µmol MOF-808	DMNP, 25 µmol	1 min	Liquid-Phase	Catalytic hydrolysis	[112]
MOF-808/bacterial cellulose sponge	1.5 µmol MOF-808 in composite	DMNP, 25 µmol	<1 min	Liquid-Phase	Catalytic hydrolysis	[113]
SA@UiO-66-NH$_2$@PAMAM hydrogel	17.6 mg	DMNP, 4 µL	7 min	Liquid-Phase	Catalytic hydrolysis	[114]
MOF-808/HIPE sponge	3.2 mg MOF-808 loading, 0.68 mol%	VX, 24 µL	<1 h	Liquid-Phase	Catalytic hydrolysis	[115]
UiO-66/DSPD Composite Films	1 × 1 × 0.015 cm^3	MPO, 25 µmol	—	Liquid-Phase	Catalytic hydrolysis	[116]

"—" = not mentioned.

5.1. Nerve Agents and Simulants

For nerve agents and simulants, Zr-MOF gels based on UIO-66 and MOF-66 have been extensively studied. As shown in Figure 14, our group and coworkers firstly reported pure macroscopic monolithic UiO-66 and UiO-66-NH$_2$ xerogels with excellent degradability for real nerve agent VX, and both of them possess a short half-life of 1.5 min and 100% conversion within 3 min. These materials can selectively catalyze the breakage of P–S during VX hydrolysis, and less toxic product breakage was obtained [105].

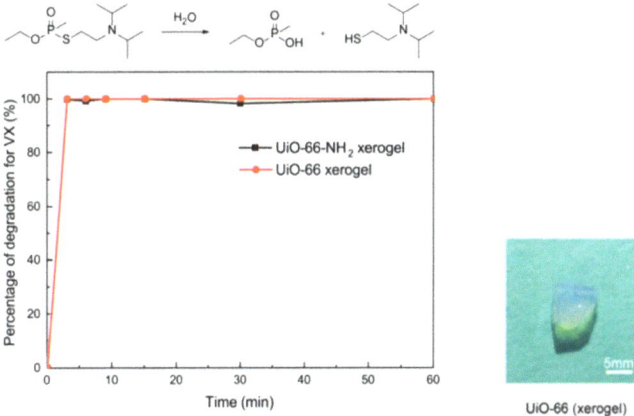

Figure 14. Degradation of VX on the UiO-66 xerogel and the UiO-66-NH$_2$ xerogel. Reproduced from Ref. [105] with permission from the Royal Society of Chemistry.

Sui and our group fabricated flexible UiO-66-NH$_2$-loaded cellulose sponge composites for rapid degradation of DMNP. The surprising hydrolysis rate with a half-life of only 9 min was attributed to the preserved catalytic activity of MOFs and the high porosity and random three-dimensional structures of the sponge [106]. UiO-66/nanocellulose aerogel composite fabricated by simple blending of UiO-66 and TEMPO-oxidized cellulose nanofibers could decompose nearly all MPO within 3 min and exhibited a 0.7 min half-life under static condition [107]. Moreover, this aerogel composite exhibits the surprising ability to dispose 53.7 g of MPO per hour with 1 m^2 of the effective area when used as the detoxification filter in continuous dynamic continuous flow systems.

A metal–organic framework-containing polymer sponge was fabricated by combining the excellent nerve agent absorption agent (styrene pHIPE) with MOF-88, which served as a hydrolysis catalyst (Figure 15) [115]. This MOF-HIPE composite can facilitate the bulk absorption, immobilization, and catalytic decomposition of P–S bonds in VX, and they rapidly hydrolyze over 80% VX in 8 h with a half-life of less than 1 h. The fibrous MOF-808 nanozyme aerogel, which was fabricated by in situ growth of MOF-808(Zr-MOF) on cellulose nanofibers, has a hierarchical macro/microporosity that provides more accessible active sites. This flexible and processable monolithic MOF composite aerogel demonstrated superior catalytic performance for hydrolysis with a very short half-life of 1 min, and DMNP was converted to nontoxic dimethyl phosphate (DMP) [112]. Superelastic cellular hierarchical metal–organic framework aerogels can be fabricated by combining functional MOFs-88 nanoparticles with structural SiO$_2$ nanofibers based on hydrogen bond-assisted interfacial coupling effect. The as-prepared MOF-808/SiO$_2$ aerogels have a preserved MOF structure, van der Waals barrier channels and minimized diffusion resistance, which all contribute to increasing the adsorption and decontamination efficiency toward CWAs. This optimized aerogel-based MOF exhibited rapid adsorption and detoxification properties for DMMP with a half-life of 5.29 min [104].

In addition, a hydrogel with polymeric networks, mechanical stability, flexibility and a high water content is a very suitable platform for the hydrolytic reaction of nerve agents, and many MOFs/hydrogel composites have been reported. For example, the inexpensive non-volatile branched polyethyleneimine hydrogel integrated with Zr-MOFs was developed for rapid degradation of organophosphorus chemicals [108]. The hydrogel possesses high amine density and plentiful water, which can regulate the micro-environment of the MOF catalytic reaction process. The obtained MOF-808 hydrogel (MOF-808/BPEIH) powder can induce near-instantaneous catalytic hydrolysis of DMNP with a short initial half-life of less than 1 min under ambient humidity, which is better than all other reported MOF-based composites. When the MOF-based composite was coated onto a textile, the as-prepared MOF-808/BPEIH/fiber composite also possessed excellent catalytic activity for DMNP with an initial half-life of 1 min and a conversion of 72% after 15 min. Regarding actual nerve agents, VX and GD, the MOF-808/BPEIH/fiber composite can degrade nearly all VX and nearly 60% of the GD after 10 min under ambient conditions, demonstrating potential for the large-scale production of protective gear in practical conditions.

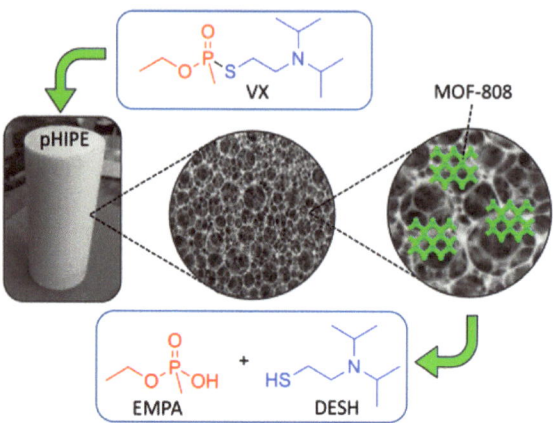

Figure 15. The structure of the MOF–HIPE composite and the degradation of VX. Reprinted with permission from [115]. Copyright 2020 American Chemical Society.

The SA@UiO-66-NH$_2$@PAMAM composite hydrogel synthesized by immobilizing UiO-66-NH$_2$ and PAMAM to the backbone of sodium alginate can rapidly degrade DMNP with a half-life as short as 7 min. This composite hydrogel easily combines with cotton fabric. Upon introducing the indicator 4-nitro-(dimethyl-tert-butyl) silica ether (P-NSE), the obtained recyclable flexible cotton not only catalyzes the hydrolysis of the nerve agent GB but also serves as a portable colorimetric platform to realize the real-time visual detection of changes in degradation [114].

To explore efficient catalysts for the destruction of nerve agents under atmospheric environments, a spontaneously super-hygroscopic MOF-gel microreactor was designed and synthesize by photoinduced integration of UiO-66-acrylamide (UiO-66-AM) and alkaline poly(dimethylaminoethyl acrylate) onto LiCl-salinized poly(N-isopropylacrylamide) gel. The resultant MOF@PDMAEA@LiCl@PNIPAM gel (MG) exhibits excellent catalytic performance for hydrolysis of DMNP with an initial half-life of ~1.9 h, and the final conversion is 95.5% [111].

5.2. Vesicant Agents and Simulants

Sulfur mustard (HD), which was first used in World War I, remains the most notorious vesicant agent. The degradation process of HD includes oxidation, dehalogenation, and hydrolysis (Figure 16), and the C–Cl of HD will be destroyed during the hydrolysis process [25]. CEES is commonly used in experiments instead of HD given its high toxicity. At present, there are many studies that focus on the detoxification of HD or simulants, which all exhibit remarkable potential in future military applications.

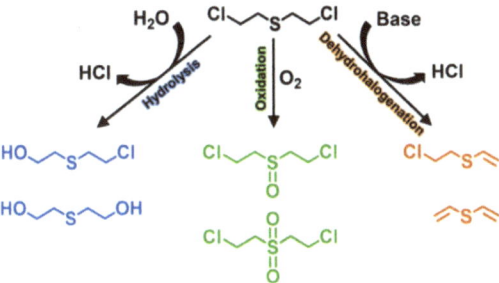

Figure 16. HD degradation pathway. Reproduced from Ref. [25] with permission from the Royal Society of Chemistry.

Pure monolithic UiO-66-NH$_2$ xerogel reported by Zhou and our group also demonstrated a fast decomposition rate of 2-CEES with a half-life of 8.2 min, higher than that of UiO-66-NH$_2$ powder ($t_{1/2}$ = 29 min) [105]. Further study showed this xerogel has a $t_{1/2}$ value of 14.4 min for the hydrolytic degradation of the real CWA sulfur mustard (HD).

Together with our collaborators, we designed and synthesized a series of defective granular UiO-66-NH$_2$ xerogels and investigated their catalytic properties for the decontamination of 2-chloroethyl ethyl sulfide (2-CEES) (Figure 17) [103]. The degradation rate increased with the increasing defect degrees and reducing the size of MOF crystals. A shortened half-life value of 7.6 min was observed, representing the best performance for MOFs reported under ambient conditions [103].

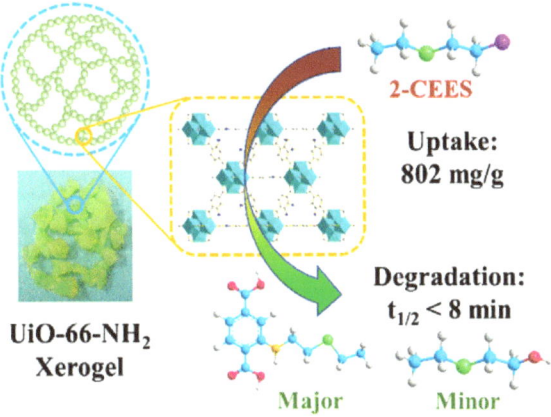

Figure 17. Structurally defective granular UiO-66-NH$_2$ xerogels and the hydrolysis of 2-CEES. Reprinted with permission from [103]. Copyright 2022 American Chemical Society.

By combining UiO-66-NH$_2$ and aramid nanofibers (ANFs), a light weight, flexible, and mechanical robust aerogel with a 3D hierarchically porous architecture was constructed. The resultant UiO-66-NH$_2$@ANF aerogels have a short half-life of 8.15 min for the detoxification of 2-chloroethyl ethyl thioether (CEES), and the removal rate is as high as 98.9%. The C-Cl bond in CEES was broken, and the fragment recombined to form BETE with low toxicity. This aerogel exhibits good mechanical stability with a recovery rate of 93.3% after 100 cycles [109].

In practical application scenarios, multiple chemical warfare agents may be used at the same time, so the ability of MOF gel to correspond to multiple toxic substances at the same time should also be explored. The monolithic UiO-66-NH$_2$ xerogel has initially demonstrated this capability and may have important application prospects.

6. Conclusions and Outlook

In summary, many studies demonstrate the potential of MOF gels and their composites as effective materials for the detoxification of chemical warfare agents (CWAs). The unique properties of MOF gels, such as their large specific surface area, hierarchical porous structures, and processability, make them highly suitable for this application. The studies have focused on nerve agents and vesicant agents, with Zr-MOF gels being the most commonly reported material.

Various approaches have been explored to enhance the catalytic performance of MOF gels, including the development of pure MOF gels, MOF-loaded composites, and MOF/hydrogel composites. The results have shown rapid and efficient degradation of CWAs, with short half-lives and high conversion rates achieved within minutes. The use of flexible and processable monolithic MOF composite aerogels has further improved the catalytic performance, enabling the disposal of significant quantities of CWAs per hour.

In addition, the combination of MOFs with different matrices, such as cellulose, graphene, and balsa wood, has expanded the functionalities and advantages of MOF gels. Silica aerogel-based MOF composites have demonstrated low densities and high specific surface areas, while wood-based aerogels have shown potential for cost-effective and continuous production. These advancements pave the way for future military applications of MOF gels in CWA detoxification.

Looking ahead, further research should focus on optimizing the performance and stability of MOF gels, exploring the potential of other MOF compositions, and investigating their efficacy against a broader range of CWAs. The coupling effects of other external conditions, such as light, microwave, ultrasound, and piezoelectric conditions, on the adsorption and catalytic degradation process should be studied, which will provide new ideas for developing efficient protecting materials. Additionally, efforts should be made to scale up the production processes and evaluate the feasibility of incorporating MOF gels into practical systems for large-scale CWA decontamination. By continuing to explore and refine the application of MOF gels for CWA detoxification, significant advancements can be made in the fields of chemical defense and military protection.

Funding: This research was funded by the National Natural Science Foundation of China (22075319, 22105223), Natural Science Foundation of Hunan Province (2021JJ40663), and the Huxiang Youth Talent Support Program (2020RC3033).

Conflicts of Interest: The authors declare no conflict of interest.

References

1. Szinicz, L. History of chemical and biological warfare agents. *Toxicology* **2005**, *214*, 167–181. [CrossRef]
2. Sidell, F.R.; Borak, J. Chemical warfare agents: II. Nerve agents. *Ann. Emerg. Med.* **1992**, *21*, 865–871. [CrossRef]
3. Mercey, G.; Verdelet, T.; Renou, J.; Kliachyna, M.; Baati, R.; Nachon, F.; Jean, L.; Renard, P.Y. Reactivators of Acetylcholinesterase Inhibited by Organophosphorus Nerve Agents. *Acc. Chem. Res.* **2012**, *45*, 756–766. [CrossRef]
4. Xu, H.; Gao, Z.; Wang, P.; Xu, B.; Zhang, Y.; Long, L.; Zong, C.; Guo, L.; Jiang, W.; Ye, Q.; et al. Biological effects of adipocytes in sulfur mustard induced toxicity. *Toxicology* **2018**, *393*, 140–149. [CrossRef] [PubMed]
5. Khateri, S.; Ghanei, M.; Keshavarz, S.; Soroush, M.; Haines, D. Incidence of lung, eye, and skin lesions as late complications in 34,000 Iranians with wartime exposure to mustard agent. *J. Occup. Environ. Med.* **2003**, *45*, 1136–1143. [CrossRef]
6. Kehe, K.; Szinicz, L. Medical aspects of sulphur mustard poisoning. *Toxicology* **2005**, *214*, 198–209. [CrossRef]
7. Jang, Y.J.; Kim, K.; Tsay, O.G.; Atwood, D.A.; Churchill, D.G. Update 1 of: Destruction and Detection of Chemical Warfare Agents. *Chem. Rev.* **2015**, *115*, PR1–PR76. [CrossRef]
8. Eubanks, L.M.; Dickerson, T.J.; Janda, K.D. Technological advancements for the detection of and protection against biological and chemical warfare agents. *Chem. Soc. Rev.* **2007**, *36*, 458–470. [CrossRef]
9. Khan, A.W.; Kotta, S.; Ansari, S.H.; Ali, J.; Sharma, R.K. Recent Advances in Decontamination of Chemical Warfare Agents. *Def. Sci. J.* **2013**, *63*, 487–496. [CrossRef]
10. Prasad, G.K.; Ramacharyulu, P.V.R.K.; Singh, B. Nanomaterials based decontaminants against chemical warfare agents. *J. Sci. Ind. Res.* **2011**, *70*, 91–104.
11. Yang, J.; Gao, M.; Zhang, M.; Zhang, Y.; Gao, M.; Wang, Z.; Xu, L.; Wang, X.; Shen, B. Advances in the adsorption and degradation of chemical warfare agents and simulants by Metal-organic frameworks. *Coord. Chem. Rev.* **2023**, *493*, 215289. [CrossRef]

12. Kiani, S.S.; Farooq, A.; Ahmad, M.; Irfan, N.; Nawaz, M.; Irshad, M.A. Impregnation on activated carbon for removal of chemical warfare agents (CWAs) and radioactive content. *Environ. Sci. Pollut. Res.* **2021**, *28*, 60477–60494. [CrossRef] [PubMed]
13. Jung, D.; Das, P.; Atilgan, A.; Li, P.; Hupp, J.T.; Islamoglu, T.; Kalow, J.A.; Farha, O.K. Reactive Porous Polymers for Detoxification of a Chemical Warfare Agent Simulant. *Chem. Mater.* **2020**, *32*, 9299–9306. [CrossRef]
14. Wang, Q.-Y.; Sun, Z.-B.; Zhang, M.; Zhao, S.-N.; Luo, P.; Gong, C.-H.; Liu, W.-X.; Zang, S.-Q. Cooperative Catalysis between Dual Copper Centers in a Metal-Organic Framework for Efficient Detoxification of Chemical Warfare Agent Simulants. *J. Am. Chem. Soc.* **2022**, *144*, 21046–21055. [CrossRef] [PubMed]
15. Liao, Y.; Song, J.; Si, Y.; Yu, J.; Ding, B. Superelastic and Photothermal RGO/Zr-Doped TiO_2 Nanofibrous Aerogels Enable the Rapid Decomposition of Chemical Warfare Agents. *Nano Lett.* **2022**, *22*, 4368–4375. [CrossRef]
16. Couzon, N.; Dhainaut, J.; Campagne, C.; Royer, S.; Loiseau, T.; Volkringer, C. Porous textile composites (PTCs) for the removal and the decomposition of chemical warfare agents (CWAs)-A review. *Coord. Chem. Rev.* **2022**, *467*, 214598. [CrossRef]
17. Kalita, P.; Paul, R.; Boruah, A.; Dao, D.Q.; Bhaumik, A.; Mondal, J. A critical review on emerging photoactive porous materials for sulfide oxidation and sulfur mustard decontamination. *Green Chem.* **2023**, *25*, 5789–5812. [CrossRef]
18. Holdren, S.; Tsyshevsky, R.; Fears, K.; Owrutsky, J.; Wu, T.; Wang, X.; Eichhorn, B.W.; Kuklja, M.M.; Zachariah, M.R. Adsorption and Destruction of the G-Series Nerve Agent Simulant Dimethyl Methylphosphonate on Zinc Oxide. *ACS Catal.* **2019**, *9*, 902–911. [CrossRef]
19. Hou, Y.J.; An, H.Y.; Zhang, Y.M.; Hu, T.; Yang, W.; Chang, S.Z. Rapid Destruction of Two Types of Chemical Warfare Agent Simulants by Hybrid Polyoxomolybdates Modified by Carboxylic Acid Ligands. *ACS Catal.* **2018**, *8*, 6062–6069. [CrossRef]
20. Snider, V.G.; Hill, C.L. Functionalized reactive polymers for the removal of chemical warfare agents: A review. *J. Hazard. Mater.* **2023**, *442*, 130015. [CrossRef]
21. Sheng, K.; Huang, X.-Q.; Wang, R.; Wang, W.-Z.; Gao, Z.-Y.; Tung, C.-H.; Sun, D. Decagram-scale synthesis of heterometallic Ag/Ti cluster as sustainable catalyst for selective oxidation of sulfides. *J. Catal.* **2023**, *417*, 185–193. [CrossRef]
22. Ma, L.; Liu, Y.; Liu, Y.; Jiang, S.; Li, P.; Hao, Y.; Shao, P.; Yin, A.; Feng, X.; Wang, B. Ferrocene-Linkage-Facilitated Charge Separation in Conjugated Microporous Polymers. *Angew. Chem.-Int. Ed.* **2019**, *58*, 4221–4226. [CrossRef] [PubMed]
23. Li, X.; Dong, J.; Liu, H.; Sun, X.; Chi, Y.; Hu, C. Recoverable amphiphilic polyoxoniobates catalyzing oxidative and hydrolytic decontamination of chemical warfare agent simulants in emulsion. *J. Hazard. Mater.* **2018**, *344*, 994–999. [CrossRef]
24. Cao, M.; Pang, R.; Wan, Q.-Y.; Han, Z.; Wang, Z.-Y.; Dong, X.-Y.; Li, S.-F.; Zang, S.-Q.; Mak, T.C.W. Porphyrinic Silver Cluster Assembled Material for Simultaneous Capture and Photocatalysis of Mustard-Gas Simulant. *J. Am. Chem. Soc.* **2019**, *141*, 14505–14509. [CrossRef] [PubMed]
25. Bobbitt, N.S.; Mendonca, M.L.; Howarth, A.J.; Islamoglu, T.; Hupp, J.T.; Farha, O.K.; Snurr, R.Q. Metal-organic frameworks for the removal of toxic industrial chemicals and chemical warfare agents. *Chem. Soc. Rev.* **2017**, *46*, 3357–3385. [CrossRef]
26. DeCoste, J.B.; Peterson, G.W. Metal-Organic Frameworks for Air Purification of Toxic Chemicals. *Chem. Rev.* **2014**, *114*, 5695–5727. [CrossRef]
27. Islamoglu, T.; Chen, Z.; Wasson, M.C.; Buru, C.T.; Kirlikovali, K.O.; Afrin, U.; Mian, M.R.; Farha, O.K. Metal-Organic Frameworks against Toxic Chemicals. *Chem. Rev.* **2020**, *120*, 8130–8160. [CrossRef]
28. Amiri, A.; Mirzaei, M. *Metal–Organic Frameworks in Analytical Chemistry*; The Royal Society of Chemistry: London, UK, 2023.
29. Khajavian, R.; Mirzaei, M.; Alizadeh, H. Current status and future prospects of metal-organic frameworks at the interface of dye-sensitized solar cells. *Dalton Trans.* **2020**, *49*, 13936–13947. [CrossRef]
30. Nazari, M.; Saljooghi, A.S.; Ramezani, M.; Alibolandi, M.; Mirzaei, M. Current status and future prospects of nanoscale metal-organic frameworks in bioimaging. *J. Mater. Chem. B* **2022**, *10*, 8824–8851. [CrossRef]
31. Bazargan, M.; Ghaemi, F.; Amiri, A.; Mirzaei, M. Metal-organic framework-based sorbents in analytical sample preparation. *Coord. Chem. Rev.* **2021**, *445*, 214107. [CrossRef]
32. Abdar, A.; Amiri, A.; Mirzaei, M. Electrospun mesh pattern of polyvinyl alcohol/zirconium-based metal-organic framework nanocomposite as a sorbent for extraction of phthalate esters. *J. Chromatography. A* **2023**, *1707*, 464295. [CrossRef]
33. Bazargan, M.; Mirzaei, M.; Amiri, A.; Ritchie, C. Efficient dispersive micro solid-phase extraction of antidepressant drugs by a robust molybdenum based coordination polymer. *Microchim. Acta* **2021**, *188*, 108. [CrossRef]
34. Abdar, A.; Amiri, A.; Mirzaei, M. Semi-automated solid-phase extraction of polycyclic aromatic hydrocarbons based on stainless steel meshes coated with metal-organic framework/graphene oxide. *Microchem. J.* **2022**, *177*, 107269. [CrossRef]
35. Hassanpoor, A.; Mirzaei, M.; Shahrak, M.N.; Majcher, A.M. Developing a magnetic metal organic framework of copper bearing a mixed azido/butane-1,4-dicarboxylate bridge: Magnetic and gas adsorption properties. *Dalton Trans.* **2018**, *47*, 13849–13860. [CrossRef]
36. Liu, Y.; Howarth, A.J.; Vermeulen, N.A.; Moon, S.-Y.; Hupp, J.T.; Farha, O.K. Catalytic degradation of chemical warfare agents and their simulants by metal-organic frameworks. *Coord. Chem. Rev.* **2017**, *346*, 101–111. [CrossRef]
37. Barton, H.F.; Jamir, J.D.; Davis, A.K.; Peterson, G.W.; Parsons, G.N. Doubly Protective MOF-Photo-Fabrics: Facile Template-Free Synthesis of PCN-222-Textiles Enables Rapid Hydrolysis, Photo-Hydrolysis and Selective Oxidation of Multiple Chemical Warfare Agents and Simulants. *Chem.-A Eur. J.* **2021**, *27*, 1465–1472. [CrossRef]
38. Zhao, H.; Tao, C.-a.; Zhao, S.; Zou, X.; Wang, F.; Wang, J. Porphyrin-Moiety-Functionalized Metal-Organic Layers Exhibiting Catalytic Capabilities for Detoxifying Nerve Agent and Blister Agent Simulants. *ACS Appl. Mater. Interfaces* **2023**, *15*, 3297–3306. [CrossRef] [PubMed]

39. Zhuang, Z.; Mai, Z.; Wang, T.; Liu, D. Strategies for conversion between metal-organic frameworks and gels. *Coord. Chem. Rev.* **2020**, *421*, 213461. [CrossRef]
40. Hou, J.; Sapnik, A.F.; Bennett, T.D. Metal-organic framework gels and monoliths. *Chem. Sci.* **2020**, *11*, 310–323. [CrossRef]
41. Chattopadhyay, P.K.; Singha, N.R. MOF and derived materials as aerogels: Structure, property, and performance relations. *Coord. Chem. Rev.* **2021**, *446*, 214125. [CrossRef]
42. Ma, S.; Xu, J.; Sohrabi, S.; Zhang, J. Metal-organic gels and their derived materials for electrochemical applications. *J. Mater. Chem. A* **2023**, *11*, 11572–11606. [CrossRef]
43. Wychowaniec, J.K.; Saini, H.; Scheibe, B.; Dubal, D.P.; Schneemann, A.; Jayaramulu, K. Hierarchical porous metal-organic gels and derived materials: From fundamentals to potential applications. *Chem. Soc. Rev.* **2022**, *51*, 9068–9126. [CrossRef]
44. Liu, G.; Li, S.; Shi, C.; Huo, M.; Lin, Y. Progress in Research and Application of Metal-Organic Gels: A Review. *Nanomaterials* **2023**, *13*, 1178. [CrossRef] [PubMed]
45. Liu, Q.; Li, S.; Yu, H.; Zeng, F.; Li, X.; Su, Z. Covalently crosslinked zirconium-based metal-organic framework aerogel monolith with ultralow-density and highly efficient Pb(II) removal. *J. Colloid Interface Sci.* **2020**, *561*, 211–219. [CrossRef]
46. Zheng, X.; Rehman, S.; Zhang, P. Room temperature synthesis of monolithic MIL-100(Fe) in aqueous solution for energy-efficient removal and recovery of aromatic volatile organic compounds. *J. Hazard. Mater.* **2023**, *442*, 129998. [CrossRef]
47. Li, L.; Xiang, S.; Cao, S.; Zhang, J.; Ouyang, G.; Chen, L.; Su, C.-Y. A synthetic route to ultralight hierarchically micro/mesoporous Al(III)-carboxylate metal-organic aerogels. *Nat. Commun.* **2013**, *4*, 1774–1779. [CrossRef]
48. Li, H.; Zhu, Y.; Zhang, J.; Chi, Z.; Chen, L.; Su, C.-Y. Luminescent metal-organic gels with tetraphenylethylene moieties: Porosity and aggregation-induced emission. *RSC Adv.* **2013**, *3*, 16340–16344. [CrossRef]
49. Wang, M.; Day, S.; Wu, Z.; Wan, X.; Ye, X.; Cheng, B. A new type of porous Zn (II) metal-organic gel designed for effective adsorption to methyl orange dye. *Colloids Surf. A-Physicochem. Eng. Asp.* **2021**, *628*, 127335. [CrossRef]
50. Tian, T.; Zeng, Z.; Vulpe, D.; Casco, M.E.; Divitini, G.; Midgley, P.A.; Silvestre-Albero, J.; Tan, J.-C.; Moghadam, P.Z.; Fairen-Jimenez, D. A sol–gel monolithic metal-organic framework with enhanced methane uptake. *Nat. Mater.* **2018**, *17*, 174–179. [CrossRef] [PubMed]
51. Tan, X.; Chen, X.; Zhang, J.; Su, C.-Y. Luminescent coordination polymer gels based on rigid terpyridyl phosphine and Ag(I). *Dalton Trans.* **2012**, *41*, 3616–3619. [CrossRef]
52. Keum, Y.; Kim, B.; Byun, A.; Park, J. Synthesis and Photocatalytic Properties of Titanium-Porphyrinic Aerogels. *Angew. Chem.-Int. Ed.* **2020**, *59*, 21591–21596. [CrossRef]
53. Qin, Z.-S.; Dong, W.-W.; Zhao, J.; Wu, Y.-P.; Zhang, Q.; Li, D.-S. A water-stable Tb(III)-based metal-organic gel (MOG) for detection of antibiotics and explosives. *Inorg. Chem. Front.* **2018**, *5*, 120–126. [CrossRef]
54. Ragon, F.; Campo, B.; Yang, Q.; Martineau, C.; Wiersum, A.D.; Lago, A.; Guillerm, V.; Hemsley, C.; Eubank, J.F.; Vishnuvarthan, M.; et al. Acid-functionalized UiO-66(Zr) MOFs and their evolution after intra-framework cross-linking: Structural features and sorption properties. *J. Mater. Chem. A* **2015**, *3*, 3294–3309. [CrossRef]
55. Ragon, F.; Horcajada, P.; Chevreau, H.; Hwang, Y.K.; Lee, U.H.; Miller, S.R.; Devic, T.; Chang, J.S.; Serre, C. In Situ Energy-Dispersive X-ray Diffraction for the Synthesis Optimization and Scale-up of the Porous Zirconium Terephthalate UiO-66. *Inorg. Chem.* **2014**, *53*, 2491–2500. [CrossRef] [PubMed]
56. Yang, Q.Y.; Vaesen, S.; Ragon, F.; Wiersum, A.D.; Wu, D.; Lago, A.; Devic, T.; Martineau, C.; Taulelle, F.; Llewellyn, P.L.; et al. A Water Stable Metal-Organic Framework with Optimal Features for CO_2 Capture. *Angew. Chem.-Int. Ed.* **2013**, *52*, 10316–10320. [CrossRef]
57. Liu, L.P.; Zhang, J.Y.; Fang, H.B.; Chen, L.P.; Su, C.Y. Metal-Organic Gel Material Based on UiO-66-NH_2 Nanoparticles for Improved Adsorption and Conversion of Carbon Dioxide. *Chem.-Asian J.* **2016**, *11*, 2278–2283. [CrossRef]
58. Bueken, B.; Van Velthoven, N.; Willhammar, T.; Stassin, T.; Stassen, I.; Keen, D.A.; Baron, G.V.; Denayer, J.F.M.; Ameloot, R.; Bals, S.; et al. Gel-based morphological design of zirconium metal–organic frameworks. *Chem. Sci.* **2017**, *8*, 3939–3948. [CrossRef]
59. Somjit, V.; Thinsoongnoen, P.; Sriphumrat, K.; Pimu, S.; Arayachukiat, S.; Kongpatpanich, K. Metal-Organic Framework Aerogel for Full pH Range Operation and Trace Adsorption of Arsenic in Water. *ACS Appl. Mater. Interfaces* **2022**, *14*, 40005–40013. [CrossRef] [PubMed]
60. Lohe, M.R.; Rose, M.; Kaskel, S. Metal–organic framework (MOF) aerogels with high micro- and macroporosity. *Chem. Commun.* **2009**, 6056–6058. [CrossRef]
61. Hu, Y.; Fan, Y.; Huang, Z.; Song, C.; Li, G. In situ fabrication of metal–organic hybrid gels in a capillary for online enrichment of trace analytes in aqueous samples. *Chem. Commun.* **2012**, *48*, 3966–3968. [CrossRef]
62. Xiang, S.; Li, L.; Zhang, J.; Tan, X.; Cui, H.; Shi, J.; Hu, Y.; Chen, L.; Su, C.-Y.; James, S.L. Porous organic–inorganic hybrid aerogels based on Cr^{3+}/Fe^{3+} and rigid bridging carboxylates. *J. Mater. Chem.* **2012**, *22*, 1862–1867. [CrossRef]
63. He, L.; Peng, Z.W.; Jiang, Z.W.; Tang, X.Q.; Huang, C.Z.; Li, Y.F. Novel Iron(III)-Based Metal-Organic Gels with Superior Catalytic Performance toward Luminol Chemiluminescence. *ACS Appl. Mater. Interfaces* **2017**, *9*, 31834–31840. [CrossRef]
64. Fan, J.; Li, L.; Rao, H.-S.; Yang, Q.-L.; Zhang, J.; Chen, H.-Y.; Chen, L.; Kuang, D.-B.; Su, C.-Y. A novel metal-organic gel based electrolyte for efficient quasi-solid-state dye-sensitized solar cells. *J. Mater. Chem. A* **2014**, *2*, 15406–15413. [CrossRef]
65. Lee, J.H.; Kang, S.; Lee, J.Y.; Jaworski, J.; Jung, J.H. Instant Visual Detection of Picogram Levels of Trinitrotoluene by Using Luminescent Metal-Organic Framework Gel-Coated Filter Paper. *Chem.-A Eur. J.* **2013**, *19*, 16665–16671. [CrossRef]

66. Cheng, Y.; Yin, M.; Ren, X.; Feng, Q.; Wang, J.; Zhou, Y. A coordination polymeric gelator based on Ag(l) and 2, 7-bis(1-imidazole)fluorene: Synthesis, characterization, gelation and antibacterial properties. *Mater. Lett.* **2015**, *139*, 141–144. [CrossRef]
67. Yang, K.; Chen, L.; Duan, X.; Song, G.; Sun, J.; Chen, A.; Xie, X. Ligand-controlled bimetallic Co/Fe MOF xerogels for CO_2 photocatalytic reduction. *Ceram. Int.* **2023**, *49*, 16061–16069. [CrossRef]
68. Lu, M.; Deng, Y.; Luo, Y.; Lv, J.; Li, T.; Xu, J.; Chen, S.-W.; Wang, J. Graphene Aerogel-Metal-Organic Framework-Based Electrochemical Method for Simultaneous Detection of Multiple Heavy-Metal Ions. *Anal. Chem.* **2019**, *91*, 888–895. [CrossRef] [PubMed]
69. Ma, X.; Lou, Y.; Chen, X.-B.; Shi, Z.; Xu, Y. Multifunctional flexible composite aerogels constructed through in-situ growth of metal-organic framework nanoparticles on bacterial cellulose. *Chem. Eng. J.* **2019**, *356*, 227–235. [CrossRef]
70. Fan, Y.; Liang, H.; Jian, M.; Liu, R.; Zhang, X.; Hu, C.; Liu, H. Removal of dimethylarsinate from water by robust NU-1000 aerogels: Impact of the aerogel materials. *Chem. Eng. J.* **2023**, *455*, 140387. [CrossRef]
71. Li, Z.; Liu, C.; Frick, J.J.; Davey, A.K.; Dods, M.N.; Carraro, C.; Senesky, D.G.; Maboudian, R. Synthesis and characterization of UiO-66-NH_2 incorporated graphene aerogel composites and their utilization for absorption of organic liquids. *Carbon* **2023**, *201*, 561–567. [CrossRef]
72. Zhu, H.; Yang, X.; Cranston, E.D.; Zhu, S.P. Flexible and Porous Nanocellulose Aerogels with High Loadings of Metal-Organic-Framework Particles for Separations Applications. *Adv. Mater.* **2016**, *28*, 7652–7657. [CrossRef] [PubMed]
73. Wang, Z.; Song, L.; Wang, Y.; Zhang, X.-F.; Hao, D.; Feng, Y.; Yao, J. Lightweight UiO-66/cellulose aerogels constructed through self-crosslinking strategy for adsorption applications. *Chem. Eng. J.* **2019**, *371*, 138–144. [CrossRef]
74. Li, H.; Ye, M.; Zhang, X.; Zhang, H.; Wang, G.; Zhang, Y. Hierarchical Porous Iron Metal-Organic Gel/Bacterial Cellulose Aerogel: Ultrafast, Scalable, Room-Temperature Aqueous Synthesis, and Efficient Arsenate Removal. *ACS Appl. Mater. Interfaces* **2021**, *13*, 47684–47695. [CrossRef]
75. Liu, L.; Yan, Y.; Cai, Z.H.; Lin, S.X.; Hu, X.B. Growth-Oriented Fe-Based MOFs Synergized with Graphene Aerogels for High-Performance Supercapacitors. *Adv. Mater. Interfaces* **2018**, *5*, 1701548. [CrossRef]
76. Jiang, G.; Jia, Y.; Wang, J.; Sun, Y.; Zhou, Y.; Ruan, Y.; Xia, Y.; Xu, T.; Xie, S.; Zhang, S.; et al. Facile preparation of novel Fe-BTC@PAN nanofibrous aerogel membranes for highly efficient continuous flow degradation of organic dyes. *Sep. Purif. Technol.* **2022**, *300*, 121753. [CrossRef]
77. He, X.; Wang, L.; Sun, S.; Yang, X.; Tian, H.; Xia, Z.; Li, X.; Yan, X.; Pu, X.; Jiao, Z. Self-assembled synthesis of recyclable g-C_3N_4/NH_2-MIL-53(Fe) aerogel for enhanced photocatalytic degradation of organic pollutants. *J. Alloys Compd.* **2023**, *946*, 169391. [CrossRef]
78. Liu, Q.; Yu, H.; Zeng, F.; Li, X.; Sun, J.; Li, C.; Lin, H.; Su, Z. HKUST-1 modified ultrastability cellulose/chitosan composite aerogel for highly efficient removal of methylene blue. *Carbohydr. Polym.* **2021**, *255*, 117402. [CrossRef]
79. Nuzhdin, A.L.; Shalygin, A.S.; Artiukha, E.A.; Chibiryaev, A.M.; Bukhtiyarova, G.A.; Martyanov, O.N. HKUST-1 silica aerogel composites: Novel materials for the separation of saturated and unsaturated hydrocarbons by conventional liquid chromatography. *RSC Adv.* **2016**, *6*, 62501–62507. [CrossRef]
80. Hu, Y.; Jiang, Y.; Ni, L.; Huang, Z.; Liu, L.; Ke, Q.; Xu, H. An elastic MOF/graphene aerogel with high photothermal efficiency for rapid removal of crude oil. *J. Hazard. Mater.* **2023**, *443*, 130339. [CrossRef]
81. Rosado, A.; Borras, A.; Fraile, J.; Navarro, J.A.R.; Suarez-Garcia, F.; Stylianou, K.C.; Lopez-Periago, A.M.; Giner Planas, J.; Domingo, C.; Yazdi, A. HKUST-1 Metal-Organic Framework Nanoparticle/Graphene Oxide Nanocomposite Aerogels for CO_2 and CH_4 Adsorption and Separation. *ACS Appl. Nano Mater.* **2021**, *4*, 12712–12725. [CrossRef]
82. Ulker, Z.; Erucar, I.; Keskin, S.; Erkey, C. Novel nanostructured composites of silica aerogels with a metal organic framework. *Microporous Mesoporous Mater.* **2013**, *170*, 352–358. [CrossRef]
83. Shalygin, A.S.; Nuzhdin, A.L.; Bukhtiyarova, G.A.; Martyanov, O.N. Preparation of HKUST-1@silica aerogel composite for continuous flow catalysis. *J. Sol-Gel Sci. Technol.* **2017**, *84*, 446–452. [CrossRef]
84. Ramasubbu, V.; Alwin, S.; Mothi, E.M.; Shajan, X.S. TiO_2 aerogel-Cu-BTC metal-organic framework composites for enhanced photon absorption. *Mater. Lett.* **2017**, *197*, 236–240. [CrossRef]
85. Zhang, Z.; Hu, J.; Tian, X.; Guo, F.; Wang, C.; Zhang, J.; Jiang, M. Facile in-situ growth of metal-organic framework layer on carboxylated nanocellulose/chitosan aerogel spheres and their high-efficient adsorption and catalytic performance. *Appl. Surf. Sci.* **2022**, *599*, 153974. [CrossRef]
86. Liu, Y.; Yu, B.; Chen, X.; Li, D.; Zhou, C.; Guo, Z.-R.; Xu, W.; Yang, S.; Zhang, J. One-Droplet Synthesis of Polysaccharide/Metal-Organic Framework Aerogels for Gas Adsorption. *Acs Appl. Polym. Mater.* **2023**, *5*, 4327–4332. [CrossRef]
87. Li, D.; Tian, X.; Wang, Z.; Guan, Z.; Li, X.; Qiao, H.; Ke, H.; Luo, L.; Wei, Q. Multifunctional adsorbent based on metal-organic framework modified bacterial cellulose/chitosan composite aerogel for high efficient removal of heavy metal ion and organic pollutant. *Chem. Eng. J.* **2020**, *383*, 123127. [CrossRef]
88. Zhou, Q.; Jin, B.; Zhao, P.; Chu, S.; Peng, R. rGO/CNQDs/ZIF-67 composite aerogel for efficient extraction of uranium in wastewater. *Chem. Eng. J.* **2021**, *419*, 129622. [CrossRef]
89. Qu, W.; Wang, Z.; Wang, X.; Wang, Z.; Yu, D.; Ji, D. High-hydrophobic ZIF-67@PLA honeycomb aerogel for efficient oil-water separation. *Colloids Surf. A-Physicochem. Eng. Asp.* **2023**, *658*, 130768. [CrossRef]
90. Song, W.; Zhu, M.; Zhu, Y.; Zhao, Y.; Yang, M.; Miao, Z.; Ren, H.; Ma, Q.; Qian, L. Zeolitic imidazolate framework-67 functionalized cellulose hybrid aerogel: An environmentally friendly candidate for dye removal. *Cellulose* **2020**, *27*, 2161–2172. [CrossRef]

91. Zhu, G.; Zhang, C.; Li, K.; Zhang, X.; Deng, S. Enhanced removal of Cu(II) ions from aqueous solution by in-situ synthesis of zeolitic imidazolate framework-67@wood aerogel composite adsorbent. *Wood Mater. Sci. Eng.* **2023**, *18*, 1–11. [CrossRef]
92. Chen, G.; He, S.; Shi, G.; Ma, Y.; Ruan, C.; Jin, X.; Chen, Q.; Liu, X.; Dai, H.; Chen, X.; et al. In-situ immobilization of ZIF 67 on wood aerogel for effective removal of tetracycline from water. *Chem. Eng. J.* **2021**, *423*, 130184. [CrossRef]
93. Yao, L.; Gu, Q.; Yu, X. Three-Dimensional MOFs@MXene Aerogel Composite Derived MXene Threaded Hollow Carbon Confined CoS Nanoparticles toward Advanced Alkali-Ion Batteries. *ACS Nano* **2021**, *15*, 3228–3240. [CrossRef]
94. Yang, Q.X.; Lu, R.; Ren, S.S.; Chen, C.T.; Chen, Z.J.; Yang, X.Y. Three dimensional reduced graphene oxide/ZIF-67 aerogel: Effective removal cationic and anionic dyes from water. *Chem. Eng. J.* **2018**, *348*, 202–211. [CrossRef]
95. Zhang, X.Q.; Liang, Q.L.; Han, Q.; Wan, W.; Ding, M.Y. Metal-organic frameworks@graphene hybrid aerogels for solid-phase extraction of nonsteroidal anti-inflammatory drugs and selective enrichment of proteins. *Analyst* **2016**, *141*, 4219–4226. [CrossRef] [PubMed]
96. Sun, T.; Hao, S.; Fan, R.; Qin, M.; Chen, W.; Wang, P.; Yang, Y. Hydrophobicity-Adjustable MOF Constructs Superhydrophobic MOF-rGO Aerogel for Efficient Oil-Water Separation. *Acs Appl. Mater. Interfaces* **2020**, *12*, 56435–56444. [CrossRef]
97. Ni, Z.; Jerrell, J.P.; Cadwallader, K.R.; Masel, R.I. Metal−Organic Frameworks as Adsorbents for Trapping and Preconcentration of Organic Phosphonates. *Anal. Chem.* **2007**, *79*, 1290–1293. [CrossRef] [PubMed]
98. Son, Y.-R.; Ryu, S.G.; Kim, H.S. Rapid adsorption and removal of sulfur mustard with zeolitic imidazolate frameworks ZIF-8 and ZIF-67. *Microporous Mesoporous Mater.* **2020**, *293*, 109819. [CrossRef]
99. Emelianova, A.; Reed, A.; Basharova, E.A.; Kolesnikov, A.L.; Gor, G.Y. Closer Look at Adsorption of Sarin and Simulants on Metal-Organic Frameworks. *ACS Appl. Mater. Interfaces* **2023**, *15*, 18559–18567. [CrossRef]
100. Plonka, A.M.; Wang, Q.; Gordon, W.O.; Balboa, A.; Troya, D.; Guo, W.; Sharp, C.H.; Senanayake, S.D.; Morris, J.R.; Hill, C.L.; et al. In Situ Probes of Capture and Decomposition of Chemical Warfare Agent Simulants by Zr-Based Metal Organic Frameworks. *J. Am. Chem. Soc.* **2017**, *139*, 599–602. [CrossRef]
101. Asha, P.; Sinha, M.; Mandal, S. Effective removal of chemical warfare agent simulants using water stable metal-organic frameworks: Mechanistic study and structure-property correlation. *RSC Adv.* **2017**, *7*, 6691–6696. [CrossRef]
102. Son, F.; Wasson, M.C.; Islamoglu, T.; Chen, Z.; Gong, X.; Hanna, S.L.; Lyu, J.; Wang, X.; Idrees, K.B.; Mahle, J.J.; et al. Uncovering the Role of Metal-Organic Framework Topology on the Capture and Reactivity of Chemical Warfare Agents. *Chem. Mater.* **2020**, *32*, 4609–4617. [CrossRef]
103. Zhou, C.; Yuan, B.; Zhang, S.; Yang, G.; Lu, L.; Li, H.; Tao, C.-a. Ultrafast Degradation and High Adsorption Capability of a Sulfur Mustard Simulant under Ambient Conditions Using Granular UiO-66-NH$_2$ Metal-Organic Gels. *Acs Appl. Mater. Interfaces* **2022**, *14*, 23383–23391. [CrossRef]
104. Yan, Z.; Liu, X.; Ding, B.; Yu, J.; Si, Y. Interfacial engineered superelastic metal-organic framework aerogels with van-der-Waals barrier channels for nerve agents decomposition. *Nat. Commun.* **2023**, *14*, 2116. [CrossRef] [PubMed]
105. Zhou, C.; Zhang, S.; Pan, H.; Yang, G.; Wang, L.; Tao, C.-a.; Li, H. Synthesis of macroscopic monolithic metal-organic gels for ultra-fast destruction of chemical warfare agents. *RSC Adv.* **2021**, *11*, 22125–22130. [CrossRef] [PubMed]
106. Shena, C.; Mao, Z.; Xua, H.; Zhang, L.; Zhonga, Y.; Wang, B.; Feng, X.; Taod, C.-a.; Suia, X. Catalytic MOF-loaded cellulose sponge for rapid degradation of chemical warfare agents simulant. *Carbohydr. Polym.* **2019**, *213*, 184–191. [CrossRef] [PubMed]
107. Seo, J.Y.; Song, Y.; Lee, J.-H.; Kim, H.; Cho, S.; Baek, K.-Y. Robust Nanocellulose/Metal-Organic Framework Aerogel Composites: Superior Performance for Static and Continuous Disposal of Chemical Warfare Agent Simulants. *ACS Appl. Mater. Interfaces* **2021**, *13*, 33516–33523. [CrossRef]
108. Ma, K.; Wasson, M.C.; Wang, X.; Zhang, X.; Idrees, K.B.; Chen, Z.; Wu, Y.; Lee, S.-J.; Cao, R.; Chen, Y.; et al. Near-instantaneous catalytic hydrolysis of organophosphorus nerve agents with zirconium-based MOF/hydrogel composites. *Chem Catal.* **2021**, *1*, 721–733. [CrossRef]
109. Jiang, N.; Liu, H.; Zhao, G.; Li, H.; Yang, S.; Xu, X.; Zhuang, X.; Cheng, B. Aramid nanofibers supported metal-organic framework aerogel for protection of chemical warfare agent. *J. Colloid Interface Sci.* **2023**, *640*, 192–198. [CrossRef]
110. Su, H.; Huang, P.; Wu, F.-Y. Visualizing the degradation of nerve agent simulants using functionalized Zr-based MOFs from solution to hydrogels. *Chem. Commun.* **2021**, *57*, 11681–11684. [CrossRef]
111. Wang, X.; Yang, J.; Zhang, M.; Hu, Q.; Li, B.-X.; Qu, Y.; Yu, Z.-Z.; Yang, D. Spontaneously Super-Hygroscopic MOF-Gel Microreactors for Efficient Detoxification of Nerve Agent Simulant in Atmospheric Environments. *Appl. Catal. B Environ.* **2023**, *328*, 122516. [CrossRef]
112. Ma, K.; Cheung, Y.H.; Kirlikovali, K.O.; Xie, H.; Idrees, K.B.; Wang, X.; Islamoglu, T.; Xin, J.H.; Farha, O.K. Fibrous Zr-MOF Nanozyme Aerogels with Macro-Nanoporous Structure for Enhanced Catalytic Hydrolysis of Organophosphate Toxins. *Adv. Mater.* **2023**, *35*, 2300951. [CrossRef]
113. Cheung, Y.H.; Ma, K.; Wasson, M.C.; Wang, X.; Idrees, K.B.; Islamoglu, T.; Mahle, J.; Peterson, G.W.; Xin, J.H.; Farha, O.K. Environmentally Benign Biosynthesis of Hierarchical MOFBacterial Cellulose Composite Sponge for Nerve Agent Protection. *Angew. Chem.* **2022**, *61*, e202202207. [CrossRef] [PubMed]
114. Shen, A.; Hao, X.; Zhang, L.; Du, M.; Li, M.; Zhao, Y.; Li, H.; Hou, L.; Duan, R.; Yang, Y. Solid-state degradation and visual detection of the nerve agent GB by SA@UiO-66-NH$_2$@PAMAM hydrogel. *Polym. Chem.* **2022**, *13*, 6205–6212. [CrossRef]

115. Kalinovskyy, Y.; Wright, A.J.; Hiscock, J.R.; Watts, T.D.; Williams, R.L.; Cooper, N.J.; Main, M.J.; Holder, S.J.; Blight, B.A. Swell and Destroy A Metal–Organic Framework-Containing Polymer Sponge That Immobilizes and Catalytically Degrades Nerve Agents. *ACS Appl. Mater. Interfaces* **2020**, *12*, 8634–8641. [CrossRef] [PubMed]
116. Long, N.H.; Park, H.-w.; Chae, G.-s.; Lee, J.H.; Bae, S.W.; Shin, S. Preparation of Peelable Coating Films with a Metal Organic Framework (UiO-66) and Self-Crosslinkable Polyurethane for the Decomposition of Methyl Paraoxon. *Polymers* **2019**, *11*, 1298. [CrossRef] [PubMed]

Disclaimer/Publisher's Note: The statements, opinions and data contained in all publications are solely those of the individual author(s) and contributor(s) and not of MDPI and/or the editor(s). MDPI and/or the editor(s) disclaim responsibility for any injury to people or property resulting from any ideas, methods, instructions or products referred to in the content.

Article

Extraction of Palladium(II) with a Magnetic Sorbent Based on Polyvinyl Alcohol Gel, Metallic Iron, and an Environmentally Friendly Polydentate Phosphazene-Containing Extractant

Pavel Yudaev [1], Irina Butorova [1], Gennady Stepanov [2], and Evgeniy Chistyakov [1,*]

[1] Mendeleev University of Chemical Technology of Russia, Miusskaya Sq. 9, 125047 Moscow, Russia
[2] State Scientific Center of the Russian Federation, Institute of Chemistry and Technology of Organoelement Compounds, 111123 Moscow, Russia
* Correspondence: ewgenijj@rambler.ru

Citation: Yudaev, P.; Butorova, I.; Stepanov, G.; Chistyakov, E. Extraction of Palladium(II) with a Magnetic Sorbent Based on Polyvinyl Alcohol Gel, Metallic Iron, and an Environmentally Friendly Polydentate Phosphazene-Containing Extractant. *Gels* **2022**, *8*, 492. https://doi.org/10.3390/gels8080492

Academic Editors: Daxin Liang, Ting Dong, Yudong Li and Caichao Wan

Received: 12 July 2022
Accepted: 6 August 2022
Published: 8 August 2022

Publisher's Note: MDPI stays neutral with regard to jurisdictional claims in published maps and institutional affiliations.

Copyright: © 2022 by the authors. Licensee MDPI, Basel, Switzerland. This article is an open access article distributed under the terms and conditions of the Creative Commons Attribution (CC BY) license (https://creativecommons.org/licenses/by/4.0/).

Abstract: In this work, a highly efficient and environmentally friendly method for extracting palladium from hydrochloric acid media was developed. The method uses a magnetic sorbent carrying an organophosphorus extractant, which is not washed from the sorbent into the aqueous phase. The extractant was characterized by ^{1}H, ^{13}C, and ^{31}P NMR spectroscopy and MALDI TOF mass spectrometry, and the palladium complex based on it was characterized by IR spectroscopy. According to an in vitro microbiological study, the extractant was non-toxic to soil microflora. It was established that the water uptake and saturation magnetization of the magnetic sorbent were sufficient for use in sorption processes. The sorption efficiency of palladium(II) with the developed sorbent can reach 71% in one cycle. After treatment of the spent sorbent with 5 M hydrochloric acid, palladium was completely extracted from the sorbent. The new sorbent is proposed for the extraction of palladium from hydrochloric acid media obtained by the leaching of electronic waste.

Keywords: phosphazene; extraction; stripping; sorption; magnetic sorbent; palladium; polyvinyl alcohol; carbonyl iron; green chemistry

1. Introduction

Palladium is a noble metal of the platinum group and is widely used in various fields of science and technology. For example, palladium is used in electronics as part of multilayer ceramic capacitors of printed circuit boards [1], in the automotive industry in catalytic converters in cars [2], as well as in jewelry [3], chemical catalysis [4,5], and hydrogen energy production [6]. However, the content of palladium in natural deposits of platinum group metals is extremely low. In particular, the average palladium content in low-sulfide platinum–palladium ores from the Kievey and North Kamennik deposits is 3.32 ppm, and that in the Fedorova Tundra deposit is 1.20 ppm [7]. Therefore, the search for methods to recover palladium from industrial waste and secondary resources, such as spent automotive catalysts or waste electrical and electronic equipment (WEEE), is a promising area of research.

To date, there is no highly efficient, selective, and simple method for extracting palladium from WEEE. Pyrometallurgical processes require very high temperatures (over 1500 °C) and generate a large amount of waste and atmospheric emissions [8]. During the hydrometallurgical treatment of WEEE, a leaching solution of WEEE is prepared in concentrated hydrochloric acid in the presence of oxidizing agents, for example, aqua regia [9], which is followed by the separation and extraction of palladium(II), platinum(IV), gold(III), silver(I), copper(II), tin(II), lead(II), nickel(II), iron(II), and zinc(II) using electrodeposition, extraction, ion exchange, membrane separation, and other techniques.

Solvent extraction with organic compounds (extractants) is the most promising method for the recovery of metals from industrial waste compared to other methods due to high productivity, economic feasibility, high speed, and simple process design [10–21]. Organophosphorus extractants are becoming increasingly important in hydrometallurgical processes [22] due to their high selectivity, good solubility of both extractants and their metal complexes in nonpolar solvents, high degree of stripping, chemical stability, acid resistance, and low cost. However, a significant disadvantage of liquid organophosphorus extractants is their high toxicity. The introduction of organophosphorus extractants into a sorbent matrix would make it possible to avoid their negative impact on the environment.

To date, the processes of metal sorption by mineral (silica gel, zeolites, bentonite, activated carbon, activated alumina, and so on) and polymeric sorbents have been studied. Mineral sorbents weakly interact with metal ions and are difficult to separate from the aqueous phase and regenerate [23]. Among polymeric sorbents for metals, the most widely used are polymers containing surface hydrophilic groups capable of coordinating metals, for example, chitosan. The benefits of chitosan as a polymeric sorbent include the lack of toxicity, biocompatibility, high density of functional groups on its surface, and ease of functionalization. The drawbacks of chitosan are its low sorption capacity, sensitivity to the pH of the aqueous phase, limited reuse, poor mechanical properties, and low stability in acidic media [24]. To increase the sorption capacity, chitosan is modified with compounds containing donor nitrogen, oxygen, and sulfur atoms, which makes the process more expensive [25–27].

To provide for easy separation of the sorbent from the aqueous phase using a permanent magnet, magnetic particles, for example, magnetite nanoparticles, are added to the polymer along with the extractant [28–30]. However, the agglomeration of magnetite nanoparticles in a polymer matrix reduces the magnetic properties [31], so it is necessary to use a finely dispersed magnetic carrier.

To improve the efficiency of palladium extraction, chelate compounds containing at least two donor atoms are used. However, many chelate complexes are soluble in water, some are toxic, and due to low content of coordination sites, they poorly bind metals. Of interest are polyfunctional compounds, aryloxycyclophosphazenes, since they are biocompatible, resistant to hydrolysis in an acidic environment, and insoluble in water. The replacement of chlorine atoms in the starting chlorophosphazene produces various structures capable of metal coordination [32,33].

Here, it is proposed to use a phosphazene-containing aminophosphonate with six coordination sites as an extractant. It is planned that an extractant introduced into a magnetic gel matrix based on polyvinyl alcohol and acid-resistant carbonyl iron will effectively and selectively extract palladium(II) from hydrochloric acid media obtained by leaching WEEE and electrical capacitors.

2. Results and Discussion

The extractant was synthesized by the Pudovik reaction from hexakis-[4-{(N-allylimino)methyl}-phenoxy]-cyclotriphosphazene (APP) and diethyl phosphite in dioxane, as shown in Figure 1.

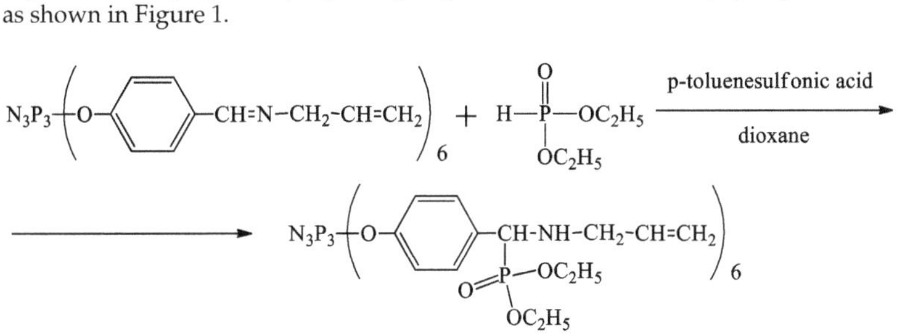

Figure 1. Scheme of extractant synthesis.

The product is a light-yellow viscous mass, soluble in most organic solvents and insoluble in water, which is important in the extraction of metals from aqueous media.

From the ^{31}P NMR spectra, it can be seen that the signal of the phosphorus nuclei of the phosphazene ring of the extractant (Figure 2B) is shifted relative to the phosphorus signal of the original APP (Figure 2A) by 0.51 ppm. This is due to a decrease in the mesomeric effect acting on the phosphorus atoms due to the disruption of conjugation between the benzene rings and the azomethine nitrogen atoms, since azomethine groups have been converted to aminophosphonate groups. The formation of aminophosphonate groups is also confirmed by the presence of a phosphorus signal at 23.21 ppm. In this case, the integrated intensity ratio of the phosphorus signals of the phosphazene ring and aminophosphonate groups is approximately 1:2, which indirectly confirms the completeness of the Pudovik reaction.

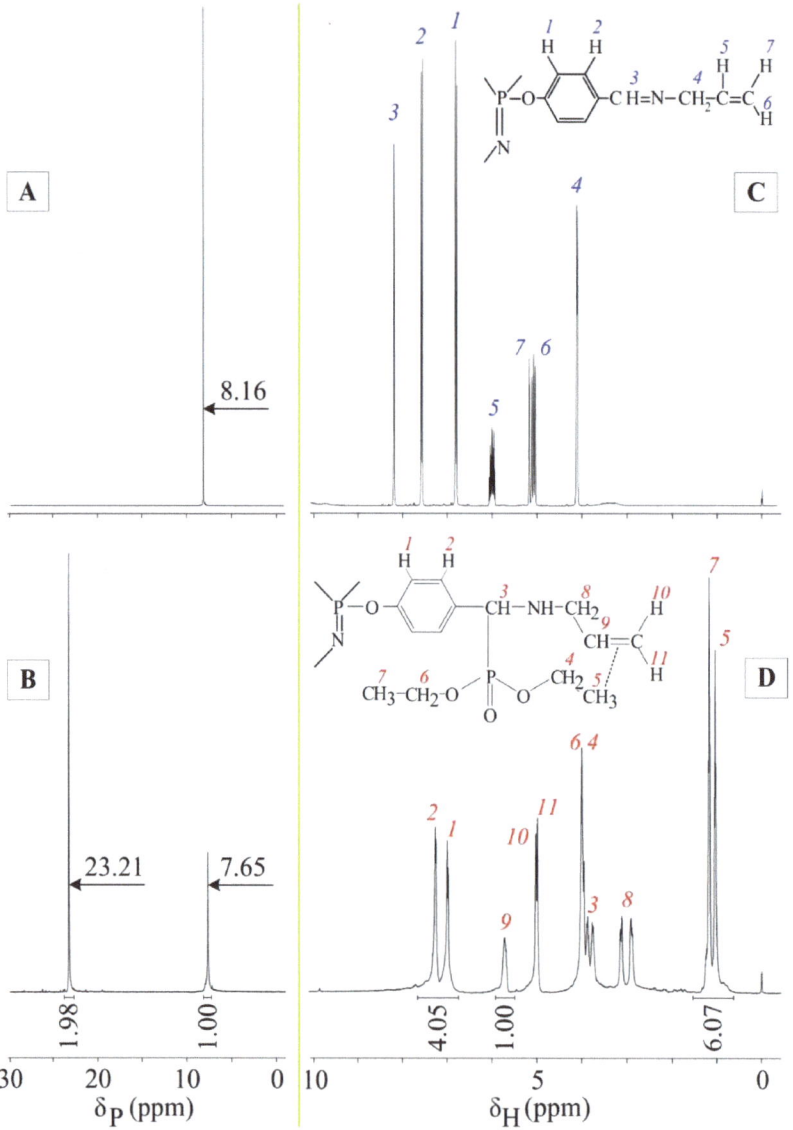

Figure 2. ^{31}P NMR spectra: APP (**A**), extractant (**B**), and ^{1}H NMR spectra: APP (**C**) and extractant (**D**).

For a more accurate assessment of the conversion of azomethine groups to aminophosphonate groups, ^{1}H NMR analysis was performed. It can be seen in the spectrum of the extractant (Figure 2D) that the proton signals of the azomethine groups at 8.2 ppm have completely disappeared (Figure 2C), while signals for the protons of the aminophosphonate CH groups (proton 3, Figure 2D) have appeared at 3.7–4 ppm. The integrated intensity ratio of the proton signals of the methylene groups in allyl radicals to the proton signals of the benzene ring is 1:2, which indicates the absence of side reactions involving azomethine groups during the synthesis. In addition, the number of protons of methyl groups in phosphonate radicals fully corresponds to the theoretical content, which confirms the formation of the target product. It is worth noting that the methyl proton signals form two triplets (0.98 and 1.12 ppm, Figure 2D) instead of one. The upfield shift of proton 5 (Figure 2D) relative to proton 7 is due to the contribution of the magnetic anisotropy of the double bond of allyl radicals. The signal shift of the methylene groups in the ethylphosphonate moieties in the ^{1}H NMR spectrum is slight, but it is clearly visible in the carbon spectrum (carbons 6 and 8, Figure 3B). On the contrary, the difference between the carbon signals of the methyl groups is less pronounced (atoms 7 and 9). The upfield shift of the carbon 5 signal from 162 ppm (Figure 3A) to 63 ppm (Figure 3B) also indicates the complete conversion of azomethine groups to aminophosphonate groups.

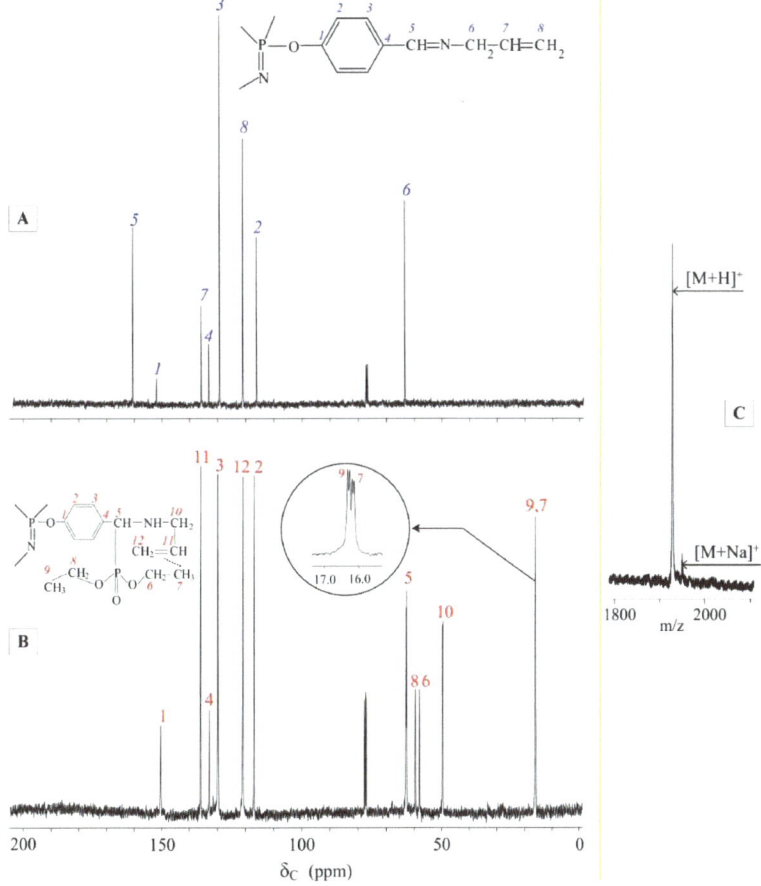

Figure 3. ^{13}C NMR spectra of APP (**A**) and extractant (**B**) and MALDI-TOF mass spectrum of extractant (**C**).

The MALDI-TOF mass spectrum of the extractant (Figure 3C) shows a molecular ion peak with a solvated matrix proton in the 1925 [M + H]$^+$ region, corresponding to the mass of the target compound, and a peak for sodium ion-solvated extractant at 1947 [M + Na]$^+$.

It was found by DSC that the extractant is amorphous with a glass transition temperature in the range of −5 to +5 °C (Figure 4).

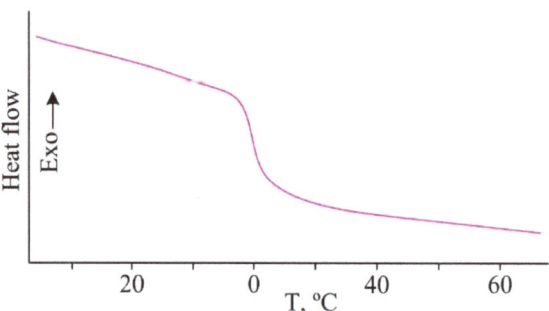

Figure 4. DSC curve of the extractant.

The extractant was tested in palladium extraction from chloride media. As a result of extraction, the corresponding complex was obtained, which turned out to be insoluble. A comparison of the IR spectra of the extractant (Figure 5A) and the palladium complex (Figure 5B) showed that palladium is coordinated by phosphoryl groups, as evidenced by a change in the shape of the P=O stretching band at 935 cm^{-1}. It was also assumed that the double bonds of allyl groups would additionally be involved in coordination; however, the vibrational band of the double bonds of allyl groups at 1501 cm^{-1} remained unchanged.

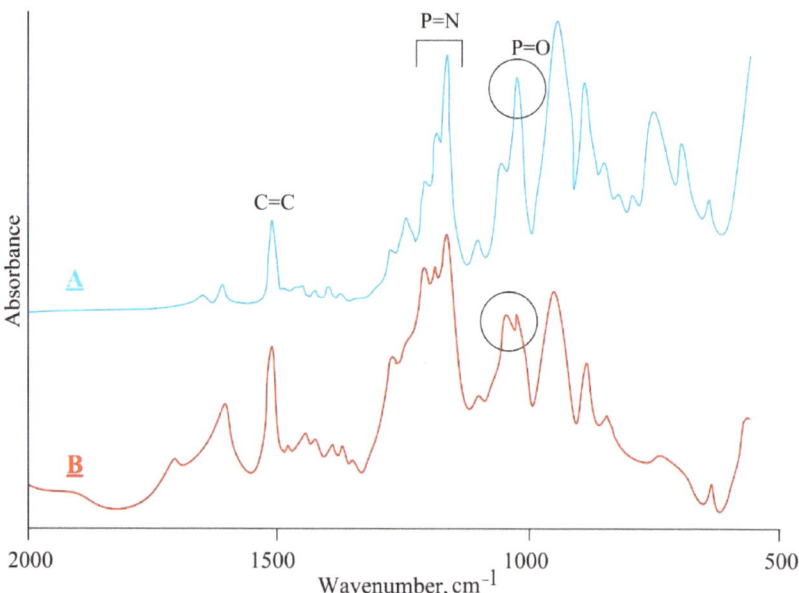

Figure 5. IR spectra of the extractant (**A**) and its palladium complex (**B**).

According to the elemental analysis of the palladium complex (Table 1), there is approximately one molecule of palladium chloride per two phosphoryl groups of the extractant. This follows from the atomic ratio of phosphorus and palladium in the obtained complex, which is 3.42:1.17; i.e., it is close to the theoretical elemental ratio for the complex with the indicated structure (3.45:1.15).

Table 1. Elemental composition of the palladium complex of the extractant, %.

Chemical Element	Actual Content		Theoretical Content	
	Weight	Atomic	Weight	Atomic
C	41.01	32.24	41.05	32.17
N	5.15	3.47	5.13	3.45
O	15.7	9.25	15.63	9.20
P	11.23	3.42	11.34	3.45
Cl	8.65	2.30	8.67	2.30
Pd	13.1	1.17	13.00	1.15
H	5.16	48.15	5.18	48.28

Since palladium(II) exists in aqueous hydrochloric acid as chloride complexes $PdCl_4^{2-}$ [34], in the case of aminophosphonates, the extraction of palladium(II) at high acid concentrations proceeds according to the outer-sphere mechanism via protonation of the aminophosphonate nitrogen atom to give the complex $\{[PdCl_4]^{2-} \cdot [HR]^+{}_x\}$, where R is the coordination sites. With a decrease in the concentration of hydrochloric acid, the coordination mainly follows the intra-sphere mechanism involving the chelation of palladium with phosphoryl groups. Moreover, in the case of synthesized phosphazene, palladium can be chelated by phosphoryl groups located both at the same and at different phosphorus atoms of the phosphazene ring. As a result, the formation of structurally diverse chelate complexes is possible (Figure 6).

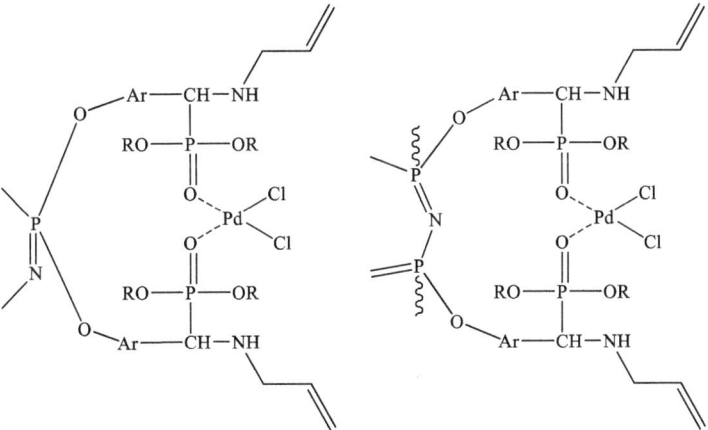

Figure 6. Palladium(II)-extractant chelate complexes with intra-sphere coordination of palladium (R = C_2H_5, Ar = p-C_6H_4).

To assess the effect of the extractant on the environment, which is important when it enters wastewater and soil, microbiological studies were carried out. It was found that when the extractant is applied to the surface of a nutrient medium inoculated with soil microflora, the extractant does not have an inhibitory effect on it. Conversely, a stimulating effect was noted compared to the control group, as evidenced by the increase in the number of microorganisms in the sample treated with the extractant (Table 2).

Table 2. Study of the effect of the extractant on soil microflora.

	Optical Density, Units	Number of Microorganisms, CFU mL^{-1}	
		Bacteria	Yeast and Fungi
Control	9.03	8.0×10^9	6.0×10^5
Sample	9.29	1.29×10^{10}	2.6×10^6

The stimulating effect of the extractant is probably due to the destruction of the phosphazene ring under the action of microbial enzymes and the formation of ammonium phosphates, which act as fertilizers.

Polymer sorbents were formed from a two-phase system: an extractant solution in THF and an aqueous solution of polyvinyl alcohol (PVA) with glutaraldehyde (GA). With rapid mixing of the components, the solutions gave a stable and relatively viscous emulsion, which further ensured a uniform distribution of the extractant in the polymer. The structuring of the system was provided by the addition of catalytic amounts of hydrochloric acid, while intermolecular crosslinking of polymer chains occurred due to the formation of acetals via the reaction of PVA hydroxyl groups and GA aldehyde groups.

When studying gelation, it was found that the gelation time and water absorption decrease with the increasing amount of GA added, and the amount of liquid displaced from the sorbent increases (Table 3), which is due to an increase in the degree of cross-linking of the polymer. From the obtained results, it follows that the best sorbent for use in sorption processes is sorbent number one, since it has the highest water uptake and does not displace water during gelation. Therefore, further studies were carried out using this sample.

Table 3. Parameters of synthesized sorbents.

No.	Gelation Time	Displaced Liquid, wt %	Water Uptake, wt %
1.	1 day	0	54.7
2.	1 day	23.4	53.4
3.	12 h	43.1	45.1
4.	7 h	71.6	39.4
5.	5 h	75.1	22.1

When conducting IR studies, it was found that the spectrum of the sorbent (Figure 7B) exhibits a vibrational band in the region of 1208–1160 cm^{-1}, which is absent in the spectrum of PVA cross-linked with glutaraldehyde (Figure 7A). This band is also observed in the spectrum of the extractant (Figure 7C) and is characteristic of the stretching vibrations of the P=N units of the phosphazene ring. This fact indicates that the extractant is present in the sorbent after washing and drying, and also that the phosphazene ring has been preserved during the synthesis and isolation of the sorbent.

The study of the extraction properties of the sorbent showed that it is effective for the sorption of palladium(II) from aqueous hydrochloric acid solutions. It was found that the extraction efficiency increases with a decrease in the acidity of the medium and reaches 57% when a 0.25 mol L^{-1} hydrochloric acid solution is used (sorbent weight 0.1 g, volume of the aqueous phase 6 mL). This value is an order of magnitude higher than that for the liquid extraction of palladium(II) from hydrochloric acid solution with commercial monodentate extractant Cyanex 923 dissolved in toluene [35] (Figure 8). When the amount of sorbent was doubled, the extraction efficiency reached 71%.

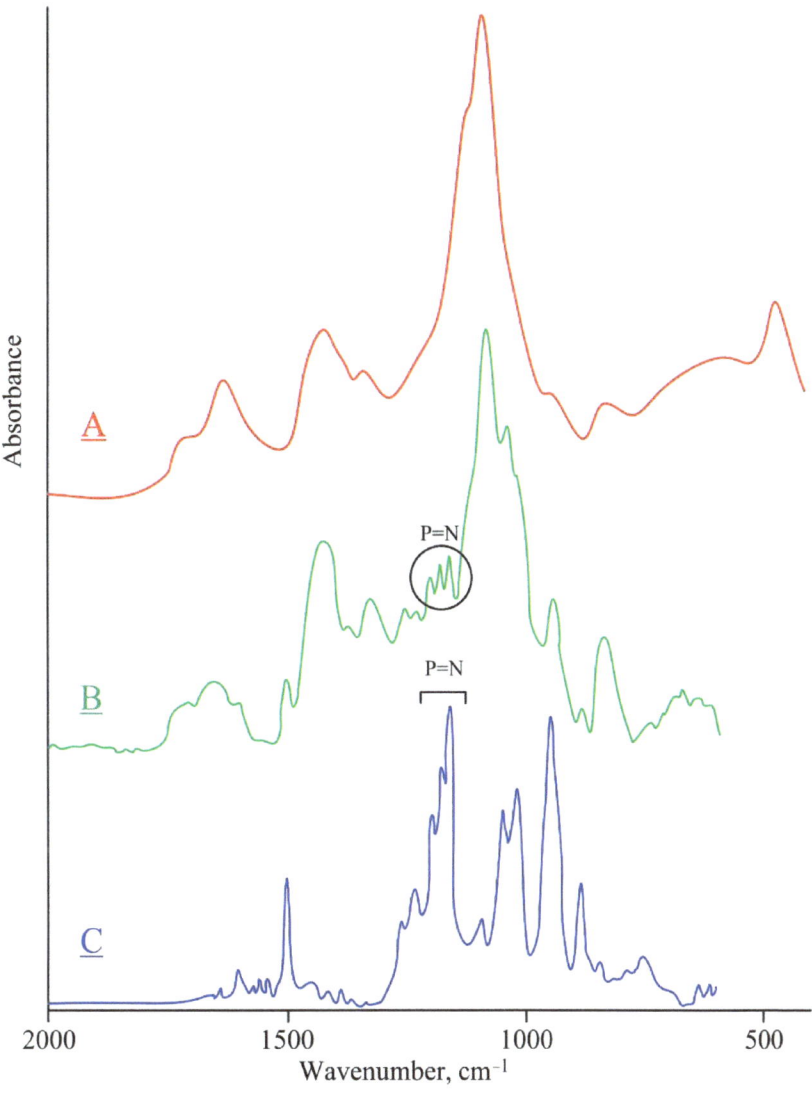

Figure 7. IR spectra of cross-linked PVA (**A**), sorbent (**B**), and extractant (**C**).

During two cycles of extraction with one portion of the sorbent (0.1 g) for each cycle, the amount of palladium recovered reached 89%.

It was also found that 100% stripping of palladium from the sorbent is accomplished in one cycle with 5 mol L^{-1} hydrochloric acid. After stripping, the sorbent can be reused without changing the extraction efficiency.

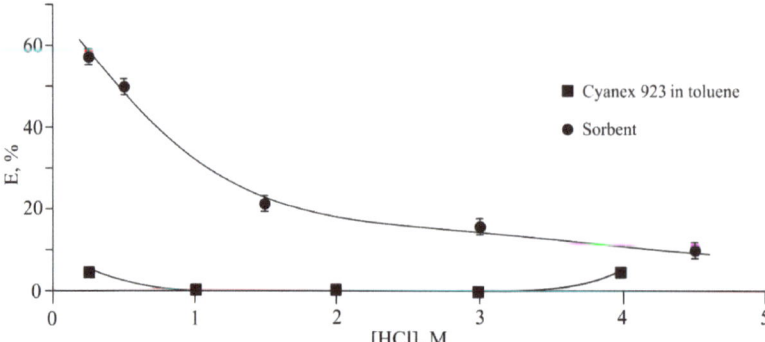

Figure 8. Extraction efficiency of palladium by the magnetic sorbent. In [35], palladium(II) was extracted from a hydrochloric acid medium using Cyanex 923 under the following initial conditions: [Pd] = 5 × 10^{-4} mol L^{-1}, [Cyanex 923] = 0.1 mol L^{-1}.

Since production wastes and secondary raw materials containing palladium contain other metals almost in all cases, it was necessary to evaluate the extraction selectivity of the developed sorbent. For example, WEEE and electrical capacitors always contain copper together with palladium. Therefore, the Pd(II) sorption was studied from a 0.25 mol L^{-1} hydrochloric acid solution in the presence of copper(II) chloride. As a result, 52% of palladium was selectively separated by the developed sorbent in one cycle, while all copper remained in the leaching solution.

Magnetic properties were imparted to the sorbent by adding encapsulated iron in the gelation stage. The stability of the dispersion was ensured by the viscosity of the system. As can be seen from the micrograph of the magnetic sorbent film (Figure 9), iron particles are evenly distributed in the gel and form small agglomerates, with their linear size not exceeding 200 µm.

Figure 9. Micrograph of a magnetic sorbent film.

According to vibrating magnetometry data (Figure 10), the saturation magnetization of the sorbent is approximately 14 emu g^{-1}. This value is sufficient for the sorbent to be separated by a magnet from water and non-magnetic particles and used in the processes of metal extraction from metallurgical waste and secondary raw materials.

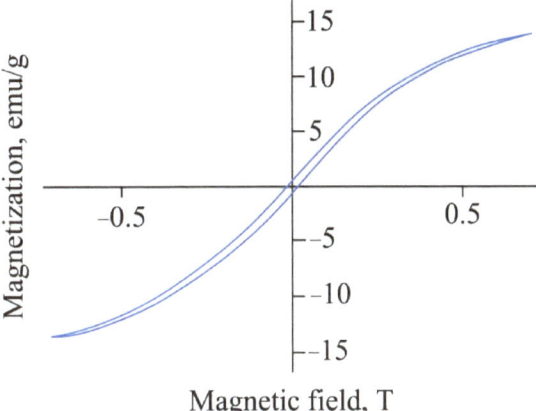

Figure 10. Magnetic properties of the sorbent with iron particles.

The study of the properties of the magnetic sorbent showed that it has similar extraction characteristics in terms of the weight of the iron-free sorbent.

3. Conclusions

The new magnetic sorbent based on polyvinyl alcohol, metallic iron, and a polydentate phosphazene-containing extractant is a promising material for the solid-phase extraction of noble metals from leaching solutions of WEEE and electrical capacitors. This is due to its acid resistance, high efficiency and selectivity, excellent sorption and magnetic properties, and environmental safety. The efficiency of sorption of palladium(II) by the developed sorbent is 57% in one cycle and 89% in two sorption cycles. The spent magnetic sorbent can also be disposed of by burial in the soil, since it does not inhibit the activity of soil microflora.

4. Materials and Methods

4.1. Materials

Polyvinyl alcohol (PVA), carbonyl iron, glutaraldehyde (GA), hydrochloric acid, diethyl phosphite, *p*-toluenesulfonic acid, dioxane, tetrahydrofuran, palladium(II) chloride, copper(II) chloride, chloroform, and potassium carbonate were products of Sigma Aldrich (Saint Louis, MO, USA). Dioxane and tetrahydrofuran were dried over sodium metal followed by distillation. The encapsulation of carbonyl iron was carried out according to the procedure described in [36].

4.2. Methods

^{1}H, ^{13}C, and ^{31}P NMR spectra were recorded on an Agilent/Varian Inova 400 spectrometer (Agilent Technologies, Santa Clara, CA, USA) at 400.02 MHz, 100.60 MHz, and 161.94 MHz, respectively. The mass spectrum was recorded on a Microflex LRF mass spectrometer (Bruker Daltonic GmbH, Leipzig, Germany). 3-Hydroxypicolinic acid was used as a matrix. IR spectra were measured on a Nicolet 380 spectrometer (Thermo Fisher Scientific, Waltham, MA, USA) in the spectral range of 4000–500 cm^{-1} with a wavenumber accuracy of 0.01 cm^{-1}. Differential scanning calorimetry (DSC) measurements were conducted using a NETZSCH STA 449F1 instrument (Erich NETZSCH GmbH & Co. Holding KG, Selb, Germany). The hysteresis loop of a magnetic composite swollen in water was recorded using a LakeShore 7407 vibrating magnetometer (LakeShore Cryotronics Inc., Westerville, OH, USA). The distribution of iron microparticles in the polymer matrix was visually assessed using a Levenhuk MED D25T optical microscope (PRC, controlled by Levenhuk, Inc., Tampa, FL, USA). The contents of Pd(II) and Cu(II) in aqueous hydrochloric acid solutions were determined using an XSeriesII ICP-MS instrument (Thermo Fisher Scientific, USA). The composition of the palladium complex of the extractant was determined on an X-Max SDD Inca Energy Dispersive spectrometer for electron probe microanalysis (Oxford Instruments, Abingdon, UK).

Studies on the effect of the extractant on the soil microflora were carried out in vitro in flasks on a liquid nutrient medium in a shaker incubator. An enrichment culture of soil microorganisms obtained by cultivating a nutrient soil on a liquid medium of the following composition was used as an inoculation material: peptone 1.0 g L^{-1}, yeast extract 0.5 g L^{-1}, NaCl 0.5 g L^{-1}, and glucose 2.0 g L^{-1}. The medium pH was 6.5. The soil to medium ratio was 1:2. The cultivation was carried out at 30 °C for 48 h with stirring at 200 rpm. The growth of microflora was evaluated spectrophotometrically by measuring the optical density at λ = 600 nm.

To determine CFU, the method of tenfold dilutions (Koch's method) was used. The obtained enrichment culture (1 mL) was added into flasks with 100 mL of liquid nutrient medium of the above composition, and then, the extractant diluted in 1 mL of acetone was added. Thus, the concentration of the extractant in the medium was 0.03314%. Then, 1 mL of sterile tap water was added to the control flask (control group).

After incubation at 30 °C for 48 h at 200 rpm, the resulting suspension was sown on a solid medium in Petri dishes. To do this, dilutions of the suspension were prepared in sterile tap water. An exact volume of dilution was added to Petri dishes with agarized nutrient medium and spread with a glass spatula over the surface of the nutrient medium, and colonies were counted after 1–15 days of incubation.

4.3. Synthesis of Hexakis-[4-{(N-allylimino)methyl}-phenoxy]-cyclotriphosphazene

Hexakis-[4-{(N-allylimino)methyl}-phenoxy]-cyclotriphosphazene (APP) was synthesized according to the procedure described in [37].

4.4. Synthesis of Hexakis-[4-{α,α-(N-allylamino)(O,O-diethylphosphoryl)methylidine}phenoxy]cyclotriphosphazene (Extractant)

A 50 mL round-bottom flask equipped with a reflux condenser and a magnetic stirrer was charged with APP (0.5 g, 0.4566 mmol), diethyl phosphite (0.59 mL, 0.4566 mmol), and *p*-toluenesulfonic acid (catalyst) (79 mg, 10 mol %), and the mixture was dissolved in 30 mL of dioxane. After complete dissolution, the reaction mixture was stirred at the boiling point of dioxane for 6 h in an argon atmosphere. Dioxane was distilled off, and the resulting liquid was dissolved in chloroform. Potassium carbonate (0.05 g) was added to the solution, and the mixture was stirred for 24 h at 25 °C. The solution was separated from the precipitate by decantation, and chloroform was distilled off on a rotary evaporator. The resulting substance was dried in an oven under vacuum at a temperature of 90 °C for 5 h. Yield: 0.70 g (80%).

4.5. Extraction of Palladium with the Developed Extractant

A solution of palladium(II) chloride (0.05 g, 0.282 mmol) in 0.5 M hydrochloric acid (3 mL) was prepared in a 10 mL glass vial. At the same time, a solution of the extractant (0.18 g, 0.0936 mmol) in chloroform (3 mL) was prepared. The extractant solution was added to the palladium(II) chloride solution and stirred at 25 °C for 48 h. The solid palladium complex formed at the interface was washed several times with distilled water and chloroform. The complex was dried under vacuum at 70 °C for 4 h.

4.6. Synthesis of Sorbents

Five solutions of PVA (0.8 g) in water (4.52 mL) containing 0.05, 0.1, 0.2, 0.4, and 0.8 mL of HA, respectively, were prepared in 10 mL glass vials. Five identical solutions of the extractant (0.1 g, 0.0520 mmol) in THF (1 mL) were prepared separately. Extractant solutions were added to the PVA solutions and vigorously stirred. Catalytic amounts of hydrochloric acid (3 drops) were added and stirred again. The emulsions were left at room temperature until gelation.

The gel was washed several times with distilled water and dried in a vacuum oven at 80 °C for 6 h.

4.7. Sorption of Palladium by the Developed Sorbent

Six solutions of palladium(II) chloride (0.012 g, 0.0677 mmol) in hydrochloric acid (6 mL) of various concentrations (0.25, 0.5, 1.5, 3, and 4.5 M) were prepared in 10 mL glass vials. Then, the sorbent (0.1 g) was placed in each vial, and the mixture was stirred for 48 h at room temperature.

4.8. Stripping of Palladium

The spent sorbent was treated with 5 M hydrochloric acid with stirring for 48 h at room temperature.

4.9. Sorption of Palladium by the Developed Gel in the Presence of Copper

A solution of palladium(II) chloride (0.012 g, 0.0677 mmol) and copper(II) chloride (0.012 g, 0.08925 mmol) in 0.25 M hydrochloric acid (6 mL) was prepared in a 10 mL glass vial. Then, the sorbent (0.1 g) was placed in the vial and stirred for 48 h at room temperature.

4.10. Synthesis of a Magnetic Sorbent Containing Acid-Resistant Iron

In a 10 mL glass vial, polyvinyl alcohol (0.8 g) was dissolved in distilled water (4.52 g). Then, encapsulated carbonyl iron powder (1 g) was introduced into the solution. After that, glutaraldehyde (0.05 mL), a solution of the extractant (0.1 g) in THF (1 mL), and 3 drops of hydrochloric acid were added. The mixture was stirred for about 7 min until the viscosity increased, after which it was left at room temperature until completely cured.

The gel was washed several times with distilled water and dried in a vacuum chamber at a temperature of 80 °C to constant weight.

Author Contributions: Conceptualization, E.C. and P.Y.; methodology, E.C., P.Y., I.B. and G.S.; validation, E.C. and P.Y.; investigation, E.C., P.Y., I.B. and G.S.; resources, P.Y.; data curation, E.C.; writing—original draft preparation, E.C. and P.Y.; writing—review and editing, E.C. and P.Y.; visualization, P.Y.; supervision, E.C.; project administration, E.C.; funding acquisition, E.C. All authors have read and agreed to the published version of the manuscript.

Funding: This research received no external funding.

Institutional Review Board Statement: Not applicable.

Informed Consent Statement: Not applicable.

Acknowledgments: Elemental analysis was performed on the equipment of the Center of Collective Use of Mendeleev University of Chemical Technology under the state contract 13.CCU.21.0009. Magnetic characteristics tested on equipment of SSC RF JSC "GNIIChTEOS".

Conflicts of Interest: The authors declare no conflict of interest.

References

1. Bourgeois, D.; Lacanau, V.; Mastretta, R.; Contino-Pépin, C.; Meyer, D. A simple process for the recovery of palladium from wastes of printed circuit boards. *Hydrometallurgy* **2019**, *191*, 105241. [CrossRef]
2. Cieszynska, A.; Wieczorek, D. Extraction and separation of palladium(II), platinum(IV), gold(III) and rhodium(III) using piperidine-based extractants. *Hydrometallurgy* **2018**, *175*, 359–366. [CrossRef]
3. Dubiella-Jackowska, A.; Namieśnik, J. Platinum Group Elements in the Environment: Emissions and Exposure. *Rev. Environ. Contam. Toxicol.* **2008**, *199*, 1–25. [CrossRef]
4. Li, W.-X.; Yang, B.-W.; Na, J.-H.; Rao, W.; Chu, X.-Q.; Shen, Z.-L. Palladium-catalyzed cross-coupling of alkylindium reagent with diaryliodonium salt. *Tetrahedron Lett.* **2022**, *95*, 153729. [CrossRef]
5. Ding, K.; Zhang, D.; Chen, J.; Han, J.; Shi, F.; Li, B.; He, X.; Wang, L.; Wang, H.; Wang, Y. Unexpected electocatalytic activity of a micron-sized carbon sphere-graphene (MS-GR) supported palladium composite catalyst for ethanol oxidation reaction (EOR). *Mater. Chem. Phys.* **2020**, *259*, 124035. [CrossRef]

6. Dergachev, Y.M. Hydrogen absorption by transition metals. *Inorg. Mater.* **2006**, *42*, 112–115. [CrossRef]
7. Groshev, N.Y.; Rundkvist, T.V.; Karykowski, B.T.; Maier, W.D.; Korchagin, A.U.; Ivanov, A.N.; Junge, M. Low-Sulfide Platinum–Palladium Deposits of the Paleoproterozoic Fedorova–Pana Layered Complex, Kola Region, Russia. *Minerals* **2019**, *9*, 764. [CrossRef]
8. Torres, R.; Lapidus-Lavine, G. Platinum, palladium and gold leaching from magnetite ore, with concentrated chloride solutions and ozone. *Hydrometallurgy* **2016**, *166*, 185–194. [CrossRef]
9. Shi, W.; Hu, Y.; Li, Q.; Lan, T.; Zhang, X.; Cao, J. Recovery of Pd(II) in chloride solutions by solvent extraction with new vinyl sulfide ploymer extractants. *Hydrometallurgy* **2021**, *204*, 105716. [CrossRef]
10. Song, L.; Wang, X.; Li, L.; Wang, Z.; Xu, H.; He, L.; Li, Q.; Ding, S. Recovery of palladium(II) from strong nitric acid solutions relevant to high-level liquid waste of PUREX process by solvent extraction with pyrazole-pyridine-based amide ligands. *Hydrometallurgy* **2022**, *211*, 105888. [CrossRef]
11. Rios-Vera, R.M.; Chagnes, A.; Hernández-Perales, L.; Martínez-Rodríguez, D.E.; Navarro-Segura, D.L.; Gaillon, L.; Sirieix-Plénet, J.; Rizzi, C.; Rollet, A.L.; Avila-Rodriguez, M.; et al. Trihexyl(tetradecyl)phosphonium bis-2,4,4-(trimethylpentyl)phosphinate micellar behavior in the extraction of Ag(I) from acidic nitrate media. *J. Mol. Liq.* **2022**, *358*, 119132. [CrossRef]
12. Junior, A.B.; Espinosa, D.; Vaughan, J.; Tenório, J. Recovery of scandium from various sources: A critical review of the state of the art and future prospects. *Miner. Eng.* **2021**, *172*, 107148. [CrossRef]
13. Zhang, S.-M.; Wu, Q.-Y.; Yuan, L.-Y.; Wang, C.-Z.; Lan, J.-H.; Chai, Z.-F.; Liu, Z.-R.; Shi, W.-Q. Theoretical study on the extraction behaviors of MoO22+ with organophosphorous extractants. *J. Mol. Liq.* **2022**, *355*, 118969. [CrossRef]
14. Sarkar, S.; Ammath, S.; Kirubananthan, S.; Suneesh, A.S. Investigation of the Phase Splitting Behaviour of U(VI) and Th(IV) loaded Trialkyl Phosphate Solvents in the Absence of Aqueous Phase. *ChemistrySelect* **2021**, *6*, 13725–13735. [CrossRef]
15. Atanassova, M.; Kukeva, R.; Stoyanova, R.; Todorova, N.; Kurteva, V. Synergistic and antagonistic effects during solvent extraction of Gd(III) ion in ionic liquids. *J. Mol. Liq.* **2022**, *353*, 118818. [CrossRef]
16. Atanassova, M.; Angelov, R.; Gerginova, D.; Karashanova, D. Neutral organophosphorus ligands as a molecular lab for simultaneous detecting of Ag(I) ions. *J. Mol. Liq.* **2021**, *335*, 116287. [CrossRef]
17. Anderson, S.; Nilsson, M.; Kalu, E.E. Electrochemical Impedance Spectroscopy (EIS) Characterization of Water/Sodium Bis(2-Ethylhexyl) Sulfosuccinate-HDEHP/n-Dodecane Reverse Micelles for Electroextraction of Neodymium. *ChemEngineering* **2017**, *1*, 3. [CrossRef]
18. Chen, J.; Zhang, H.; Zeng, Z.; Gao, Y.; Liu, C.; Sun, X. Separation of lithium and transition metals from the leachate of spent lithium-ion battery by extraction-precipitation with p-tert-butylphenoxy acetic acid. *Hydrometallurgy* **2021**, *206*, 105768. [CrossRef]
19. Nguyen, V.N.H.; Song, S.J.; Lee, M.S. Separation of palladium and platinum metals by selective and simultaneous leaching and extraction with aqueous/non-aqueous solutions. *Hydrometallurgy* **2022**, *208*, 105814. [CrossRef]
20. Liu, T.; Chen, J.; Li, H.; Li, K.; Li, D. Further improvement for separation of heavy rare earths by mixtures of acidic organophosphorus extractants. *Hydrometallurgy* **2019**, *188*, 73–80. [CrossRef]
21. Junior, A.B.B.; Espinosa, D.C.R.; Tenório, J.A.S. Selective separation of Sc(III) and Zr(IV) from the leaching of bauxite residue using trialkylphosphine acids, tertiary amine, tri-butyl phosphate and their mixtures. *Sep. Purif. Technol.* **2021**, *279*, 119798. [CrossRef]
22. Kukkonen, E.; Virtanen, E.J.; Moilanen, J.O. α-Aminophosphonates, -Phosphinates, and -Phosphine Oxides as Extraction and Precipitation Agents for Rare Earth Metals, Thorium, and Uranium: A Review. *Molecules* **2022**, *27*, 3465. [CrossRef] [PubMed]
23. Samiey, B.; Cheng, C.-H.; Wu, J. Organic-Inorganic Hybrid Polymers as Adsorbents for Removal of Heavy Metal Ions from Solutions: A Review. *Materials* **2014**, *7*, 673–726. [CrossRef] [PubMed]
24. Omer, A.M.; Dey, R.; Eltaweil, A.S.; El-Monaem, E.M.A.; Ziora, Z.M. Insights into recent advances of chitosan-based adsorbents for sustainable removal of heavy metals and anions. *Arab. J. Chem.* **2021**, *15*, 103543. [CrossRef]
25. Kong, D.; Foley, S.R.; Wilson, L.D. An Overview of Modified Chitosan Adsorbents for the Removal of Precious Metals Species from Aqueous Media. *Molecules* **2022**, *27*, 978. [CrossRef]
26. Surgutskaia, N.S.; Di Martino, A.; Zednik, J.; Ozaltin, K.; Lovecká, L.; Bergerová, E.D.; Kimmer, D.; Svoboda, J.; Sedlarik, V. Efficient Cu^{2+}, Pb^{2+} and Ni^{2+} ion removal from wastewater using electrospun DTPA-modified chitosan/polyethylene oxide nanofibers. *Sep. Purif. Technol.* **2020**, *247*, 116914. [CrossRef]
27. Wang, S.; Liu, Y.; Yang, A.; Zhu, Q.; Sun, H.; Sun, P.; Yao, B.; Zang, Y.; Du, X.; Dong, L. Xanthate-Modified Magnetic $Fe_3O_4@SiO_2$-Based Polyvinyl Alcohol/Chitosan Composite Material for Efficient Removal of Heavy Metal Ions from Water. *Polymers* **2022**, *14*, 1107. [CrossRef]
28. Jakavula, S.; Biata, N.R.; Dimpe, K.M.; Pakade, V.E.; Nomngongo, P.N. Magnetic Ion Imprinted Polymers (MIIPs) for Selective Extraction and Preconcentration of Sb(III) from Environmental Matrices. *Polymers* **2021**, *14*, 21. [CrossRef]
29. Han, Q.; Du, M.; Guan, Y.; Luo, G.; Zhang, Z.; Li, T.; Ji, Y. Removal of simulated radioactive cerium (III) based on innovative magnetic trioctylamine-polystyrene composite microspheres. *Chem. Phys. Lett.* **2020**, *741*, 137092. [CrossRef]
30. Yudaev, P.A.; Chistyakov, E.M. Ionic Liquids as Components of Systems for Metal Extraction. *ChemEngineering* **2022**, *6*, 6. [CrossRef]
31. Yu, M.; Wang, L.; Hu, L.; Li, Y.; Luo, D.; Mei, S. Recent applications of magnetic composites as extraction adsorbents for determination of environmental pollutants. *TrAC Trends Anal. Chem.* **2019**, *119*, 115611. [CrossRef]
32. Ozay, O.; Ozay, H. Novel hexacentered phosphazene compound as selective Fe^{3+} ions sensor with high quantum yield: Synthesis and application. *Phosphorus Sulfur. Silicon Relat. Elem.* **2018**, *194*, 221–228. [CrossRef]

33. Çıralı, D.E.; Dayan, O. Synthesis of Tetranuclear Ruthenium (Ii) Complex of Pyridyloxy-Substituted 2,2′-Dioxybiphenyl-Cyclotriphosphazene Platform and its Catalytic Application in the Transfer Hydrogenation of Ketones. *Phosphorus Sulfur. Silicon Relat. Elem.* **2015**, *190*, 1100–1107. [CrossRef]
34. Cherkasov, R.A.; Garifzyanov, A.R.; Zakharov, S.V.; Vinokurov, A.V.; Galkin, V.I. Liquid extraction of noble metal ions with bis(α-aminophosphonates). *Russ. J. Gen. Chem.* **2006**, *76*, 417–420. [CrossRef]
35. Gupta, B.; Singh, I. Extraction and separation of platinum, palladium and rhodium using Cyanex 923 and their recovery from real samples. *Hydrometallurgy* **2013**, *134–135*, 11–18. [CrossRef]
36. Chistyakov, E.; Kolpinskaya, N.; Yudaev, P. Magnetic polymer granules based on metallic iron and polyvinyl alcohol. *Int. Multidiscip. Sci. Geoconf. SGEM* **2020**, *20*, 507–514. [CrossRef]
37. Chistyakov, E.; Yudaev, P.; Nelyubina, Y. Crystallization of Nano-Sized Macromolecules by the Example of Hexakis-[4-{(N-Allylimino)methyl}phenoxy]cyclotriphosphazene. *Nanomaterials* **2022**, *12*, 2268. [CrossRef]

Article

The Use of Hydrogels in the Treatment of Metal Cultural Heritage Objects

Elodie Guilminot

Arc'Antique Conservation and Research Laboratory, 26 Rue de la Haute Forêt, 44300 Nantes, France; elodie.guilminot@loire-atlantique.fr

Abstract: Currently gels are widely used in the restoration of paintings, graphic arts, stuccowork and stonework, but their use in metal restoration is less widespread. In this study, several polysaccharide-based hydrogels (agar, gellan and xanthan gum) were selected for use in metal treatments. The use of hydrogels allows to localize a chemical or electrochemical treatment. This paper presents several examples of treatment of metal objects of cultural heritage, i.e., historical or archaeological objects. The advantages, disadvantages and limits of hydrogel treatments are discussed. The best results are obtained for the cleaning of copper alloys via associating an agar gel with a chelating agent (EDTA (ethylenediaminetetraacetic acid) or TAC (tri-ammonium citrate)). The hot application allows to obtain a peelable gel, particularly adapted for historical objects. Electrochemical treatments using hydrogels have been successful for the cleaning of silver and for the dechlorination of ferrous or copper alloys. The use of hydrogels for the cleaning of painted aluminum alloys is possible but it has to be coupled with mechanical cleaning. However, for the cleaning of archaeological lead, the cleaning using hydrogels was not very effective. This paper shows the new possibilities of using hydrogels for the treatment of metal cultural heritage objects: agar is the most promising hydrogel.

Keywords: hydrogel; agar; conservation-restoration; cleaning treatment; stabilization; iron; copper alloys; lead; Al alloys; silver

Citation: Guilminot, E. The Use of Hydrogels in the Treatment of Metal Cultural Heritage Objects. *Gels* **2023**, *9*, 191. https://doi.org/10.3390/gels9030191

Academic Editors: Daxin Liang, Ting Dong, Yudong Li and Caichao Wan

Received: 14 February 2023
Revised: 28 February 2023
Accepted: 28 February 2023
Published: 2 March 2023

Copyright: © 2023 by the author. Licensee MDPI, Basel, Switzerland. This article is an open access article distributed under the terms and conditions of the Creative Commons Attribution (CC BY) license (https:// creativecommons.org/licenses/by/ 4.0/).

1. Introduction

When a metal object is discarded, the metal tends to revert to its natural mineral state and corrosion takes place. The resulting corrosion products form layers that vary considerably depending on the metal and its exposure to environment: iron corrosion products are often voluminous and can be highly reactive, especially in the presence of chlorides, while Pb objects tend to form a highly protective, thin corrosion layer, except when organic acids are present. A conservators' objective is to locate the original surface of the object, which is often obscured by the various layers of corrosion.

Work carried out by conservators must be the least invasive possible while ensuring the stability of the object and restoring its legibility. Ideally, a cleaning treatment would act on a single element (generally corrosion products, sediments or dust) without affecting the layer(s) to be preserved. A range of stabilization and cleaning techniques has been developed and their effectiveness proven: these techniques may be mechanical, chemical or electrochemical [1]. The most common mechanical cleaning methods involve removing corrosion products with a scalpel, brush or compressed air, or by sandblasting. These mechanical techniques are widely used but can be very time consuming and are unsuitable for some objects. Chemical cleaning involves the use of chelating agents or acids to selectively remove corrosion products. Electrochemical treatments are used to remove chlorides or reduce corrosion products. However, chemical and electrochemical treatments require the object to be immersed in the treatment solution. If the object is composed of more than one material, the use of this type of treatment is largely ruled out. Moreover, if chemical solutions penetrate too far into the object, this can result in a weakening of the layers underlying the original surface.

Conservators often have to combine different techniques to obtain a satisfactory result and are always looking for new protocols that allow them better control. For this reason, the Arc'Antique laboratory has been testing and developing gel-applied treatments for metals for several years. In the early 1990s, Wolbers came up with an innovative methodology to treat paintings: the application of liquid cleaning agents embedded in gel matrices [2,3]. In conservation, the term "gels" is often linked with other words, such as packs, pastes, poultices, compresses and pads, to indicate that localized cleaning is intended. The gel acts as a vehicle for applying the treatment solution to the surface to be cleaned. Gelled formulations are used to lengthen solution retention time and to improve control over the cleaning process. Many formulations are common in conservation–restoration [4,5]: polyacrylic (Pemulen® or Carbopol®), cellulose-based (CMC (carboxymethylcellulose) and cellulose ether), polysaccharides (agar, gellan gum or xanthan gum), emulsions, and poly(vinyl alcohol) (PVA-borax and gels developed by the University of Florence (Center for Colloid and Surface Science: CSGI)).

Interactions between polymer chains create a 3D structural network in gels. These interactions can be in the form of a hydrophobic, electrostatic or van-der-Waals interactions or hydrogen bonds: the system is then called a "physical gel" [6]. If connections between different polymer chains are due to covalent bonds, the gel is known as a "chemical gel". Gels can also be classified according to the nature of the solution embedded in the gel: In the case of a liquid phase composed of an organic solvent, they are referred to as organogels; if the solvent is water, they are called hydrogels. For the restoration of cultural heritage metal, physical hydrogels are preferred for various reasons: low cost, accessibility, ease of processing and good compatibility with the active agents most commonly used by restorers.

In this article, the study will focus on physical hydrogels based on polysaccharides: agar, gellan gum and xanthan gum [7,8]. Normally, physical gels contain reversible bonds formed by temporary associations between chains. These associations have finite lifespans and are continuously breaking up and reforming. These weak physical bonds are often hydrogen bonds, but can also be a formation of block copolymer micelles and ionic associations [9]. However, some physical hydrogels form more stable bonds. The strong physical bonds between polymer chains are effectively permanent within a given set of experimental conditions. Examples of strong physical bonds are lamellar microcrystals, glassy nodules, and double and triple helices. Hence, these physical hydrogels are analogous to chemical gels. Agar can form a semi-rigid network. It is composed of a mixture of agarose and agaropectin in variable proportions depending on the type of algae used and the manufacturing process [10]. The linear chains of agarose arrange themselves into double helix structures that aggregate to form "suprafibers" comprising anywhere up to 10^4 helices. Agar gels have the remarkable property of reversibility. They simply melt on heating and solidify again upon cooling. These transformations can be repeated indefinitely in the absence of aggressive substances. Generally, 1–5% wt agar is mixed in aqueous solution and heated up to 85 °C. It then develops a random coil structure which has the ability to progressively rearrange itself and gel during the cooling stage, at below 40 °C [11,12]. The gelling temperature of agar is 38–42 °C. So, when the hot agar gel is cooled on a surface, the film adheres to the surface and is easy to remove. These gels are called peelable gels [13]. It was shown that the agar concentration strongly influences the water state within the gel network: agar at 1% wt releases far more water than gels at 3% and 5% wt, acting as a "free water reservoir" [14]. Conservators can adapt the rigidity of the gel film to suit the surface to be treated and adapt its sensitivity to water.

The use of agar gels in the cleaning of artworks has been studied for more than a decade and its application concerns a variety of substrates with different physico–mechanical features: from stone [15,16] and plaster surfaces [17] with compact soiling, to the most fragile, such as paper [18,19] or paintings [11,20,21]. Several research studies have recently been launched with the aim of understanding the cleaning mechanism of agar and optimizing its use in the restoration of heritage building materials [13,14,16,17,22]. These studies have shown the possibility of successfully using chelating agents combined with agar gels to remove corrosion product stains from marble.

Gellan gum has similar characteristics to a peelable gel. It is a linear, anionic heteropolysaccharide produced by a microorganism (Sphingomonas elodea) [23]. The molecules are transformed into an ordered double-helical conformation upon cooling, followed by associations between the helices through weak interactions, such as hydrogen bonds and van-der-Waals forces. In the presence of cations (especially Ca^{2+}), gellan gum forms hard, brittle gels that are able to promote and stabilize the ordered "crystalline-like" structure. However, the addition of cations is not necessary for the formation of high-acyl gellan gum gels [24], that have a higher gelling temperature than agar gels: about 70 °C. Another difference with agar is that gellan gum is less thermo-reversible: once the film is formed, conservators cannot reuse it (or any excess gel prepared). Gellan gum is mainly used to clean paper because it keeps water damage to a minimum, and thus more effectively preserves the integrity of the original, historical paper [25,26]. Sometimes gellan gum is not used as a rigid peelable gel, it can be prepared cold and form a viscous gel. For example, if biopatin is applied, the integration of microorganisms requires that the gel must not be heated during the preparation process [27].

Unlike agar or gellan gum, xanthan gum has two conformations: an ordered conformation (helix) and a disordered one (random coil) [28]. This double conformation excludes the possibility of obtaining a peelable gel, as is possible with agar or gellan gum gels. Xanthan gum is an extracellular polysaccharide composed of glucose and secreted by Xanthomonas Campestris bacteria. Xanthan gels obtained using dissolution at ambient temperatures tend to be highly viscous. While the stability of peelable gels (agar or gellan gum) is strongly pH-dependent, as the double helix structure is maintained in neutral pH, xanthan gels maintain a high viscosity in a wide pH range [29]. The main disadvantage of these viscous gels is residue [30].

In this paper, different physical hydrogels based on polysaccharides (agar, gellan gum and xanthan gum) were tested for the cleaning and stabilization of metal cultural heritage objects. In each case study, the main research issue is presented by detailing the objectives of the restoration, the cleaning protocols tested and the relevant results.

2. Results
2.1. Choice of Treatments

Choice of treatment was determined by the selection of the active agent needed to clean or stabilize the object. Conservators know which active agents are suitable for cleaning relevant metals in chemical treatments. Generally, to remove copper corrosion products, solutions with chelating agents are used, such as EDTA (ethylenediaminetetraacetic acid) or TAC (tri-ammonium citrate). In the case of lead corrosion products (based on lead carbonates), acidic solutions are also recommended. For cleaning silver, electrolytic reduction yields good results. Electrolysis is also used for the stabilization of ferrous or copper objects.

The use of gels allows the quantity of active agents to be reduced and their degree of penetration into the object limited. A gel treatment can also be used to treat a surface locally, enabling the whole object to be preserved. Previous studies have shown the compatibility of different gels with the active agents most widely used by conservators [7,31]. Physical hydrogels based on polysaccharides with a helix structure (agar and gellan gum) remain stable in a neutral pH range (about 5 to 8). If the treatment solutions are more acidic or basic, it is possible to form gel films with demineralized water and then immerse the film in the treatment solution for 2 h. The gel film maintains its mechanical strength throughout the application (treatment time of less than 2 h); however, contact with the object's surface is less effective than when the gel is applied hot. Xanthan gels do not form rigid gels and can be used in a wider pH range (between 4 and 10).

Table 1 summarizes the compatibility between treatment solutions and polysaccharide-based hydrogels (agar, gellan gum and xanthan gum).

Table 1. Compatibility of treatment solutions with physical polysaccharide-based hydrogels (agar, gellan gum and xanthan gum).

Hydrogels	Treatment Solutions	Compatibility
Agar	Nitric acid (10^{-4} M)	Possible with pre-made film
	Citric acid (5 wt%)	Possible with pre-made film
	Citric acid (0.055 M)	Compatible (hot application)
	Oxalic acid (5 wt%)	Possible with pre-made film
	KNO_3 (1 wt%)	Compatible (hot application)
	TAC (2.5 or 5 wt%)	Compatible (hot application)
	di-EDTA (0.5–2.5 or 5 wt%)	Compatible (hot application)
	tetra-EDTA (5 wt%)	Compatible (hot application)
	Metasilicate sodium (2 wt%)	Incompatible
	NaOH (2 wt%)	Incompatible
Gellan gum	Nitric acid (10^{-4} M)	Incompatible
	Citric acid (5 wt%)	Incompatible
	Citric acid (0.055 M)	Compatible (hot application)
	Oxalic acid (5 wt%)	Incompatible
	KNO_3 (1 wt%)	Compatible (hot application)
	TAC (2.5 or 5 wt%)	Compatible (hot application)
	di-EDTA (0.5–2.5 or 5 wt%)	Compatible (hot application)
	tetra-EDTA (5 wt%)	Compatible (hot application)
	Metasilicate sodium (2 wt%)	Incompatible
	NaOH (2 wt%)	Incompatible
Xanthan gum	Nitric acid (10^{-4} M)	Possible but very viscous
	Citric acid (5 wt%)	Possible but very viscous
	Citric acid (0.055 M)	Compatible
	Oxalic acid (5 wt%)	Possible but very viscous
	KNO_3 (1 wt%)	Compatible
	TAC (2.5 or 5 wt%)	Compatible
	di-EDTA (0.5–2.5 or 5 wt%)	Compatible
	tetra-EDTA (5 wt%)	Compatible
	Metasilicate sodium (2 wt%)	Possible but very viscous
	NaOH (2 wt%)	Incompatible

2.2. Case Study

2.2.1. Cleaning of Copper Alloys—Historical Artifact

Our first case study concerns the cleaning of objects from the Islamic art collection of the French author, Pierre Loti, whose former home is now a French museum. Located in Rochefort, the author's collection of art objects is remarkable for its scope and features a wealth of various typologies, creative techniques and materials. The case of the gun (Figure 1) illustrates the problems involved in cleaning a tarnished copper alloy. The object also typifies the complexity of treating composite objects because it is constituted of a wooden butt covered with copper alloy plates decorated with semi-precious gems and corals. The copper alloy plates displayed uniform, light corrosion, and were stained with grease deposits. The multi-material nature of the object made chemical treatment by immersion impossible. Mechanical cleaning was possible but would have been long and difficult because of the decoration. Gel cleaning provided an attractive alternative. After initial cleaning with a mixture of water/ethanol applied using a cotton swab, the surface was de-greased with acetone. The different gels (agar, gellan gum and xanthan gum) were tested with a TAC solution at 2.5 wt% on the copper alloy plates. The best results were achieved using agar gel applied hot. Application time can vary between 10 min and 1 h, depending on the conservator's cleaning objective. Gel treatment successfully removed most of the copper corrosion products. The surface was rinsed with demineralized water applied with a cotton swab. The cleaning resulted in a homogeneous surface, as shine could be controlled.

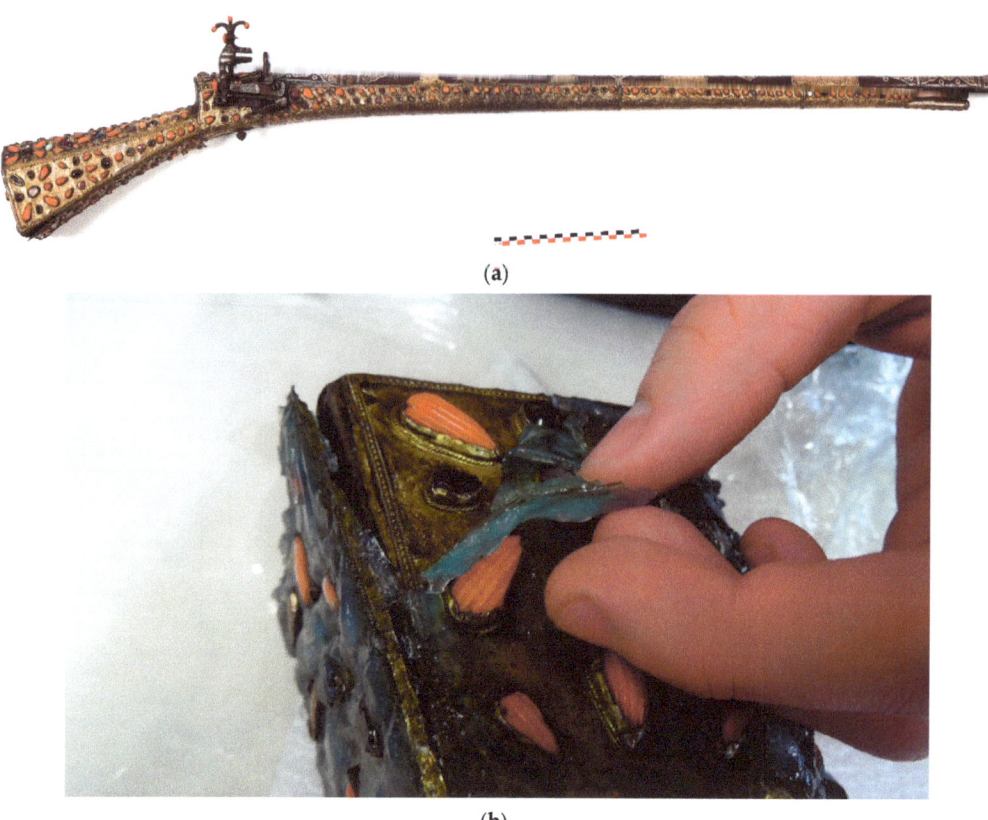

Figure 1. The gun from the Pierre Loti Museum collection (**a**), removal of the agar gel (**b**) (© J.G. Aubert/Arc'Antique—Grand Patrimoine de Loire Atlantique, Nantes, France).

Gellan gum gels also gave good results but processing was more delicate. The gel rapidly rigidified because of its higher gelation temperature than agar. Xanthan gum gels also removed corrosion products but left a lot of residues on the surface.

These treatments were also carried out by applying the gel as a film: Agar gel films were pre-made with demineralized water and immersed in a TAC solution before application. The cleaning was effective due to good contact between the gel film and the object. However, if an air bubble forms between the gel film and the object, the area remains tarnished.

2.2.2. Cleaning of Gilded Copper Alloys—Historical Artifact

The second case study also concerns a historical object: an Armenian censer from the Dobrée Museum collection (Figure 2). The object is made of gilded copper with a set of chased, embossed and openwork decorations. The copper showed thick corrosion masking the gilding (sometimes with gaps). The surface of the object was also covered with a thick, dirty varnish. An application of agar gel with tetrasodium EDTA (5 wt%) for 5 min removed dirt, varnish and part of the corrosion products. Cleaning was completed by the application of an agar gel with di-EDTA (2.5 wt%) for 5 min. Then, the surface was rinsed with demineralized water using a cotton swab. Agar gel cleaning produced a result that conservators consider to be highly satisfactory.

Figure 2. The Armenian censer from the Dobrée Museum collection during treatment (© L. Preud'homme/Arc'Antique—Grand Patrimoine de Loire Atlantique, Nantes, France).

2.2.3. Cleaning of Copper Alloys—Archaeological Artifact

The first two case studies showed that gel cleanings are effective in removing corrosion from copper. The third case study presents a similar problem: the removal of copper corrosion products from silver-plated copper alloy coins. In this case, however, the coins were archaeological objects, from the Cléons treasure (Haute Goulaine, Pays de Loire, France) dating from Antiquity (between 271 and 274 CE). They are in the effigy of various Roman emperors: Volusianus, Valerianus I, Gallienus and Saloninus. The Cléons treasure was discovered in 1901 and is conserved in the Dobrée museum collection. These coins were used in a study comparing cleaning with a range of gels, published by Giraud et al. [7]. In this article, we will focus on the tests performed with hydrogels: agar, gellan gum and xanthan gum (Figure 3). The objective of the cleaning was to reveal the significant surface which corresponds to the silver plating. The silver plating was covered with a mixture of corrosion products and sediments. The corrosion products above the silver plating were mainly composed of copper carbonates. During the burial period, a layer of corrosion products containing mainly copper oxides had developed below the silver plating. The treatment selected was an application of disodium EDTA gels at different concentrations (0.5, 2 or 5 wt%). It was applied to one half of the surface of a coin for 20 min, four or five times. The coins had not undergone mechanical cleaning before the first application of the gel. The treated surface was rinsed using a cotton swab impregnated with acetone diluted to 50% v. The coins were photographed prior to treatment and after each application of gel. The results achieved after the final application are presented in Figure 3. All treatments with 0.5 wt% di-EDTA were ineffective. Gel treatment with 2 wt% di-EDTA partially removed corrosion products but was not complete after five applications. Gel treatment with 5 wt% di-EDTA ensured effective cleaning after four or five applications. Peelable hydrogels (agar gels or gellan gum) gave the best results: they ensured good contact between the gel and the surface of the object; the treatment was effective when the solution contained a sufficient concentration of di-EDTA. They were easily removed and left little visible residue on the surface of the coin; post-treatment rinsing was therefore limited and the result of the treatment was homogeneous. The application of these hydrogels is precise and allows for the treatment of a well-defined surface. Viscous hydrogels (xanthan gum) also effectively cleaned the surface, but after treatment the surface was less well defined and gel removal was difficult. Xanthan gum left a large amount of residue that required copious rinsing. Despite thorough rinsing, it is possible that residue may have penetrated into the cracks and remains on the surface of the archaeological object.

Figure 3. Coins from the Cléons treasure (only half of the coin's surface was treated in each case) before or after treatments: before (**a**) and after 5 applications of agar with di-EDTA 0.5 wt% (**b**), before (**c**) and after 5 applications of agar with di-EDTA 2 wt% (**d**), before (**e**) and after 5 applications of agar with di-EDTA 5 wt% (**f**), before (**g**) and after 5 applications of gellan with di-EDTA 0.5 wt% (**h**), before (**i**) and after 5 applications of gellan with di-EDTA 2 wt% (**j**), before (**k**) and after 4 applications of gellan with di-EDTA 5 wt% (**l**), before (**m**) and after 5 applications of xanthan with di-EDTA 0.5 wt% (**n**), before (**o**) and after 5 applications of xanthan with di-EDTA 2 wt% (**p**), before (**q**) and after 5 applications of xanthan with di-EDTA 5 wt% (**r**) (© C. Colonnier/Arc'Antique—Grand Patrimoine de Loire Atlantique, Nantes, France).

2.2.4. Cleaning of Lead Artefacts

The fourth case study concerns another archaeological object: a curse tablet from the collection of the Medals and Antiques Department of the National Library of France. Most of the tablets from this collection come from North Africa, date from the early centuries AD, and were acquired in the 19th century. Such tablets were used as magical objects and some of them are inscribed [31]. Their condition had degraded as a result of being conserved in oak coin cabinets and exposed to acetic acid vapors. The inscriptions were either completely covered over by hard, thick corrosion products, or lightly veiled with whitish corrosion products. Layers containing the incisions sometimes either lacked adherence and displayed extensive cracking or, inversely, were compact and solid. The cleaning objective was to remove corrosion products so that the inscriptions could be read. With regard to chemical treatments, tests by conservators revealed that only acidic solutions could eliminate corrosion products from lead. Peelable hydrogels were not very compatible with acidic solutions. A film of water agar was first formed on the tablet and took on the shape of the surface. It was then immersed for 2 h in an oxalic acid solution. Next, it was placed on the relevant surface for 2 h. The operation was repeated three times. The photo in Figure 3b shows that the gel shrank after immersion in the acidic solution, and although it retained its mechanical strength, it was degraded. The contact between the gel film and the surface to be treated was not as good as when the agar gel was applied hot. The cleaning resulting from the use of oxalic acid in agar gel was not completely successful (Figure 4): The layer of corrosion products was reduced but only in the areas in direct contact with the gel. In addition, a whitish veil formed on the surface and increased with the number of gel applications. Xanthan gels were more compatible with acidic solutions. The cleaning test was performed with citric acid (5 wt%) in xanthan gel. The gel was applied for 1 h and the operation was repeated twice. The surface was rinsed using a cotton swab impregnated with water, but the gel proved very difficult to remove completely. Figure 4 shows the detail of the cleaned surface: corrosion products are still largely present and the treatment was not really effective. Gel treatments are therefore not really suitable for this type of archaeological object. Mechanical cleaning (micro-sandblasting with plant-based abrasive or with cationic exchange resins) yielded better results.

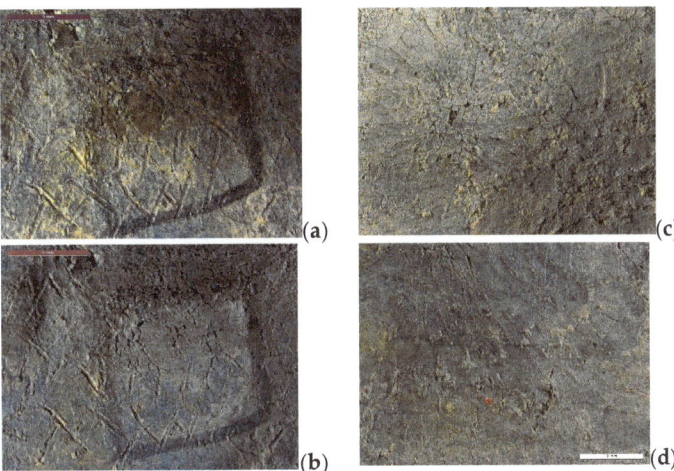

Figure 4. Results of gel treatments on curse tablet: before (**a**) and after application of agar with oxalic acid (**b**), before (**c**) and after application of xanthan with citric acid (**d**), (© L. Rossetti/Arc'Antique—Grand Patrimoine de Loire Atlantique, Nantes, France).

2.2.5. Cleaning of Painted Aluminum Alloys

In this case study, cleaning tests were carried out on painted surfaces from two authentic WWII aircraft wrecks:

A plate fragment from a Messerschmitt Bf109, which crashed during WWII at Le Rheu (west of Rennes, France) (Figure 5a). The fragment is in a good state of conservation. It is a copper-based aluminum alloy with cladding (Alclad: layer of pure aluminum above the alloy), it has a layer of paint with good cohesion with the metal surface. The object's surface was covered with black deposits. The cleaning objective was to remove these black deposits without damaging the paint.

A wing part from a Spitfire Mk VII MB887, which crashed on 1st June 1944, off the Saint Brieuc coast (France) (Figure 5b). The wing part is in relatively good condition: It is also a copper-based aluminum alloy with cladding coated with remains of the original paint. On the object's surface, there were aluminum corrosion products, some concretions and some traces of iron corrosion products. The cleaning objective was to remove the corrosion products and concretions without damaging the paint remains.

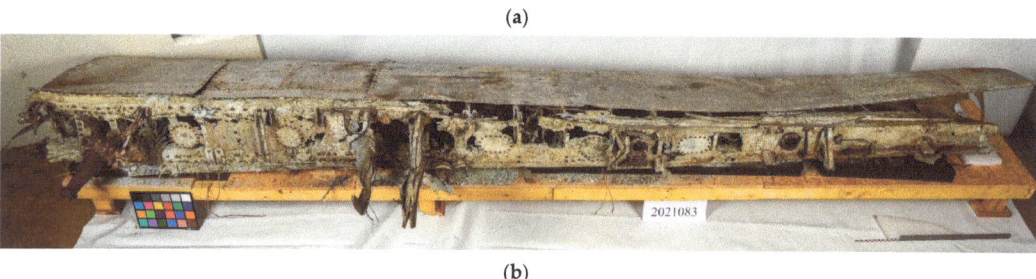

Figure 5. A small plate fragment from a Messerschmitt Bf109 (**a**) and a wing part from a Spitfire Mk VII MB887 (**b**) (© L. Preud'homme/Arc'Antique—Grand Patrimoine de Loire Atlantique, Nantes, France).

For Al alloys, chemical cleaning protocols are still experimental and limited data are available. The main treatments were developed in Australia and use solutions of citric acid [32], sodium metasilicates or tetrasodium EDTA [33]. Tri-ammonium citrate (TAC) solutions were also tested as they are used on painted surfaces [34]. For sodium metasilicate and tri-sodium EDTA solutions, xanthan gels were applied, agar gels were used for citric acid (0.055 M) and TAC. The results for the Messerschmitt Bf109 fragment are shown in Figure 5. The best results were obtained with citric acid or sodium metasilicate solutions: black deposits were removed with no damage to the paint. The ammonium citrate solution removed most of the black deposits, but it also eliminated the thinner parts of the painted layer. The EDTA solution was effective: all the black deposits disappeared but the paint was damaged by the cleaning. Different treatment solutions were also tested on the Spitfire wing part (Figure 6). Metasilicate sodium had little effect on corrosion products even after

60 min of treatment: this solution was therefore not suitable for cleaning the Spitfire wing. TAC or EDTA were more effective solutions because of their action on corrosion products. When tetra- and disodium EDTA solutions were applied using gel, they partially removed the corrosion products and had a slight effect on concretions but they also caused the paint remains to deteriorate. The TAC gel solution offered the best compromise for removing corrosion products without damaging the traces of paintwork but did not completely clean the object. To clean the wing part, agar gel with TAC was applied with a spray gun, in line with accepted procedures for large surfaces [21]. Three 20 min applications were necessary for a satisfactory result. Dirt was removed and rust stains reduced. Then, to homogenize the surface of those parts with a larger quantity of corrosion products and concretions, they were cleaned by sandblasting with vegetal abrasive.

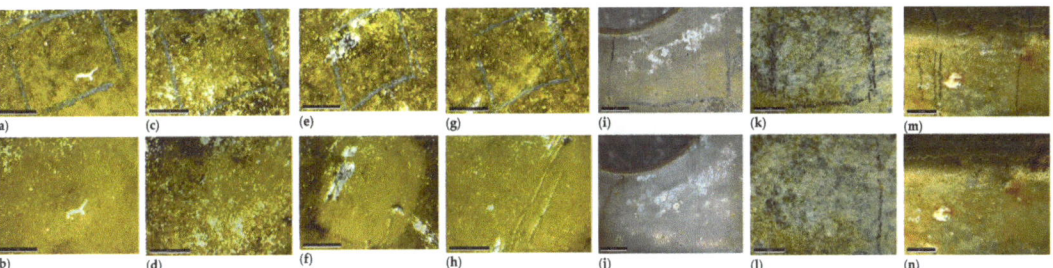

Figure 6. Results of gel treatments on the fragment from a Messerschmitt Bf109: before (**a**) and after treatment using agar with citric acid (0.055 M) (**b**), before (**c**) and after treatment using agar with TAC (**d**), before (**e**) and after treatment using xanthan with metasilicate sodium (**f**), before (**g**) and after treatment using xanthan with tetra-EDTA (**h**) (© S. Bampitzaris/Arc'Antique—Grand Patrimoine de Loire Atlantique, Nantes, France). Results of gel treatments on the wing part from a Spitfire Mk VII MB887: before (**i**) and after treatment using agar with TAC (**j**), before (**k**) and after treatment using xanthan with metasilicate sodium (**l**), before (**m**) and after treatment using xanthan with tetra-EDTA (**n**) (© E. Paillaux/Arc'Antique—Grand Patrimoine de Loire Atlantique, Nantes, France).

2.2.6. Cleaning of Vermeil Gold

This new case study concerns the cleaning of a further object from the Islamic art collection of the French author, Pierre Loti. It is a dagger in a good state of conservation (Figure 7a), composed of a steel blade and a walrus ivory handle with silver and vermeil plating. The silver elements displayed fine, homogeneous silver sulfide corrosion (Figure 7b). Tarnished silver objects are often cleaned mechanically but the relief of the decoration made it difficult to apply a mechanical treatment. Electrochemical treatments thus provided a good alternative [35]: First, the silver sulfides were reduced to silver at −0.89 V/SCE (saturated calomel electrode), then the reduced silver deposited on the gilding was oxidized at +0.66 V/SCE to reveal the gilding. However, since the dagger is a composite object, the ivory parts could not be immersed. Electrochemical treatment was therefore applied locally using an agar gel with KNO_3 at 1 wt%. Cleaning was completed by rinsing the surface with a water/ethanol mixture applied with a soft brush (Figure 7c). In this case study, the gel treatment was successful.

2.2.7. Stabilization of Copper Alloys or Iron

As seen in the previous case, local electrochemical treatments offer a very satisfactory treatment alternative for composite objects. Other electrochemical treatments have been developed for the treatment of metals, in particular dechlorination treatments for iron [36] or copper alloys [37]. The electrochemical setup is a classic three-electrode electrolysis setup. The object to be treated acts as the working electrode. The gel (agar with KNO_3 1 wt%) acts as an electrolytic solution and the counter electrode is a stainless-steel grid. The reference electrode is placed in an extension containing a conductive solution (KNO_3

1 wt%) and inserted in the gel (Figure 8). The reference electrode is a saturated calomel electrode (SCE). Cathodic potential is applied to the object to be treated (working electrode) to allow chloride ions to migrate into the gel: the potential is −1.04 V/SCE for iron objects or −0.19 V/SCE for copper alloys. A previous study has shown the effectiveness of this treatment [38]. For it to be effective, contact between the corrosion layers and the gel must be very good; a drop of the electrolyte (KNO_3 at 1 wt%) must be placed on the surface to be treated. The gel must be changed every 30 min. The difficulty of these electrochemical treatments using gel is to determine the end of treatment. The amount of extracted chloride can be monitored by measuring chloride concentration in the gel after treatment by X-ray fluorescence (XRF) [38]. When all chlorides have been extracted, the object is stable. The absence of active corrosion can also be checked by measuring the oxygen consumption of the object in a leakproof pocket [39].

Figure 7. A knife from the collection of the Pierre Loti Museum (**a**), and of a detail of the handle before (**b**) and after treatment (**c**) (© J.G. Aubert/Arc'Antique—Grand Patrimoine de Loire Atlantique, Nantes, France).

Figure 8. The electrolytic assembly for the stabilization using gel of an archaeological iron object (© C. Fontaine/Arc'Antique—Grand Patrimoine de Loire Atlantique, Nantes, France).

3. Discussion

These first studies on the use of gels for the treatment of metals show that gels can provide good treatment alternatives but that they are not suitable in every case. Treatments using agar gels applied hot yielded the best results. The limits of hot applied agar gels are their compatibility with treatment solutions. Not all active agents commonly used in chemical treatments are compatible with agar. The treatment solutions must have a pH close to the neutral range, otherwise the helix network of the agar gel cannot form and the gel is not peelable. Agar gel treatments can incorporate solutions of TAC, EDTA, dilute citric acid or a neutral conductive solution such as KNO_3. The case studies showed that gel treatments work better on historical than archaeological objects. This is due to difference in the surface state of the objects. Generally, historical objects have a homogeneous surface whereas archaeological objects have cracked, heterogeneous corrosion layers. Agar gels applied hot to an archaeological object tend to penetrate the cracks and porosities of the corrosion layers. When the gel is removed, parts of the surface may become detached and residues may remain in the corrosion layers. Post-treatment gel residue has not yet been quantified. An initial study evaluated the presence of residues by integrating a fluorescent marker, fluorescein (at 10^{-4} M), in the gels. The technique was not precise enough to determine the quantity of the residues, but it enabled a comparison to be made with residues due to agar or xanthan gels, respectively, on a historical silver object with a surface relief (Figure 9). After removal of the gel, the surface was rinsed with demineralized water using a cotton swab. The green color of the gel residues showed up under ultraviolet light (UV). Agar residues were very localized in cavities and represented about 0.7% of the treated surface while xanthan residues formed a veil on the surface and represented about

20% of the treated surface. Although xanthan gels are more compatible with chemical agents than agar gels, they form a viscous gel that is difficult to rinse off. Xanthan residues are significant and are the main reason that conservators avoid using it for the treatment of metal objects.

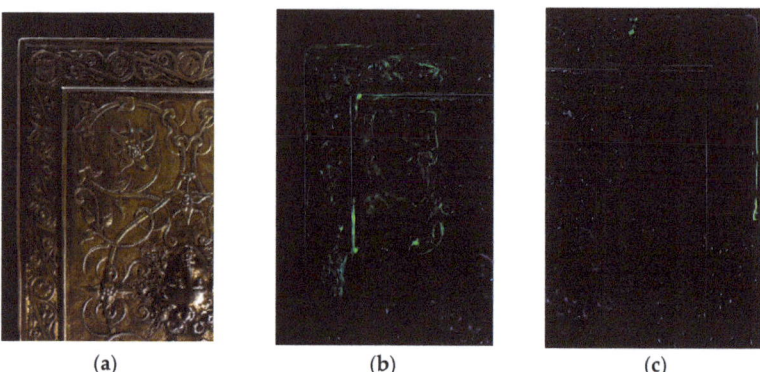

Figure 9. A historical silver object with surface relief: (**a**) under normal lighting, (**b**) under UV light after removal of a xanthan gel containing 10^{-4} M fluorescein, (**c**) under UV light after removal of an agar gel containing 10^{-4} M fluorescein (© E. Guibert-Martin/Arc'Antique—Grand Patrimoine de Loire Atlantique, Nantes, France).

Gellan gels and agar gels form peelable gels and have similar compatibility with chemical agents. The advantage of agar gels is their thermo-reversibility: they can be reheated several times, which is not the case for gellan gels. Moreover, the low cost of agar gels makes them easy for conservators to access. Gellan gels solidify very quickly because they have a gelling temperature of 70 °C while agar gels gel at around 40 °C. Agar gels may seem more practical to use but in hot weather they may become unusable and unable to solidify. In this case, the use of gellan gels is to be preferred.

Another advantage of using gels is the possibility of localized treatment. They are particularly suited to the treatment of composite objects that cannot withstand immersion in a chemical bath. The use of gel also allows the quantity of solution and active agent to be limited. This is often cited as an ecological benefit of gel treatments. Gel treatment as a new green method is currently being researched by Edith Joseph at the University of Neuchatel [8]. The use of siderophores for the cleaning of historical ferrous metals yields very promising results [40].

These different studies are shared among French conservators who are participating in the "Gels Métaux" collaborative project [41]. Through a series of exchange days and workshops, scientists and conservators present their respective studies and treatment examples as well as the success and limits of their research.

4. Conclusions and Perspectives

For several decades, the use of gels has greatly modified the practices of painting and graphic art conservators. Progressively, they are now becoming more commonplace in the restoration of metals. The use of cleaning gels for historical objects is increasingly becoming an integral part of conservation practice. The best results were obtained using agar gels associated with a chelating agent (EDTA or TAC) applied hot. These gel treatments were used to remove dirt and corrosion products from historical copper and iron objects. They are particularly suitable for composite objects. For objects featuring decoration (gilding, silver plating, painting), gel treatments can be an interesting alternative. Gel treatments allow for the local application of a chemical or electrochemical treatment. Peelable gels are preferable on surfaces in good condition, such as those of historical objects. The hot application of these peelable gels (agar or gellan gum) ensures good contact between the gel and the surface to be treated. The use of gel films is generally unsuitable for the treatment of metal objects because the geometry of the objects prevents good contact between the gel film and the treated surface. Viscous gels, such as xanthan gum, adhere well to the surface of the object and have a good level of compatibility with many active agents. However, xanthan gels are very difficult to remove, even after copious rinsing. On a surface in good condition—as is generally the case for historical objects—even after thorough rinsing a large quantity of xanthan gel residue remains in the form of a thin layer spread over the surface.

For archaeological objects, the use of gels is more problematic. Gels can penetrate corrosion layers and cause peeling or/and leave residues after treatment. The question of residues is an issue that remains to be investigated. Gel residues must be quantified and their impact on the conservation of objects must be determined. It should also be possible to limit their presence by optimizing gel application protocols.

5. Material and Methods

5.1. Preparation Protocol of Gels

To prepare agar gel, first a solution was heated to 50 °C and mixed with 3 wt% agar (AgarArt® provided by CTS Conservation, France). It was then heated to 90 °C until the mixture became homogeneous and translucent. It was allowed to cool for 24 h before being heated a second time to 90 °C. The preparation of gellan gum (Kelcogel® provided by CTS or Phytagel® from Sigma Aldrich), requires only a single heating. When peeling gels (agar or gellan gum) was applied hot, the gel was mixed directly with the treatment solution and applied to the object with a syringe (hot application). For cold application, films (2 or 3 mm in thickness) were formed with demineralized water. The films were then immersed in the treatment solution for 2 h at room temperature, impregnating the gel with the treatment solution.

Xanthan gum preparations (from Kremer, or Vanzan® by CTS Conservation) were made at room temperature 24 h prior to use. A concentration of 2 wt% (Kremer) or 5 wt% (CTS Conservation) xanthan gum was added to the treatment solution, and after 24 h the mixture became homogeneous. The gel can be applied cold. Unused gel must be refrigerated (24 h) to avoid the development of mold.

5.2. Treatment Solutions

The active agents selected are used in chemical or electrochemical treatments in metal restoration. The 4 types of selected active agents tested were: acidic solutions to eliminate corrosion products, sediments and concretions; neutral or alkaline conductive solutions to carry out electrolysis, and chelating agent solutions to remove corrosion products. The list of selected active agents is detailed in Table 2.

Table 2. List of treatment solutions.

Type	Composition	Concentration	pH
Acidic solution	Nitric acid	10^{-4} M	3
	Citric acid	5 wt%	3
	Citric acid	0.055M	5.4
	Oxalic acid	5 wt%	2
Neutral conductive solution	KNO_3	1 wt%	5.6
Alkaline solution	Metasilicate Sodium	2 wt%	9
	NaOH	2 wt%	13.5
Chelating agent	TAC (tri-ammonium citrate)	2.5 wt%	7
	TAC (tri-ammonium citrate)	5 wt%	7.5
	Disodium EDTA (ethylenediaminetetraacetic acid)	0.5 wt%	4.7
	Disodium EDTA (ethylenediaminetetraacetic acid)	2 wt%	4.5
	Disodium EDTA (ethylenediaminetetraacetic acid)	5 wt%	4.4
	Tetrasodium EDTA (ethylenediaminetetraacetic acid)	5 wt%	10

Funding: Research for the Metal Gels project was supported by DRAC Pays de Loire, and research on aluminum was supported by JPI-CH, within the framework of the "Procraft" project.

Data Availability Statement: The data of these studies are kept only at the Arc'Antique laboratory.

Acknowledgments: The author would like to thank all the students who contributed to the gel studies, and especially Andrea Dupke, Alban Gomez, Tiffanie Giraud, Clémence Fontaine, Erwan Guibert-Martin, Sotiris Bampitzaris and Eve Paillaux. The author wishes to express her gratitude to the conservators, and more particularly to Loretta Rossetti and Aymeric Raimon, as well as to Charlène Pelé-Meziani for the analyses. The author really appreciates everything the co-founders of the gel metals project, Manuel Leroux and Aymeric Raimon, have done. Many thanks to Edith Joseph for having associated the author to the Helix project and to Luana Cuvillier for her high-quality work in this project.

Conflicts of Interest: The author declares no conflict of interest.

References

1. Turner-Walker, G. The conservation of metals. In *A Practical Guide to the Care and Conservation of Metals*; Wang, X., Ed.; Art and Design Agency: London, UK, 2008; pp. 61–76.
2. Stulik, D.; Miller, D.; Khanjian, H.; Khandekar, N.; Wolbers, R.; Carlson, J.; Petersen, C. *Solvent Gels for the Cleaning of Works of Art: The Residue Question*; Dorge, V., Ed.; The Getty Conservation Institute: Los Angeles, CA, USA, 2004; 160p.
3. Khandekar, N. A survey of the conservation literature relating to the development of aqueous gel cleaning on painted and varnished surfaces. *Stud. Conserv.* **2000**, *45*, 10–20. [CrossRef]
4. Angelova, L.; Ormsby, B.; Townsend, J.; Wolbers, R. *Gels in the Conservation of Art*; Archetype Publications: London, UK, 2017; 400p.
5. Baglioni, P.; Berti, D.; Bonini, M.; Carretti, E.; Dei, L.; Fratini, E.; Giorgi, R. Micelle, microemulsions, and gels for the conservation of cultural heritage. *Adv. Colloid Interface Sci.* **2014**, *205*, 361–371. [CrossRef] [PubMed]
6. Bonelli, N.; Chelazzi, D.; Baglioni, M.; Giorgi, R.; Baglioni, P. Confined Aqueous Media for the Cleaning of Cultural Heritage: Innovative Gels and Amphiphile-Based Nanofluids. In *Nanoscience and Cultural Heritage*; Dillmann, P., Bellot-Gurlet, L., Nenner, I., Eds.; Atlantis Press: Paris, France, 2016; pp. 283–311.
7. Giraud, T.; Gomez, A.; Lemoine, S.; Pelé-Meziani, C.; Raimon, A.; Guilminot, E. Use of gels for the cleaning of archaeological metals. Case study of silver-plated copper alloy coins. *J. Cult. Herit.* **2021**, *52*, 73–83. [CrossRef]
8. Passaretti, A.; Cuvillier, L.; Sciutto, G.; Guilminot, E.; Joseph, E. Biologically Derived Gels for the Cleaning of Historical and Artistic Metal Heritage. *Appl. Sci.* **2021**, *11*, 3405. [CrossRef]
9. Gulrez, S.; Al-Assaf, S.; Phillips, G. Hydrogels: Methods of preparation, characterization and applications. In *Progress in Molecular and Environmental Bioengineering—From Analysis and Modeling to Technology Applications*; InTech: Vienna, Austria, 2011; Chapter 5; pp. 117–151.
10. Armisen, R.; Galatas, F. Agar. In *Handbook of Hydrocolloïds*; Woodhead Plushing Limited: Sawston, UK; CRC Press: Boca Raton, FL, USA, 2009; pp. 82–107.
11. Cremonesi, P. Rigid gels and enzyme cleaning. In *Cleaning 2010: New Insights into the Cleaning of Paintings*; Smithsonian Institution Scholarly Press: Washington, DC, USA, 2013; pp. 179–184.
12. Cremonesi, P. Surface cleaning? Yes, freshly grated Agar gel, please. *Stud. Conserv.* **2016**, *61*, 362–367. [CrossRef]
13. Bertasa, M.; Poli, T.; Riedo, C.; Di Tullio, V.; Capitani, D.; Proietti, N.; Canevali, C.; Sansonetti, A.; Scalarone, D. A study of non-bounded/bounded water and water mobility in different agar gels. *Microchem. J.* **2018**, *139*, 306–314. [CrossRef]
14. Bertasa, M.; Canevali, C.; Sansonetti, A.; Lazzari, M.; Malandrino, M.; Simonutti, R.; Scalarone, D. An in-depth study on the agar gel effectiveness for built heritagecleaning. *J. Cult. Herit.* **2021**, *47*, 12–20. [CrossRef]
15. Gullotta, D.; Saviello, D.; Gherardi, F.; Toniolo, L.; Anzani, M.; Rabbolini, A.; Goidanich, S. Setup of a sustainable indoor cleaning methodology for thesculpted stone surfaces of the Duomo of Milan. *Herit. Sci.* **2014**, *2*, 6. [CrossRef]
16. Canevali, C.; Fasoli, M.; Bertasa, M.; Botteon, A.; Colombo, A.; Di Tullio, V.; Capitani, D.; Proietti, N.; Scalarone, D.; Sansonetti, A. A multi-analytical approach for thestudy of copper stain removal by agar gels. *Microchem. J.* **2016**, *129*, 249–258. [CrossRef]
17. Bertasa, M.; Bandini, F.; Felici, A.; Lanfranchi, M.R.; Negrotti, R.; Riminesi, C.; Scalarone, D.; Sansonetti, A. A soluble salts extraction with different thickeners:monitoring of the effects on plaster. *IOP Conf. Ser. Mater. Sci. Eng.* **2018**, *364*, 012076. [CrossRef]
18. Casoli, A.; Cremonesi, P.; Isca, C.; Groppetti, R.; Pini, S.; Senin, N. Evaluation of the effect of cleaning on the morphological properties of ancient paper surfaces. *Cellulose* **2013**, *20*, 2027–2043. [CrossRef]
19. Sullivan, M.; Duncan, V.; Berrie, B.; Richard, W.; Angelova, L.; Ormsby, B.; Townsend, J.; Weiss, R. Rigid polysaccharide gels for paper conservation: A residue study. In *Gels in the Conservation of Art*; Archetype Publications: London, UK, 2017; pp. 42–50.
20. Barkovic, M.; Diamond, O.; Cross, M. The use of agar gel for treating water stains on an acrylic canvas. In *Gels in the Conservation of Art*; Archetype Publications: London, UK, 2017; pp. 51–56.
21. Giordano, A.; Cremonesi, P. New Methods of Applying Rigid Agar Gels: From Tiny to Large-scale Surface Areas. *Stud. Conserv.* **2021**, *66*, 437–448. [CrossRef]

22. Sansonetti, A.; Bertasa, M.; Canevali, C.; Rabbolini, A.; Anzani, M.; Scalarone, D. A review in using agar gels for cleaning art surfaces. *J. Cult. Herit.* **2020**, *44*, 285–296. [CrossRef]
23. Iannuccelli, S.; Sotgiu, S. Wet treatments of works of art on paper with rigid Gellan gels. In Proceedings of the AIC's 38th Annual Meeting, Milwaukee, WI, USA, 11–14 May 2010; pp. 25–39.
24. Sworn, G. Gellan gum. In *Handbook of Hydrocolloïds*; Woodhead Plushing Limited: Sawston, UK; CRC Press: Boca Raton, FL, USA, 2009; pp. 204–227.
25. De Filpo, G.; Palermo, A.; Tolmino, R.; Formoso, P.; Nicoletta, F. Gellan gum hybrid hydrogels for the cleaning of paper artworks contaminated with Aspergillus versicolor. *Cellulose* **2016**, *23*, 3265–3279. [CrossRef]
26. Mazzuca, C.; Micheli, L.; Carbone, M.; Basoli, F.; Cervelli, E.; Iannuccelli, S.; Sotgiu, S.; Palleschi, A. Gellan hydrogel as a powerful tool in paper cleaning process: A detailed study. *J. Colloid Interface Sci.* **2014**, *416*, 205–211. [CrossRef] [PubMed]
27. Albini, M.; Letardi, P.; Mathys, L.; Brambilla, L.; Schröter, J.; Junier, P.; Joseph, E. Comparison of a Bio-Based Corrosion Inhibitor versus Benzotriazole on Corroded Copper Surfaces. *Corros. Sci.* **2018**, *143*, 84–92. [CrossRef]
28. Garcia-Ochoa, F.; Santos, V.E.; Casas, J.A.; Gomez, E.E. Xanthan gum: Production, recovery, and properties. *Biotechnol. Adv.* **2000**, *18*, 549–579. [CrossRef]
29. Sworn, G. Xanthan gum. In *Handbook of Hydrocolloïds*; Woodhead Plushing Limited: Sawston, UK; CRC Press: Boca Raton, FL, USA, 2009; pp. 186–203.
30. Casoli, A.; Di Diego, Z.; Isca, C.; Zaira, D.; Clelia, I. Cleaning painted surfaces: Evaluation of leaching phenomenon induced by solvents applied for the removal of gel residues. *Environ. Sci. Pollut. Res.* **2014**, *21*, 3252–13263. [CrossRef]
31. Rossetti, L.; Stephant, N.; Aubert, J.G.; Mélard, N.; Guilminot, E. Study and Conservation of Lead Curse Tablets. In Proceedings of the Metal 2019 Proceedings of the Interim Meeting of the ICOM-CC Metals Working Group, Neuchâtel, Switzerland, 2–6 September 2019; Chemello, C., Brambilla, L., Joseph, E., Eds.; International Councils of Museums—Committee for Conservation: Neuchâtel, Switzerland, 2019; pp. 329–337.
32. Bailey, G.T. Stabilization of a wrecked and corroded aluminium aircraft. In Proceedings of the ICOM Metal 2004, Canberra, Australia, 4–8 October 2004; pp. 453–464.
33. MacLeod, I.; Kelly, D. *The Effects of Chloride Ions on the Corrosion Aluminium Alloys Used in the Construction of Australia II*; AICCM Bulletin: Moonah, Australia, 2001; pp. 10–19.
34. Morrison, R.; Bagley-Young, A.; Burnstock, A.; Jan van den Berg, K.; Van Keulen, H. An Investigation of Parameters for the Use of Citrate Solutions for Surface Cleaning Unvarnished Paintings. *Stud. Conserv.* **2007**, *52*, 255–270. [CrossRef]
35. Degrigny, C.; Wéry, M.; Vescoli, V.; Blengino, J.M. Altération et nettoyage de pièces en argent doré. *Stud. Conserv.* **1996**, *41*, 170–178.
36. Guilminot, E.; Neff, D.; Rémazeilles, C.; Reguer, S.; Kergourlay, F.; Pelé, C.; Dillmann, P.; Refait, P.; Nicot, F.; Mielcarek, F.; et al. Influence of crucial parameters on the dechlorination treatments of ferrous objects from seawater. *Stud. Conserv.* **2012**, *57*, 227–236. [CrossRef]
37. Païn, S.; Bertholon, R.; Lacoudre, N. La déchloruration des alliages cuivreux par électrolyse à faible polarisation dans le sesquicarbonate de sodium. *Stud. Conserv.* **1991**, *36*, 33–43. [CrossRef]
38. Fontaine, C.; Lemoine, S.; Pelé-Meziani, C.; Guilminot, E. The use of gels in localized dechlorination treatments of metallic cultural heritage objects. *Herit. Sci.* **2022**, *10*, 117. [CrossRef]
39. Matthiesen, H. Oxygen monitoring in the corrosion and preservation of metallic heritage artefacts. *Corros. Conserv. Cult. Herit. Met. Artefacts* **2013**, *65*, 368–391.
40. Cuvillier, L.; Passaretti, A.; Raimon, A.; Dupuy, V.; Guilminot, E.; Joseph, E. Exploiting Biologically Synthetized Chelators in Conservation: Gel-based Bio-cleaning of Corroded Iron Heritage Objects. In Proceedings of the Metal 2022 Proceedings of the Interim Meeting of the ICOM-CC Metals Working Group, Helsinki, Finland, 5–9 September 2022; Mardikian, P., Näsänen, L., Arponen, A., Eds.; International Councils of Museums—Committee for Conservation: Helsinki, Finland, 2022; pp. 25–34.
41. Guilminot, E.; Gomez, A.; Raimon, A.; Leroux, M. Use of Gels for the treatment of Metals. In Proceedings of the Metal 2019 Proceedings of the Interim Meeting of the ICOM-CC Metals Working Group, Neuchâtel, Switzerland, 2–6 September 2019; Chemello, C., Brambilla, L., Joseph, E., Eds.; International Councils of Museums—Committee for Conservation: Neuchâtel, Switzerland, 2019; p. 473.

Disclaimer/Publisher's Note: The statements, opinions and data contained in all publications are solely those of the individual author(s) and contributor(s) and not of MDPI and/or the editor(s). MDPI and/or the editor(s) disclaim responsibility for any injury to people or property resulting from any ideas, methods, instructions or products referred to in the content.

Article

Bamboo Nanocellulose/Montmorillonite Nanosheets/Polyethyleneimine Gel Adsorbent for Methylene Blue and Cu(II) Removal from Aqueous Solutions

Xuelun Zhang [1,†], Feng Li [2,†], Xiyu Zhao [1], Jiwen Cao [1,3], Shuai Liu [1,4,5], You Zhang [1], Zihui Yuan [1], Xiaobo Huang [1], Cornelis F. De Hoop [6], Xiaopeng Peng [3,4,5,*] and Xingyan Huang [1,7,*]

1. College of Forestry, Sichuan Agricultural University, Chengdu 611130, China
2. Research Institute of Characteristic Flowers and Trees, Chengdu Agricultural College, Chengdu 611130, China
3. State Key Laboratory of Tree Genetics and Breeding, Key Laboratory of Tree Breeding and Cultivation of the National Forestry and Grassland Administration, Research Institute of Forestry, Chinese Academy of Forestry, Beijing 100091, China
4. Key Laboratory of Pulp and Paper Science & Technology of Ministry of Education, Qilu University of Technology (Shandong Academy of Sciences), Jinan 250353, China
5. National Forestry and Grassland Administration Key Laboratory of Plant Fiber Functional Materials, Fuzhou 350108, China
6. School of Renewable Natural Resources, Louisiana State University Agricultural Center, Baton Rouge, LA 70803, USA
7. Wood Industry and Furniture Engineering Key Laboratory of Sichuan Provincial Department of Education, Chengdu 611130, China

* Correspondence: xp@caf.ac.cn (X.P.); hxy@sicau.edu.cn (X.H.)
† These authors contributed equally to this work.

Citation: Zhang, X.; Li, F.; Zhao, X.; Cao, J.; Liu, S.; Zhang, Y.; Yuan, Z.; Huang, X.; De Hoop, C.F.; Peng, X.; et al. Bamboo Nanocellulose/ Montmorillonite Nanosheets/ Polyethyleneimine Gel Adsorbent for Methylene Blue and Cu(II) Removal from Aqueous Solutions. *Gels* 2023, 9, 40. https://doi.org/10.3390/gels9010040

Academic Editor: Annarosa Gugliuzza

Received: 15 November 2022
Revised: 30 November 2022
Accepted: 1 December 2022
Published: 4 January 2023

Copyright: © 2023 by the authors. Licensee MDPI, Basel, Switzerland. This article is an open access article distributed under the terms and conditions of the Creative Commons Attribution (CC BY) license (https:// creativecommons.org/licenses/by/ 4.0/).

Abstract: In recent years, the scarcity of pure water resources has received a lot of attention from society because of the increasing amount of pollution from industrial waste. It is very important to use low-cost adsorbents with high-adsorption performance to reduce water pollution. In this work, a gel adsorbent with a high-adsorption performance on methylene blue (MB) and Cu(II) was prepared from bamboo nanocellulose (BCNF) (derived from waste bamboo paper) and montmorillonite nanosheet (MMTNS) cross-linked by polyethyleneimine (PEI). The resulting gel adsorbent was characterized by Fourier transform infrared spectroscopy (FTIR), field emission scanning electron microscopy (SEM), X-ray photoelectron spectroscopic (XPS), etc. The results indicated that the MB and Cu(II) adsorption capacities of the resulting gel adsorbent increased with the solution pH, contact time, initial concentration, and temperature before equilibrium. The adsorption processes of MB and Cu(II) fitted well with the fractal-like pseudo-second-order model. The maximal adsorption capacities on MB and Cu(II) calculated by the Sips model were 361.9 and 254.6 mg/g, respectively. The removal of MB and Cu(II) from aqueous solutions mainly included electrostatic attraction, ion exchange, hydrogen bonding interaction, etc. These results suggest that the resulting gel adsorbent is an ideal material for the removal of MB and Cu(II) from aqueous solutions.

Keywords: gel; adsorption; nanocellulose; montmorillonite; methylene blue; heavy metal

1. Introduction

In recent years, many pollutants, such as nitrates [1,2], phosphorus [3], antibiotics [4], dyes [5], and heavy metals [6], have been discharged into the water as a result of rapid industrial development and human activities. Dyes and heavy metals are two of the most typical pollutants in industrial wastewater. They have serious impacts on human living environments [7,8]. The dyes in the water bodies can absorb light and hinder the penetration of light, thereby reducing the photosynthetic activities of aquatic plants and microorganisms, and inhibiting their growth [9]. Methylene blue (MB), a cationic dye, is widely used in textile, paper, coating, and printing [10,11]. The residual MB in wastewater has a serious influence on the environment and human health due to its non-biodegradable feature [12]. Cu(II), a common used heavy metal, is produced during copper smelting, processing, and electroplating [13]. Due to its toxicity and non-degradability in water, it could gradually accumulate into food chains, endangering human health [14]. As a result, Cu(II) is one of the priority pollutants classified by the US EPA [15]. Therefore, MB and Cu(II) in wastewater must be eliminated before being discharged into the environment.

The removal methods of MB and Cu(II) include ion exchange [16], chemical precipitation [17], electrodialysis [18], flocculation [19], membrane technology [20], adsorption [21], electrodialysis, etc. Chemical precipitation and flocculation are simple to operate, but they usually produce sludge and are difficult to remove; membrane technology and ion exchange introduce new chemicals; electrodialysis is also difficult to put into practical use given the high costs involved [22]. Adsorption is one of the most useful strategies in wastewater treatment because of its convenience, effectiveness, low cost, and no harmful products [23]. Many adsorbent materials, such as activated carbon, carbon nanotube, graphene, and metal-organic framework, have been applied to remove dyes and heavy metals in wastewater [24–27]. It is worth noting that some natural adsorbents, such as loofah sponges and sugarcane bagasse, can also remove pollution in wastewater, even though their adsorption capacities are not satisfactory [28,29]. The development of a natural, cheap, efficient, and biodegradable adsorbent is still a major concern in the study of wastewater treatment [30].

Cellulose is a naturally acquirable material with good biodegradability and renewability [31]. Nanoscale cellulose, i.e., nanocellulose (CNF) and cellulose nanocrystal (CNC) were developed from cellulose-containing materials [32]. The high aspect ratio, abundant hydroxyl groups, and structural flexibility of CNF are beneficial to removing dyes and heavy metal ions from wastewater. For example, CNF gel absorbent produced from TEMPO-oxidized CNF has a good adsorption effect on cationic dye (malachite green) [33]. Moreover, the adsorption efficiency of CNF gel toward Cu(II) could be enhanced via polyethyleneimine (PEI) [34].

Montmorillonite (MMT) is a clay mineral in nature. It has been demonstrated that MMT could be used as a reinforcing agent in the cellulose framework [35]. MMT has a three-layer sheet structure, consisting of a middle layer of aluminum oxide octahedron (Al^{3+}) between two layers of silicon-oxygen tetrahedrons (Si^{4+}) [36]. The Al^{3+} and Si^{4+} in MMT will be replaced by other cations with lower charges, resulting in a negatively charged MMT [37]. To eliminate the charge imbalance in the crystal lattice, a large number of exchangeable cations, such as Na^+ and Ca^{2+}, will be generated between layers [38]. The overall negative charge and cation exchange in MMT will contribute to its good adsorption performance on cationic pollutants in wastewater [5]. In addition, the lattice structure of MMT is maintained by weak electrostatic force and van der Waals force. It is easy to exfoliate MMT into nanosheets [39]. It will boost its surface areas, resulting in the full exposure of adsorption sites and functional groups [40,41]. Therefore, MMT nanosheets (MMTNS) have great potential to be used as adsorbents.

The application of gel in the removal of dyes and heavy metal ions from wastewater has received a lot of attention [16,42]. The gel structure is built from hydrophilic groups, for example, –OH, –COOH, –NH_2, –$CONH_2$, and –SO_3H, or hydrated polymer networks under aqueous conditions [43]. Many materials have been used to prepare gels, such as chitosan, alginate, and cellulose. However, these biopolymers have many drawbacks, such as low stability and limited potential to remove dye molecules. Cross-linking is the most critical strategy to prepare the gel with a stable 3D structure for improving its physical and mechanical properties [44,45]. For example, gel beads prepared from sodium alginate and carboxymethyl cellulose by blending and cross-linking have adsorption capacities for lead ions due to their hydroxyl and carboxyl groups [46].

Polyethyleneimine (PEI) has a large number of amino groups that have good adsorption capacities in heavy metal ions and they are able to form cross-linking sites with other functional groups to enhance the structure stability [47]. In this study, PEI with abundant amino groups was used as a cross-linking agent to reinforce the connection between bamboo nanocellulose and MMTNS for producing a stable gel adsorbent. Similar work on CNF/MMT/PEI had similar raw materials to ours [35]. The authors focused on the adsorption of anionic dye through amino groups from PEI, while MMT was overall negatively charged, it could be easily exfoliated into nanosheets to increase its surface areas and adsorption sites for positively charged cationic dyes.

Thereby, the performance of the resulting gel adsorbent in this work to remove MB and Cu(II) from the aqueous solutions was investigated. The novelty of this work was the utilization of waste bamboo paper, montmorillonite, and polyethyleneimine as raw materials, for their low costs and great potential adsorption abilities after combining, and to prepare bio-based gel adsorbents, which had good adsorption capacities for MB and Cu(II). The adsorption processes and maximum adsorption capacities were determined by a kinetic study and adsorption isotherm, respectively. Furthermore, SEM and XPS analyses were carried out to study the adsorption mechanism.

2. Results and Discussion

2.1. Characterization of the BMP Gel Adsorbent

2.1.1. FTIR Spectra

The FTIR spectra of BCNF, MMTNS, and BMP gel are shown in Figure 1. The remarkable BCNF bands at 3334 and 1601 cm^{-1} corresponded to O–H stretching, and C=O stretching, respectively [48]. The bands observed at approximately 2903 cm^{-1} were attributed to the symmetrical stretching vibration of the C–H bond [49]. The band at 1431 cm^{-1} corresponded to CH_2 deformation; 1375–1315 cm^{-1} was assigned to C–H and C–OH deformation, and 1160–896 cm^{-1} corresponded to the C–O-stretched backbone vibrations and glycosidic linkages between sugar units [50,51]. A strong band at 3633 cm^{-1}, assigned to Al–OH vibration, was observed from MMTNS [52]. The main bands at 1640, 930, and 795 cm^{-1} corresponded to the stretching vibration of the hydroxyl groups, O–Si–O stretching, and the Si–O–Al stretching in MMTNS, respectively [53,54]. Moreover, it was found that the band of hydrogen bonding in the BMP gel was well retained at 3334 cm^{-1}. These findings indicated that strong hydrogen bonding interactions were formed between BCNF and MMTNS [55]. Meanwhile, a secondary amide shoulder from PEI was observed at 1649 cm^{-1} of the BMP gel adsorbent, indicating that PEI was involved successfully [56]. Moreover, a strong stretching vibration band of carboxylic acid anion was found at 1560 cm^{-1} on the IR spectrum of the BMP gel adsorbent [57]. These findings suggested the formation of a cross-linked structure between –COOH (BCNF) and -NH_2 (PEI) through an electrostatic attraction (charged salt groups formed–NH_3(+)/–NH_2–(+) and (−)OOC– interactions). The proposed preparation mechanism of the BCNF/MMTNS/PEI gel adsorbent is shown in Figure 2a.

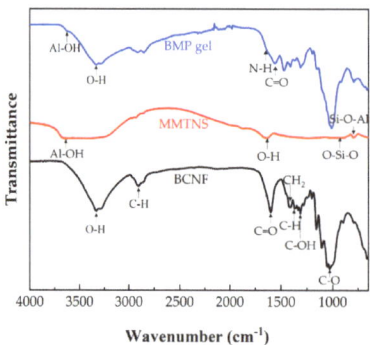

Figure 1. FTIR spectra of BCNF, MMTNS, and BMP gel adsorbent.

2.1.2. SEM

The surface morphology and structure of the MMTNS and BMP gel adsorbent are presented in Figure 2. There were many nanosheets on the surface of MMTNS (Figure 2b), suggesting that MMT was successfully exfoliated by ultrasonic separation. It can be observed from Figure 2c that the gel adsorbent had a three-dimensional layered structure with a large interlayer space. The porosity of the BMP gel adsorbent was calculated to be 46.9% using Image J. The resulting gel adsorbent prepared from BCNF and MMTNS had a porous structure. It could provide channels for the adsorbate to pass through [58]. The cell wall of the BMP gel adsorbent (Figure 2c) showed a dense, smooth, and non-porous surface, suggesting that BCNF and MMTNS dispersed uniformly and connected tightly after cross-linking. It is worth noting that N elements were observed on the EDS image of the BMP gel, indicating that PEI was successfully involved.

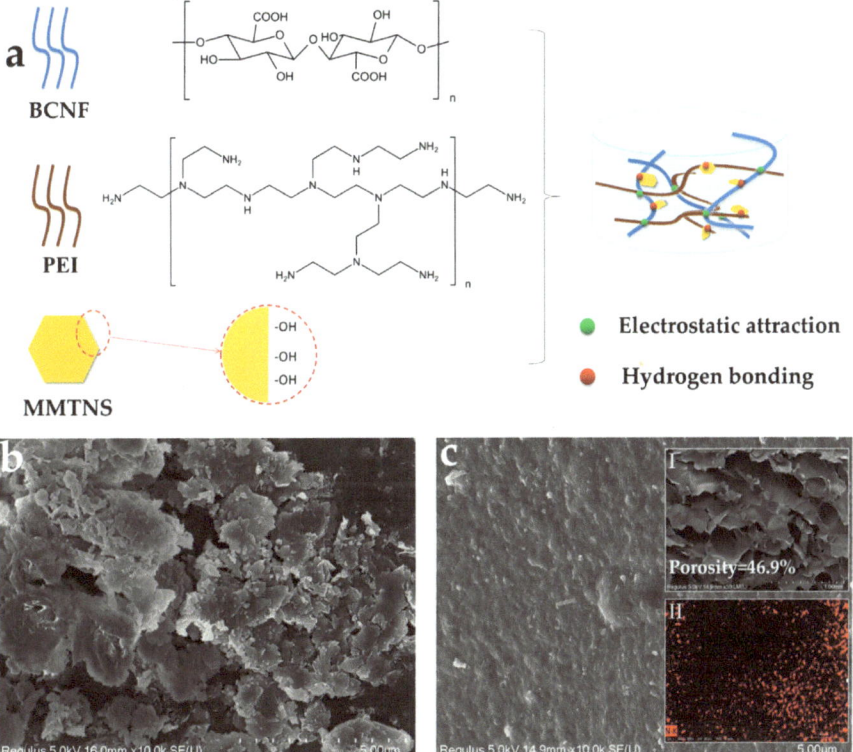

Figure 2. Proposed preparation mechanism of the BMP gel adsorbent (**a**); SEM images of MMTNS (**b**) and the BMP gel adsorbent (**c**).

2.2. MB and Cu(II) Adsorption

2.2.1. Effect of Initial pH

The solution pH is an important factor affecting the adsorption process [59]. The point of zero charge (pH_{pzc}) is the pH at which the positive and negative charges are balanced (dissociation into the liquid by H^+ and OH^- ions) [60]. Thus, pH_{pzc} is a useful measurement for assessing the surface acidity of the BMP gel and characterizing functional groups on its surface. The pH_{pzc} of the BMP gel was 8.71, as shown in Figure 3a. It indicated that the surface of the BMP gel was negatively charged at the pH solution> 8.71, and vice versa.

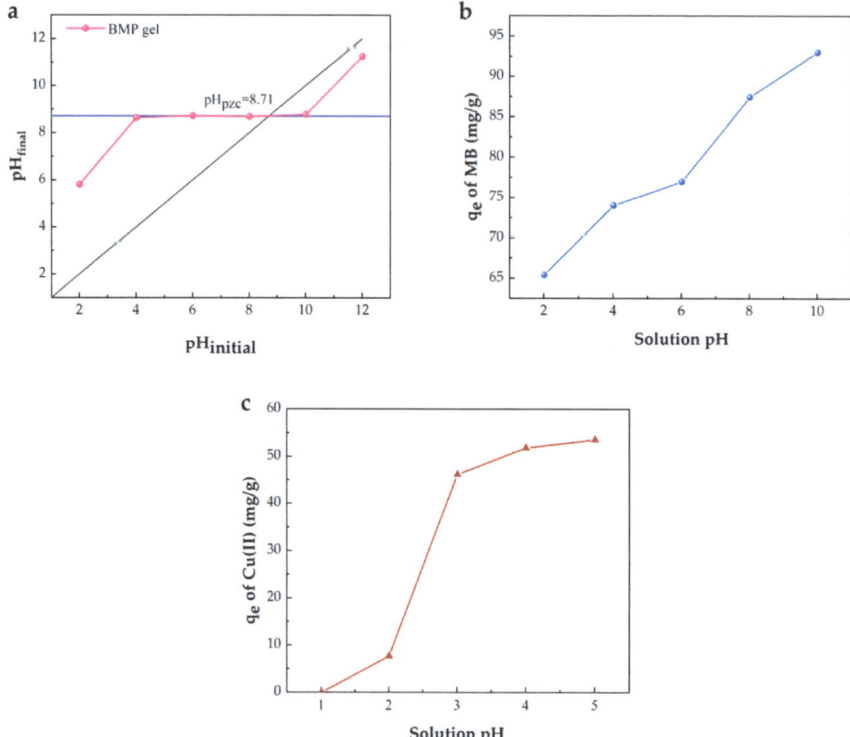

Figure 3. pH$_{pzc}$ of the BMP gel adsorbent (**a**); effect of the solution pH on the adsorption of MB (**b**) and Cu(II) (**c**) (C_0 = 100 mg/L, t = 24 h, T = 25 °C).

MB is a cationic dye in a wide pH range from 0 to 14. It was observed that the adsorption capacity of MB increased sharply with an increase of the solution pH, and the maximum adsorption capacity was 93.0 mg/g at pH = 10 (Figure 3b). MB could be easily adsorbed on the external surface and pores of the resulting gel adsorbent that is larger than its molecular size [50]. Moreover, MB is either mono- (HMB^{2+}) or di-protonated (H$_2$MB^{3+}) at acidic pH conditions [61]. The low MB adsorption at pH < pH$_{pzc}$ was due to the competition between H$^+$ and MB. With the increasing solution pH, most of the carboxyl groups in the BMP gel adsorbent were ionized into carboxylate anions (–COO$^-$); thus, the strong electrostatic adsorption between the negative surface charge and the cationic dye molecule increased [62]. Furthermore, the hydrogen bond interaction between the imine group (RCH = NR) of the MB molecule and the –OH group of gel adsorbent could enhance the adsorption of MB [63].

Cu(II) in solution is easily converted into copper hydroxide precipitation when pH is above 5. Therefore, the effect of pH on the adsorption of Cu(II) was studied in acid environments (from 1 to 5) at 25 °C in this work (Figure 3c). The maximum adsorption capacity of Cu(II) on the BMP gel was 53.6 mg/g at pH = 5. The adsorption of Cu(II) improved with the increase of solution pH. The protonation of the amino groups on PEI was dominant (as pH ≤ 2). As a result, there was an electrostatic repulsion between the BMP gel adsorbent and Cu(II) [64]. As the solution pH increased, the electrostatic force between MMTNS and the original interlayer cation gradually strengthened and the competition from H$^+$ tended to weaken [65]. Meanwhile, amino groups were deprotonated, thereby improving the adsorption capacity of Cu(II) [66].

2.2.2. Adsorption Kinetics and Isotherms

Figure 4 displays the adsorption kinetics and isotherms of MB and Cu(II) into the BMP gel adsorbent. Table 1 lists the fitting parameters. MB and Cu(II) were adsorbed to reach adsorption equilibrium within 6 and 10 h, respectively. The adsorption rate slowed down with time, and the adsorption capacity tended to be stable. It suggested that the adsorption rate was a time-dependent factor. The physical meaning of time dependence is that the reaction path changes with time [67]. It can be observed from Figure 4a,b that the adsorption processes of MB and Cu(II) were fitted better by the fractal-like pseudo-second-order model than the other kinetic models. The correlation coefficients (R_3^2) were 0.99 and 0.97, respectively. Moreover, the low reduced Chi-Sqr value also verified that the fractal-like pseudo-second-order model was the most suitable one to fit MB and Cu(II) adsorption processes [68].

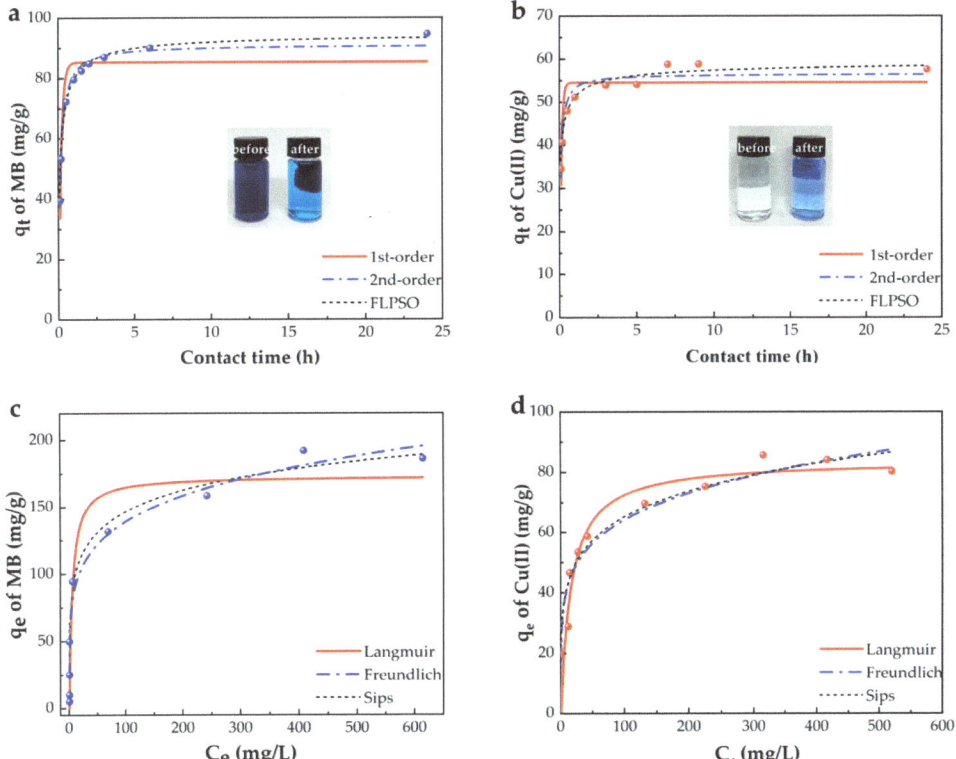

Figure 4. Kinetic models for MB adsorption (**a**) (pH = 10, C_0 = 100 mg/L, T = 25 °C) and Cu(II) adsorption (**b**) (pH = 5, C_0 = 100 mg/L, T = 25 °C); (**c**) isothermal models for MB adsorption (pH = 10, t = 24 h, T = 25 °C) and Cu(II) adsorption (**d**) (pH = 5, t = 24 h, T = 25 °C).

Table 1. Kinetic and isothermal parameters for adsorption of MB and Cu(II).

Model	Parameters	Adsorbate MB	Adsorbate Cu(II)
Pseudo-first-order	q_e (mg/g)	85.3679	54.593
	k_1 (g/(mg·h))	6.0452	9.9634
	R_1^2	0.9024	0.7876
	Reduced Chi-Sqr	37.1686	17.3686
Pseudo-second-order	q_e (mg/g)	91.0366	56.5531
	k_2 (g/(mg·h))	0.0923	0.2877
	R_2^2	0.989	0.9386
	Reduced Chi-Sqr	4.1878	5.0132
Fractal-like pseudo-second-order	q_e (mg/g)	94.8239	60.043
	k (g/(mg·h))	0.0548	0.0961
	a	0.7888	0.5773
	R_3^2	0.9983	0.978
	Reduced Chi-Sqr	0.7763	2.0980
Langmuir	q_m (mg/g)	173.3430	83.8704
	K_L (L/mg)	0.1935	0.0645
	R_4^2	0.8954	0.8824
	R_L (mg/L)	$0 < R_L < 1$	$0 < R_L < 1$
	Reduced Chi-Sqr	684.1315	70.1838
Freundlich	n	5.3604	5.3692
	K_F (mg/g)/(L/mg)	59.0696	27.2935
	R_5^2	0.9637	0.9326
	Reduced Chi-Sqr	235.3060	40.2597
Sips	q_m (mg/g)	361.8749	254.6286
	K_S (L/mg)	0.2042	0.1118
	n	3.8214	4.0920
	R_6^2	0.9683	0.9345
	Reduced Chi-Sqr	239.8241	44.7173

The concentration dependence adsorptions of MB and Cu(II) by the BMP gel adsorbent are shown in Figure 4c,d. The obtained high correlation coefficient (R^2) and low reduced Chi-Sqr values from these isothermal Langmuir, Freundlich, and Sips models are represented in Table 1. The R_6^2 values (MB: 0.97; Cu(II): 0.93) calculated from the Sips isotherm model were greater than the other models, indicating that the Sips isotherm better described the adsorptions of MB and Cu(II) by the adsorbent BMP gel than the other ones. This model indicated that the adsorption processes of MB and Cu(II) were followed by a combined model: monomolecular (at high concentration) and diffuse (at low concentration) [69]. The maximal adsorption capacities of MB and Cu(II) were 361.9 and 254.6 mg/g, respectively, calculated by the Sips isotherm model, and were higher than most of the previously reported works (Table 2). In addition, the adsorptions of MB and Cu(II) were fitted well by the Freundlich model. The n^{-1} in Freundlich indicates the advantage of the adsorption process. If $n^{-1} < 1$, the adsorption intensity is favorable over the entire range of the concentration studied. If $n^{-1} > 1$, it means that the adsorption capacity is desirable at a high concentration but much less so at a lower concentration [14]. The value of n^{-1} is 0.19 for both MB and Cu(II) (Table 1), indicating favorable adsorptions over the entire concentration ranges for both MB and Cu(II).

Table 2. Comparison of the maximum removal capacities of MB and Cu(II) with other biosorbents.

Adsorbent	Adsorbate	pH	T (°C)	Q_{max} (mg/g)	Ref.
Graphene oxide/CNF aerogel	MB	-	-	111.2	[70]
PVA/chitosan/MMT hydrogel	MB	8	30	132.2	[71]
Cellulose-derived carbon/MMT	MB	8	25	138.1	[50]
Sugarcane bagasse	MB	-	45	9.41	[53]
BMP gel adsorbent	MB	10	25	361.9	This work
CNF aerogel	Cu(II)	6	29.85	30.0	[14]
Cellulose/acrylonitrile/methacrylic acid	Cu(II)	5.5	25	76.8	[64]
TEMPO-oxidized CNF	Cu(II)	5	30	52.3	[66]
BMP gel adsorbent	Cu(II)	5	25	254.6	This work

2.2.3. Effect of Interfering Ions

Actual wastewater contains several common ions, such as K^+, Na^+, Mg^{2+}, Ca^{2+}, etc. To check the effects of these ions, we studied the adsorption of MB into BMP in the presence of 100 mg/L of a dye solution with 10 mM aqueous solutions of salt. The results are shown in Figure 5. The adsorption of MB into BMP was 94.6% in the blank group. By adding these interfering ions, the adsorptions of MB slightly decreased in the order of Na^+ (93.7%) > K^+ (92.3%) > Ca^{2+} (89.5%) > Mg^{2+} (80.8%). The decrease in the removal was due to differences in the radii of the hydrated ions. Ions with smaller hydrodynamic radii can compete with larger-sized contaminants and are easily absorbed, resulting in a decrease in MB removal [72]. It is worth noting that BMP can effectively remove MB from aqueous media even in the presence of interfering ions.

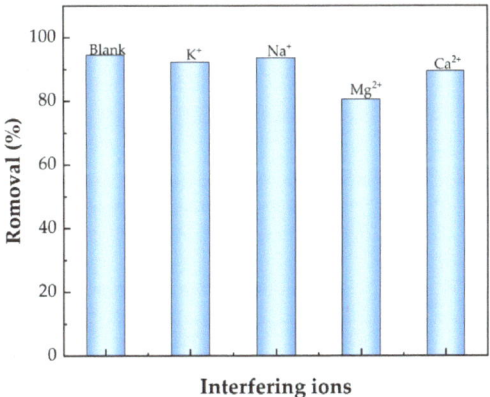

Figure 5. Effect of interfering ions on the MB adsorption by the BMP.

2.2.4. Adsorption Thermodynamics and Adsorbent Reusability

As the temperature increased from 25 to 45 °C (Figure 6a,b), the adsorption capacities of MB and Cu(II) by the BMP gel adsorbent increased from 131.8 to 147.6 mg/g and from 58.7 to 63.5 mg/g, respectively. Meanwhile, the MB and Cu(II) removal efficiencies increased from 65.9% to 73.8% and from 58.7% to 63.5%, respectively, with increasing temperatures. The thermodynamic parameters of the MB and Cu(II) adsorptions by the BMP gel adsorbent are presented in Table 3. The positive $\Delta H°$ elucidated that the adsorption mechanism was endothermic, which implied that a large amount of heat was required to transfer dyes and metal ions from the aqueous phase to the solid phase. The negative value of Gibbs free energy ($\Delta G°$) indicated that the adsorptions of MB and Cu(II) on the adsorbent were spontaneous [72,73]. The positive $\Delta S°$ values suggested an increase in the degrees of randomness at the interface between the solid and liquid during the adsorption of MB and Cu(II) on the resulting BMP gel adsorbent [53].

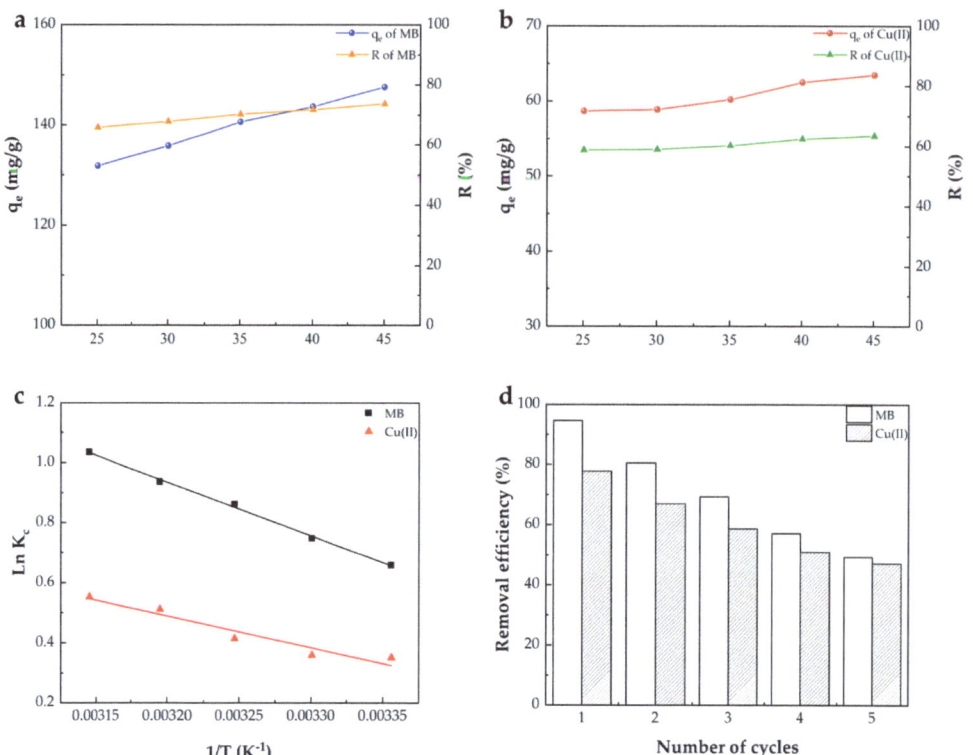

Figure 6. Effect of temperature on the adsorption of (**a**) MB (C_0 = 200 mg/L; the pH solution = 10) and (**b**) Cu(II) (C_0 = 100 mg/L; the pH solution = 5); the plot of LnK_c versus 1/T (**c**); reusability of the BMP gel adsorbent for MB and Cu(II) (**d**).

Table 3. Thermodynamic parameters of the adsorption of MB and Cu(II).

Adsorbate	T (K)	K_c	$\Delta G°$ (kJ/mol)	$\Delta H°$ (kJ/mol)	$\Delta S°$ (Jmol/K)
MB	298	1.9332	−1.6331	14.8247	6.6408
	303	2.1144	−1.8863		
	308	2.3684	−2.2079		
	313	2.5517	−2.4377		
	318	2.8165	−2.7377		
Cu(II)	298	1.4213	−0.8711	8.7197	3.8453
	303	1.4331	−0.9065		
	308	1.5145	−1.0628		
	313	1.6681	−1.3315		
	318	1.7390	−1.4628		

The regeneration of the adsorbent was also studied to provide a basis for its practical application. The reusability of the BMP gel adsorbent was tested by repeating five cycles of the adsorption–desorption process (Figure 6d). After five cycles, the removal efficiencies for MB and Cu(II) remained at 49.3% and 47.1%, respectively, indicating that they had acceptable reusability. The reduction of removal efficiency could be attributed to the incomplete desorption of MB and Cu(II) [74]. It is easy to remove the gel from the suspension for its solid shape, thereby, it has good potential to be used as an industrial adsorbent.

2.2.5. Adsorption Mechanism Analysis

Figure 7 shows the SEM images of the gel adsorbent after the adsorptions of MB and Cu(II). Compared with the smooth surface before the adsorption (Figure 2c), the gel adsorbent surface became rougher after adsorbing MB and Cu(II) (Figure 5a,b). Furthermore, as shown in Figure 5d, the blue points (denoted Cu elements) are evenly distributed on the surface of the adsorbed BMP gel adsorbent, suggesting that Cu was evenly covered on the surface of the BMP gel adsorbent. These results indicate the successful adsorptions of MB and Cu(II), and the uniform distribution of the adsorption sites on the adsorbent. These SEM images suggest that the BMP gel adsorbent had great potential to be used as an adsorbent candidate.

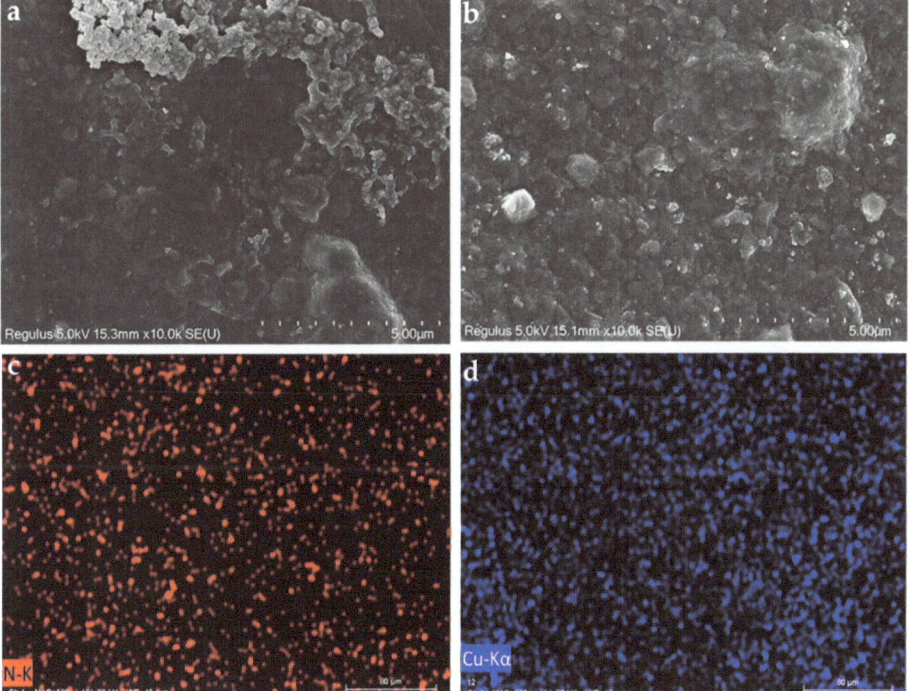

Figure 7. SEM images of the gel adsorbent after the adsorption of MB (**a**) and Cu(II) (**b**); EDS images of gel adsorbent after the adsorption of MB (**c**) and Cu (II) (**d**).

The chemical compositions of the BMP gel adsorbents before and after adsorption were characterized by XPS. As shown in Figure 8a, there was an obvious decrease of Na1s in the BMP gel adsorbent after MB and Cu(II) adsorptions. The wide scan spectra of pristine BMP gel adsorbent did not have any signal in the Cu2p region, but binding energy peaks at around 934.36 eV appeared after Cu(II) adsorption. These findings suggest that the exchangeable cations, Na^+, existed in MMTNS [65]. In the C1s plot (Figure 8b), the binding energies of C–O and O–C=O were at 285.79 and 287.50 eV before adsorption, respectively [75]. They shifted to 286.03 and 287.20 eV after MB adsorption and shifted to 286.32 and 288.32 eV after Cu(II) adsorption, owing to the combination of groups, such as -OH and -COOH on the BMP gel adsorbent with MB or Cu(II) [76]. The O1s initial peak could be divided into two regions at 530.57 and 531.85 eV, being consistent with the oxygen of C=O and C–O (Figure 8c) [77]. They shifted to 531.48 and 532.28 eV after the MB adsorption and shifted to 531.31 and 532.73 eV after the Cu(II) adsorption, which suggested that oxygen atoms in carboxyl and hydroxyl groups were involved in the adsorption process [21]. The N1s spectra of the BMP gel adsorbent before adsorption had peaks at 398.64, 399.49, and 401.15 eV, which were assigned to -NH_2, -NH-, and -NH_3^+, respectively (Figure 8d) [48,78]. After MB was adsorbed, the binding energy peaks of N1s shifted to 399.08, 400.04, and 401.67 eV. After Cu(II) was adsorbed, the binding energy peaks of N1s shifted to 399.18, 400.49, and 402.25 eV. Simultaneously, there was an obvious decrease in the N1s intensities. These findings confirmed that all three kinds (primary amine, secondary amine, and tertiary amine groups) of amino groups from PEI were involved in the removal of MB and Cu(II) [77].

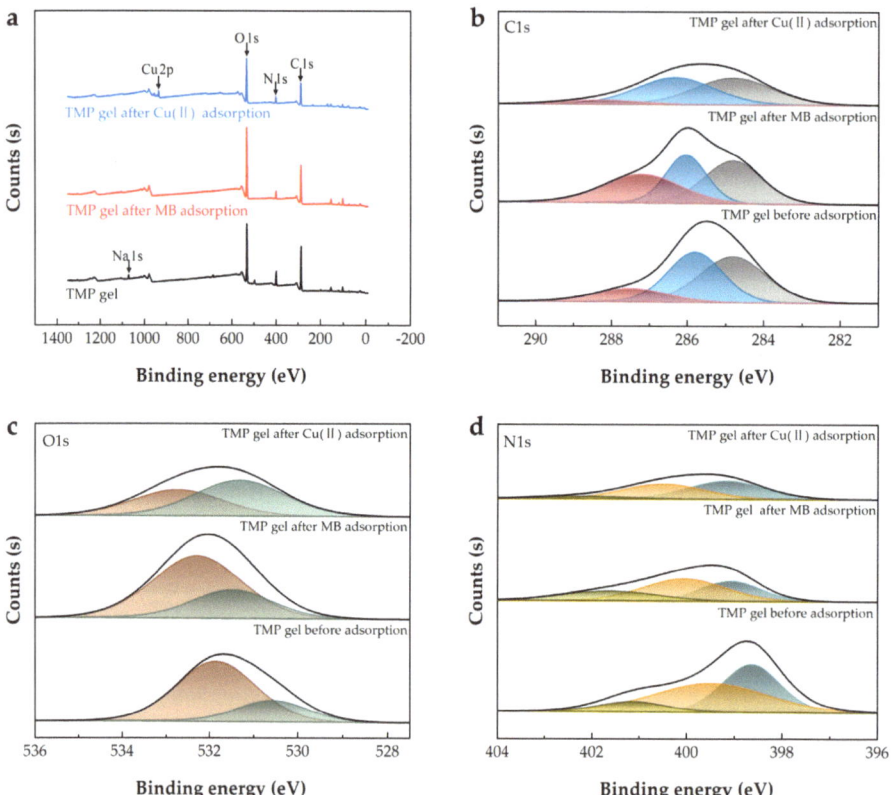

Figure 8. XPS spectra of the BMP gel adsorbent before and after adsorption on MB and Cu(II) (**a**); C1s region (**b**); O1s region (**c**); N1s region (**d**).

Under the alkaline condition, the cationic group of MB could be easily attracted by the deprotonated carboxyl group in the BMP gel adsorbent through the electrostatic interaction. A large number of hydroxyl groups in the BMP gel adsorbent could form hydrogen bonds with the imine groups (RCH=NR) of MB molecules, which also enhanced adsorption. In addition, the van der Waals force may play an important role in the MB adsorption process [79]. As for the adsorption of Cu(II), the partially ionized carboxyl groups in the BMP gel adsorbent could form electrostatic adsorption with Cu(II). The PEI grafted on the gel adsorbent had a large number of amino groups and could also be connected with Cu(II) to promote adsorption [34]. It is also worth noting that MB molecules and Cu(II) ions could exchange cations with ions between the MMTNS layers [16]. Comprehensively, BCNF was rich in hydroxyl and carboxyl groups, which help form a stable structure of the BMP gel adsorbent. It could adsorb positively charged MB and Cu(II). MMTNS could be used as a reinforcement agent for cellulose framework, and the cations between the MMTNS layers could exchange ions with MB and Cu(II). PEI, as a cross-linking agent with a large number of amino groups, could increase the structural stability and boost the removal efficiencies of MB and Cu(II). From the above, the proposed adsorption mechanisms of MB and Cu(II) by the BMP gel adsorbent are shown in Figure 9.

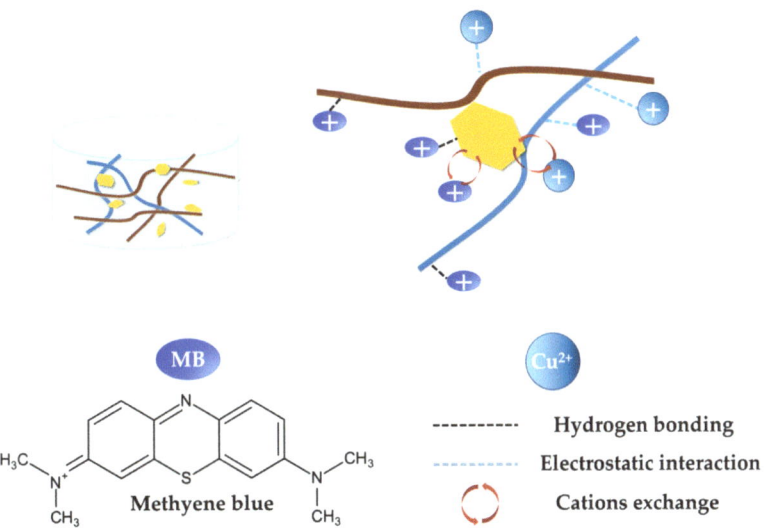

Figure 9. The proposed adsorption mechanism of MB and Cu(II) by the BMP gel adsorbent.

3. Conclusions

In this study, the BCNF/MMTNS/PEI (BMP) gel adsorbent was successfully prepared from bamboo nanocellulose and montmorillonite nanosheets, cross-linked by polyethyleneimine, and used as an adsorbent to remove the cationic dye, methylene blue (MB), heavy metal, and Cu(II) from aqueous solutions. The FTIR and SEM results showed that the resulting gel adsorbent had abundant hydroxyl, carboxyl, and amino groups, as well as a porous structure. The kinetics study of adsorption processes on MB and Cu(II) was well-fitted by a fractal-like pseudo-second-order model. According to the Sips isotherm model, the calculated maximum adsorption capacities of MB and Cu(II) were 361.9 mg/g and 254.6 mg/g, respectively. The adsorption mechanisms mainly included electrostatic attraction, ion exchange, hydrogen bond interactions, etc. These results suggest that the adsorbent had great potential to be used to remove MB and Cu(II) from aqueous solutions.

4. Materials and Methods

4.1. Materials

The waste bamboo paper was collected from the laboratory. Montmorillonite (MMT) was bought from Zhejiang Fenghong New Materials co., Ltd. (Huzhou, China). Methylene blue (MB), copper sulfate pentahydrate, and 2,2,6,6-Tetramethylpiperidine (TEMPO) were obtained from Sinopharm Chemical Regent Co., Ltd. (Shanghai, China). Polyethyleneimine (PEI, 50% aqueous solution) was purchased from Aladdin Reagent Co., Ltd. (Shanghai, China). Sodium hypochlorite (NaClO), glacial acetic acid (CH_3COOH), sodium bromide (NaBr), sodium hydroxide (NaOH), and hydrochloric acid (HCl) were obtained from

Chengdu Kelong Chemical Co., Ltd. (Chengdu, China). All chemicals were of analytical grade and used without further purification.

4.2. Preparation of Bamboo Nanocellulose (BCNF)

BCNF was prepared from waste bamboo paper in accordance with the previously reported methods [80]. Briefly, cellulose was obtained from the waste bamboo paper after bleaching and alkali treatment. Then 10 g of cellulose was mixed with 1000 mL of deionized water, 0.36 g of TEMPO, and 37.5 mL of NaClO. Then, the pH was adjusted to 10 by HCl and NaOH. Next, the suspension was treated with an ultrasonic homogenizer (Scientz, China) for 1 h. Finally, it was freeze-dried to obtain BCNF.

4.3. Preparation of MMTNS

The 5 wt% of MMT suspension was centrifuged at a speed of 1000 r/min for 2 min to remove large particles. Then 7 wt% of MMT was exfoliated by an ultrasonic processor (Scientz, China) at 400 w of power for 15 min. Finally, the exfoliated MMTNS sample was dried at 60 °C.

4.4. Preparation of BCNF/MMTNS/PEI (BMP) Gel Adsorbent

The preparation process is shown in Figure 10. A total of 0.6 g of BCNF and 0.6 g of MMTNS were evenly dispersed in 100 mL of deionized water by an ultrasonic treatment. Then, PEI was added to produce the hydrogel. Finally, it was pre-frozen at −20 °C and then freeze-dried at −50 °C for 72 h to obtain the resulting BMP gel adsorbent.

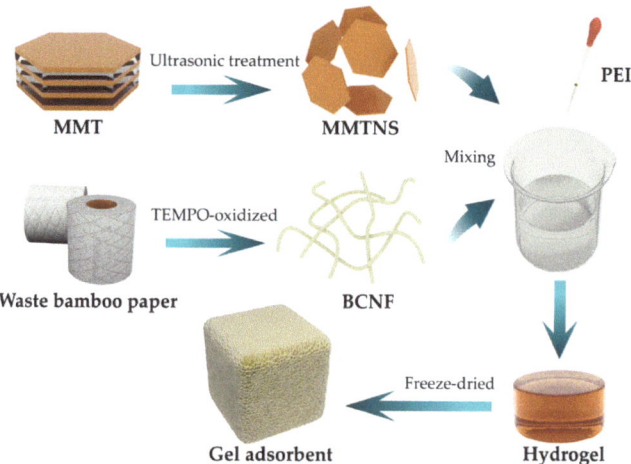

Figure 10. Schematic diagram of the preparation of the BMP gel adsorbent.

4.5. Adsorbent Characterization

Fourier transform infrared spectroscopy (FTIR) (Nicolet iS50, Thermo Fisher, Waltham, MA, USA) was carried out to detect the change of chemical groups among samples. The field emission scanning electron microscopy (SEM, SU 8010, Hitachi, Japan) equipped with energy dispersive spectroscopy (EDS) was used to analyze the morphology and elemental presence of the BMP gel adsorbent. X-ray photoelectron spectroscopic (XPS) was conducted using the Theta Probe Angle-Resolved XPS System, Thermo Fisher Scientific (UK), with an Al Ka X-ray source. The point of zero charge (pH_{pzc}) of the BMP gel was determined according to the pH drift method [60]. The porosity of the BMP gel adsorbent was calculated by Image J.

4.6. Adsorption Experiment

The resulting gel adsorbent was applied to remove MB and Cu(II) in the batch adsorption experiment. Briefly, a 30 mg sample was added to a 30 mL adsorbate solution. The mixture was shaken at 170 rpm with a mechanical shaker for 24 h. The effects of the pH solution (MB: 2–10, Cu(II): 1–5), contact time (0–24 h), initial concentration (MB: 5–800 mg/L, Cu(II): 20–600 mg/L), and temperature (25–45 °C) on the adsorption efficiency were studied. The adsorption capacity of gel adsorbent on MB was estimated by calculating the change between the initial and residual MB concentrations using a UV spectrophotometer (UV-4802H, Unico, Shanghai, China) at 664 nm. The concentration of Cu(II) was determined using an AA-6300 atomic absorption spectrophotometer with an air acetylene burner (AAS, Shimadzu, Japan). The adsorption capacity (q) and removal efficiency (R) were calculated by Equations (1) and (2), respectively.

$$q = \frac{C_0 - C_e}{m} \times V \tag{1}$$

$$R(\%) = \frac{C_0 - C_e}{C_0} \times 100\% \tag{2}$$

where q represents the adsorption capacity (mg/g); C_0 and C_e are the initial concentration and equilibrium concentrations (mg/L) of the MB or Cu(II) solution, respectively; V represents the volume of the adsorbent solution (L); m is the weight of the dried adsorbent (g).

The pseudo-first-order in Equation (3), pseudo-second-order in Equation (4), and fractal-like pseudo-second-order model in Equation (5) were fitted to analyze the adsorption process.

$$q_t = q_e(1 - e^{-k_1 t}) \tag{3}$$

$$q_t = \frac{k_2 q_e^2 t}{1 + k_2 q_e t} \tag{4}$$

$$q_t = \frac{k q_e^2 t^a}{1 + k q_e t^a} \tag{5}$$

where t is the contact time (h); q_e and q_t are the adsorption capacities at equilibrium and time t, respectively (mg/g); k, k_1, and k_2 are the rate constants (g/(mg·h)); and a is the fractal exponent.

Langmuir adsorption isotherm (Equation (6)), Freundlich isotherm (Equation (7)), and the Sips isotherm model (Equation (8)) were used to analyze the adsorption isotherm.

$$q_e = \frac{q_m K_L C_e}{1 + K_L C_e} \tag{6}$$

$$q_e = K_F C_e^{1/n} \tag{7}$$

$$q_e = \frac{q_m K_S C_e^{1/n}}{1 + K_S C_e^{1/n}} \tag{8}$$

where C_e is the concentration of MB or Cu(II) at equilibrium (mg/L); q_e and q_m are the adsorption capacity of the adsorbent at equilibrium and the maximum adsorption capacity at saturation (mg/g), respectively; n is the exponent (dimensionless); K_F, K_L, and K_S represent Freundlich, Langmuir, and Sips adsorption constants.

The separation factor (R_L) was used to describe the favorable degree of the adsorption process using Equation (9):

$$R_L = \frac{1}{1 + K_L \times C_0} \tag{9}$$

where R_L is a dimensionless equilibrium parameter or the separation factor; C_0 is the initial pollutant concentration (mg/L); $R_L > 1$ indicates unfavorable adsorption; $R_L = 1$

corresponds to a linear adsorption process; $0 < R_L < 1$ indicates favorable adsorption, and $R_L = 0$ means irreversible adsorption.

To investigate the thermodynamic adsorption behaviors of MB and Cu(II), the thermodynamic parameters ($\Delta G°$, $\Delta H°$, and $\Delta S°$) were obtained using Equations (10)–(12):

$$\Delta G° = \Delta H° - T\Delta S° \tag{10}$$

$$K_c = \frac{q_e}{C_e} \tag{11}$$

$$lnK_c = \frac{\Delta S°}{R} - \frac{\Delta H°}{R} \times \frac{1}{T} \tag{12}$$

where $\Delta G°$ is the Gibbs free energy change (kJ/mol); $\Delta H°$ is the enthalpy change (kJ/mol) and $\Delta S°$ is the entropy change (kJ/mol); K_c is the distribution coefficient; R is the universal gas constant (8.314 J/mol K); T is the absolute temperature (K).

4.7. Reusability Experiment

A total of 30 mg of the BMP gel adsorbent was added to the 30 mL MB (100 mg/L) and Cu(II) solution (60 mg/L) for 24 h. Then, the adsorbed adsorbent was washed for 24 h with ethanol and the NaOH solution for MB and Cu(II), respectively, followed by freeze-drying at -50 °C for the next cycle. It was repeated 5 times to study the reusability.

Author Contributions: Conceptualization, X.Z. (Xuelun Zhang)and F.L.; methodology, Y.Z.; software, X.H. (Xiaobo Huang) and Z.Y.; validation, J.C., X.Z. (Xiyu Zhao), and S.L.; writing—original draft preparation, X.Z. (Xuelun Zhang); writing—review and editing, F.L.; visualization, C.F.D.H.; supervision, X.H. (Xingang Huang); project administration, X.P.; funding acquisition, X.P. and X.H. (Xingyan Huang). All authors have read and agreed to the published version of the manuscript.

Funding: This work was supported by the Fundamental Research Funds for the Central Non-Profit Research Institution of Chinese Academy of Forestry (CAFYBB2022SY003), the National Natural Science Foundation of China (32101598), the Foundation of Key Laboratory of Pulp and Paper Science and Technology of Ministry of Education of China (KF201909), the National Forestry and Grassland Administration Key Laboratory of Plant Fiber Functional Materials (2022KFJJ02), and the Project of Nanhai Series of Talent Cultivation Program.

Institutional Review Board Statement: Not applicable.

Informed Consent Statement: Not applicable.

Data Availability Statement: Not applicable.

Conflicts of Interest: The authors declare no conflict of interest.

References

1. Wang, Z.; Dai, L.; Yao, J.; Guo, T.; Hrynsphan, D.; Tatsiana, S.; Chen, J. Improvement of *Alcaligenes* sp.TB Performance by Fe-Pd/Multi-Walled Carbon Nanotubes: Enriched Denitrification Pathways and Accelerated Electron Transport. *Bioresour. Technol.* **2021**, *327*, 124785. [CrossRef] [PubMed]
2. Lin, X.; Lu, K.; Hardison, A.K.; Liu, Z.; Xu, X.; Gao, D.; Gong, J.; Gardner, W.S. Membrane Inlet Mass Spectrometry Method (REOX/MIMS) to Measure 15N-Nitrate in Isotope-Enrichment Experiments. *Ecol. Indic.* **2021**, *126*, 107639. [CrossRef]
3. Guan, Q.; Zeng, G.; Song, J.; Liu, C.; Wang, Z.; Wu, S. Ultrasonic Power Combined with Seed Materials for Recovery of Phosphorus from Swine Wastewater via Struvite Crystallization Process. *J. Environ. Manag.* **2021**, *293*, 112961. [CrossRef]
4. Chen, F.; Ma, J.; Zhu, Y.; Li, X.; Yu, H.; Sun, Y. Biodegradation Performance and Anti-Fouling Mechanism of an ICME/Electro-Biocarriers-MBR System in Livestock Wastewater (Antibiotic-Containing) Treatment. *J. Hazard. Mater.* **2022**, *426*, 128064. [CrossRef] [PubMed]
5. Zhao, Y.; Kang, S.; Qin, L.; Wang, W.; Zhang, T.; Song, S.; Komarneni, S. Self-Assembled Gels of Fe-Chitosan/Montmorillonite Nanosheets: Dye Degradation by the Synergistic Effect of Adsorption and Photo-Fenton Reaction. *Chem. Eng. J.* **2020**, *379*, 122322. [CrossRef]
6. Liu, W.; Huang, F.; Liao, Y.; Zhang, J.; Ren, G.; Zhuang, Z.; Zhen, J.; Lin, Z.; Wang, C. Treatment of Cr^{VI}-Containing $Mg(OH)_2$ Nanowaste. *Angew. Chem. Int. Ed.* **2008**, *47*, 5619–5622. [CrossRef]

7. Kabir, S.F.; Cueto, R.; Balamurugan, S.; Romeo, L.D.; Kuttruff, J.T.; Marx, B.D.; Negulescu, I.I. Removal of Acid Dyes from Textile Wastewaters Using Fish Scales by Absorption Process. *Clean Technol.* **2019**, *1*, 311–324. [CrossRef]
8. Bai, B.; Bai, F.; Li, X.; Nie, Q.; Jia, X.; Wu, H. The Remediation Efficiency of Heavy Metal Pollutants in Water by Industrial Red Mud Particle Waste. *Environ. Technol. Innov.* **2022**, *28*, 102944. [CrossRef]
9. Salleh, M.A.M.; Mahmoud, D.K.; Karim, W.A.W.A.; Idris, A. Cationic and Anionic Dye Adsorption by Agricultural Solid Wastes: A Comprehensive Review. *Desalination* **2011**, *280*, 1–13. [CrossRef]
10. Gao, J.; Li, Z.; Wang, Z.; Chen, T.; Hu, G.; Zhao, Y.; Han, X. Facile Synthesis of Sustainable Tannin/Sodium Alginate Composite Hydrogel Beads for Efficient Removal of Methylene Blue. *Gels* **2022**, *8*, 486. [CrossRef]
11. Ning, F.; Zhang, J.; Kang, M.; Ma, C.; Li, H.; Qiu, Z. Hydroxyethyl Cellulose Hydrogel Modified with Tannic Acid as Methylene Blue Adsorbent. *J. Appl. Polym. Sci.* **2021**, *138*, 49880. [CrossRef]
12. He, Y.; Jiang, D.B.; Chen, J.; Jiang, D.Y.; Zhang, Y.X. Synthesis of MnO_2 Nanosheets on Montmorillonite for Oxidative Degradation and Adsorption of Methylene Blue. *J. Colloid Interface Sci.* **2018**, *510*, 207–220. [CrossRef] [PubMed]
13. Kong, W.; Li, Q.; Li, X.; Su, Y.; Yue, Q.; Zhou, W.; Gao, B. Removal of Copper Ions from Aqueous Solutions by Adsorption onto Wheat Straw Cellulose-based Polymeric Composites. *J. Appl. Polym. Sci.* **2018**, *135*, 46680. [CrossRef]
14. She, J.; Tian, C.; Wu, Y.; Li, X.; Luo, S.; Qing, Y.; Jiang, Z. Cellulose Nanofibrils Aerogel Cross-Linked by Poly(Vinyl Alcohol) and Acrylic Acid for Efficient and Recycled Adsorption with Heavy Metal Ions. *J. Nanosci. Nanotechnol.* **2018**, *18*, 4167–4175. [CrossRef] [PubMed]
15. Yang, G.-X.; Jiang, H. Amino Modification of Biochar for Enhanced Adsorption of Copper Ions from Synthetic Wastewater. *Water Res.* **2014**, *48*, 396–405. [CrossRef]
16. Sun, X.-F.; Hao, Y.; Cao, Y.; Zeng, Q. Superadsorbent Hydrogel Based on Lignin and Montmorillonite for Cu(II) Ions Removal from Aqueous Solution. *Int. J. Biol. Macromol.* **2019**, *127*, 511–519. [CrossRef]
17. Huang, H.; Liu, J.; Zhang, P.; Zhang, D.; Gao, F. Investigation on the Simultaneous Removal of Fluoride, Ammonia Nitrogen and Phosphate from Semiconductor Wastewater Using Chemical Precipitation. *Chem. Eng. J.* **2017**, *307*, 696–706. [CrossRef]
18. Wang, Q.; Jiang, C.; Wang, Y.; Yang, Z.; Xu, T. The Reclamation of Aniline Wastewater and CO_2 Capture Using Bipolar Membrane Electrodialysis. *ACS Sustain. Chem. Eng.* **2016**, *4*, 5743–5751. [CrossRef]
19. Teh, C.Y.; Budiman, P.M.; Shak, K.P.Y.; Wu, T.Y. Recent Advancement of Coagulation–Flocculation and Its Application in Wastewater Treatment. *Ind. Eng. Chem. Res.* **2016**, *55*, 4363–4389. [CrossRef]
20. Wang, M.; Sun, F.; Zeng, H.; Su, X.; Zhou, G.; Liu, H.; Xing, D. Modified Polyethersulfone Ultrafiltration Membrane for Enhanced Antifouling Capacity and Dye Catalytic Degradation Efficiency. *Separations* **2022**, *9*, 92. [CrossRef]
21. Zeng, H.; Hao, H.; Wang, X.; Shao, Z. Chitosan-Based Composite Film Adsorbents Reinforced with Nanocellulose for Removal of Cu(II) Ion from Wastewater: Preparation, Characterization, and Adsorption Mechanism. *Int. J. Biol. Macromol.* **2022**, *213*, 369–380. [CrossRef] [PubMed]
22. Zhou, Y.; Lu, J.; Zhou, Y.; Liu, Y. Recent Advances for Dyes Removal Using Novel Adsorbents: A Review. *Environ. Pollut.* **2019**, *252*, 352–365. [CrossRef] [PubMed]
23. Yu, H.; Hong, H.-J.; Kim, S.M.; Ko, H.C.; Jeong, H.S. Mechanically Enhanced Graphene Oxide/Carboxymethyl Cellulose Nanofibril Composite Fiber as a Scalable Adsorbent for Heavy Metal Removal. *Carbohydr. Polym.* **2020**, *240*, 116348. [CrossRef] [PubMed]
24. Wang, C.; Yang, S.; Ma, Q.; Jia, X.; Ma, P.-C. Preparation of Carbon Nanotubes/Graphene Hybrid Aerogel and Its Application for the Adsorption of Organic Compounds. *Carbon* **2017**, *118*, 765–771. [CrossRef]
25. Azaman, S.A.H.; Afandi, A.; Hameed, B.H.; Din, A.T.M. Removal of Malachite Green from Aqueous Phase Using Coconut Shell Activated Carbon: Adsorption, Desorption, and Reusability Studies. *J. Appl. Sci. Eng.* **2018**, *21*, 317–330. [CrossRef]
26. Yu, Z.; Wei, L.; Lu, L.; Shen, Y.; Zhang, Y.; Wang, J.; Tan, X. Structural Manipulation of 3D Graphene-Based Macrostructures for Water Purification. *Gels* **2022**, *8*, 622. [CrossRef]
27. Zhu, L.; Zong, L.; Wu, X.; Li, M.; Wang, H.; You, J.; Li, C. Shapeable Fibrous Aerogels of Metal–Organic-Frameworks Templated with Nanocellulose for Rapid and Large-Capacity Adsorption. *ACS Nano* **2018**, *12*, 4462–4468. [CrossRef]
28. Xiao, W.-D.; Xiao, L.-P.; Xiao, W.-Z.; Liu, K.; Zhang, Y.; Zhang, H.-Y.; Sun, R.-C. Cellulose-Based Bio-Adsorbent from TEMPO-Oxidized Natural Loofah for Effective Removal of Pb(II) and Methylene Blue. *Int. J. Biol. Macromol.* **2022**, *218*, 285–294. [CrossRef]
29. Andrade Siqueira, T.C.; Zanette da Silva, I.; Rubio, A.J.; Bergamasco, R.; Gasparotto, F.; Aparecida de Souza Paccola, E.; Ueda Yamaguchi, N. Sugarcane Bagasse as an Efficient Biosorbent for Methylene Blue Removal: Kinetics, Isotherms and Thermodynamics. *Int. J. Environ. Res. Public Health* **2020**, *17*, 526. [CrossRef]
30. Liu, C.; Omer, A.M.; Ouyang, X. Adsorptive Removal of Cationic Methylene Blue Dye Using Carboxymethyl Cellulose/k-Carrageenan/Activated Montmorillonite Composite Beads: Isotherm and Kinetic Studies. *Int. J. Biol. Macromol.* **2018**, *106*, 823–833. [CrossRef]
31. Gu, H.; Zhou, X.; Lyu, S.; Pan, D.; Dong, M.; Wu, S.; Ding, T.; Wei, X.; Seok, I.; Wei, S.; et al. Magnetic Nanocellulose-Magnetite Aerogel for Easy Oil Adsorption. *J. Colloid Interface Sci.* **2020**, *560*, 849–856. [CrossRef] [PubMed]
32. Voisin, H.; Bergström, L.; Liu, P.; Mathew, A. Nanocellulose-Based Materials for Water Purification. *Nanomaterials* **2017**, *7*, 57. [CrossRef] [PubMed]

33. Jiang, F.; Dinh, D.M.; Hsieh, Y.-L. Adsorption and Desorption of Cationic Malachite Green Dye on Cellulose Nanofibril Aerogels. *Carbohydr. Polym.* **2017**, *173*, 286–294. [CrossRef]
34. Hong, H.-J.; Ban, G.; Kim, H.S.; Jeong, H.S.; Park, M.S. Fabrication of Cylindrical 3D Cellulose Nanofibril(CNF) Aerogel for Continuous Removal of Copper(Cu2+) from Wastewater. *Chemosphere* **2021**, *278*, 130288. [CrossRef] [PubMed]
35. Fan, K.; Zhang, T.; Xiao, S.; He, H.; Yang, J.; Qin, Z. Preparation and Adsorption Performance of Functionalization Cellulose-Based Composite Aerogel. *Int. J. Biol. Macromol.* **2022**, *211*, 1–14. [CrossRef] [PubMed]
36. Ma, J.; Lei, Y.; Khan, M.A.; Wang, F.; Chu, Y.; Lei, W.; Xia, M.; Zhu, S. Adsorption Properties, Kinetics & Thermodynamics of Tetracycline on Carboxymethyl-Chitosan Reformed Montmorillonite. *Int. J. Biol. Macromol.* **2019**, *124*, 557–567. [CrossRef] [PubMed]
37. Yi, H.; Jia, F.; Zhao, Y.; Wang, W.; Song, S.; Li, H.; Liu, C. Surface Wettability of Montmorillonite (0 0 1) Surface as Affected by Surface Charge and Exchangeable Cations: A Molecular Dynamic Study. *Appl. Surf. Sci.* **2018**, *459*, 148–154. [CrossRef]
38. Brigatti, M.F.; Galan, E.; Theng, B.K.G. Chapter 2 Structures and Mineralogy of Clay Minerals. In *Developments in Clay Science*; Elsevier: Amsterdam, The Netherlands, 2006; Volume 1, pp. 19–86. ISBN 978-0-08-044183-2.
39. Chen, T.; Yuan, Y.; Zhao, Y.; Rao, F.; Song, S. Preparation of Montmorillonite Nanosheets through Freezing/Thawing and Ultrasonic Exfoliation. *Langmuir* **2019**, *35*, 2368–2374. [CrossRef]
40. Wang, W.; Wang, J.; Zhao, Y.; Bai, H.; Huang, M.; Zhang, T.; Song, S. High-Performance Two-Dimensional Montmorillonite Supported-Poly(Acrylamide-Co-Acrylic Acid) Hydrogel for Dye Removal. *Environ. Pollut.* **2020**, *257*, 113574. [CrossRef]
41. Wang, W.; Ni, J.; Chen, L.; Ai, Z.; Zhao, Y.; Song, S. Synthesis of Carboxymethyl Cellulose-Chitosan-Montmorillonite Nanosheets Composite Hydrogel for Dye Effluent Remediation. *Int. J. Biol. Macromol.* **2020**, *165*, 1–10. [CrossRef]
42. Melo, B.C.; Paulino, F.A.A.; Cardoso, V.A.; Pereira, A.G.B.; Fajardo, A.R.; Rodrigues, F.H.A. Cellulose Nanowhiskers Improve the Methylene Blue Adsorption Capacity of Chitosan-g-Poly(Acrylic Acid) Hydrogel. *Carbohydr. Polym.* **2018**, *181*, 358–367. [CrossRef] [PubMed]
43. Pandey, S.; Do, J.Y.; Kim, J.; Kang, M. Fast and Highly Efficient Removal of Dye from Aqueous Solution Using Natural Locust Bean Gum Based Hydrogels as Adsorbent. *Int. J. Biol. Macromol.* **2020**, *143*, 60–75. [CrossRef] [PubMed]
44. Zainal, S.H.; Mohd, N.H.; Suhaili, N.; Anuar, F.H.; Lazim, A.M.; Othaman, R. Preparation of Cellulose-Based Hydrogel: A Review. *J. Mater. Res. Technol.* **2021**, *10*, 935–952. [CrossRef]
45. Chen, Y.; Li, J.; Lu, J.; Ding, M.; Chen, Y. Synthesis and Properties of Poly(Vinyl Alcohol) Hydrogels with High Strength and Toughness. *Polym. Test.* **2022**, *108*, 107516. [CrossRef]
46. Ren, H.; Gao, Z.; Wu, D.; Jiang, J.; Sun, Y.; Luo, C. Efficient Pb(II) Removal Using Sodium Alginate–Carboxymethyl Cellulose Gel Beads: Preparation, Characterization, and Adsorption Mechanism. *Carbohydr. Polym.* **2016**, *137*, 402–409. [CrossRef] [PubMed]
47. Ghriga, M.A.; Grassl, B.; Gareche, M.; Khodja, M.; Lebouachera, S.E.I.; Andreu, N.; Drouiche, N. Review of Recent Advances in Polyethylenimine Crosslinked Polymer Gels Used for Conformance Control Applications. *Polym. Bull.* **2019**, *76*, 6001–6029. [CrossRef]
48. Rong, N.; Chen, C.; Ouyang, K.; Zhang, K.; Wang, X.; Xu, Z. Adsorption Characteristics of Directional Cellulose Nanofiber/Chitosan/Montmorillonite Aerogel as Adsorbent for Wastewater Treatment. *Sep. Purif. Technol.* **2021**, *274*, 119120. [CrossRef]
49. Mokhtari, A.; Sabzi, M.; Azimi, H. 3D Porous Bioadsorbents Based on Chitosan/Alginate/Cellulose Nanofibers as Efficient and Recyclable Adsorbents of Anionic Dye. *Carbohydr. Polym.* **2021**, *265*, 118075. [CrossRef]
50. Tong, D.S.; Wu, C.W.; Adebajo, M.O.; Jin, G.C.; Yu, W.H.; Ji, S.F.; Zhou, C.H. Adsorption of Methylene Blue from Aqueous Solution onto Porous Cellulose-Derived Carbon/Montmorillonite Nanocomposites. *Appl. Clay Sci.* **2018**, *161*, 256–264. [CrossRef]
51. Wu, L.M.; Tong, D.S.; Zhao, L.Z.; Yu, W.H.; Zhou, C.H.; Wang, H. Fourier Transform Infrared Spectroscopy Analysis for Hydrothermal Transformation of Microcrystalline Cellulose on Montmorillonite. *Appl. Clay Sci.* **2014**, *95*, 74–82. [CrossRef]
52. Xie, H.; Pan, Y.; Xiao, H.; Liu, H. Preparation and Characterization of Amphoteric Cellulose–Montmorillonite Composite Beads with a Controllable Porous Structure. *J. Appl. Polym. Sci.* **2019**, *136*, 47941. [CrossRef]
53. Shehap, A.M.; Nasr, R.A.; Mahfouz, M.A.; Ismail, A.M. Preparation and Characterizations of High Doping Chitosan/MMT Nanocomposites Films for Removing Iron from Ground Water. *J. Environ. Chem. Eng.* **2021**, *9*, 104700. [CrossRef]
54. Dao, T.B.T.; Ha, T.T.L.; Nguyen, T.D.; Le, H.N.; Ha-Thuc, C.N.; Nguyen, T.M.L.; Perre, P.; Nguyen, D.M. Effectiveness of Photocatalysis of MMT-Supported TiO_2 and TiO_2 Nanotubes for Rhodamine B Degradation. *Chemosphere* **2021**, *280*, 130802. [CrossRef] [PubMed]
55. Long, L.-Y.; Li, F.-F.; Weng, Y.-X.; Wang, Y.-Z. Effects of Sodium Montmorillonite on the Preparation and Properties of Cellulose Aerogels. *Polymers* **2019**, *11*, 415. [CrossRef] [PubMed]
56. Parker, F.S. Amides and Amines. In *Applications of Infrared Spectroscopy in Biochemistry, Biology, and Medicine*; Springer US: Boston, MA, USA, 1971; pp. 165–172. ISBN 978-1-4684-1874-3.
57. Singha, N.R.; Mahapatra, M.; Karmakar, M.; Dutta, A.; Mondal, H.; Chattopadhyay, P.K. Synthesis of Guar Gum-g-(Acrylic Acid-Co-Acrylamide-Co-3-Acrylamido Propanoic Acid) IPN via in Situ Attachment of Acrylamido Propanoic Acid for Analyzing Superadsorption Mechanism of Pb(II)/Cd(II)/Cu(II)/MB/MV. *Polym. Chem.* **2017**, *8*, 6750–6777. [CrossRef]
58. Wang, W.; Zhao, Y.; Yi, H.; Chen, T.; Kang, S.; Li, H.; Song, S. Preparation and Characterization of Self-Assembly Hydrogels with Exfoliated Montmorillonite Nanosheets and Chitosan. *Nanotechnology* **2018**, *29*, 025605. [CrossRef]

59. Akter, M.; Bhattacharjee, M.; Dhar, A.K.; Rahman, F.B.A.; Haque, S.; Rashid, T.U.; Kabir, S.M.F. Cellulose-Based Hydrogels for Wastewater Treatment: A Concise Review. *Gels* **2021**, *7*, 30. [CrossRef]
60. Mok, C.F.; Ching, Y.C.; Osman, N.A.A.; Muhamad, F.; Hai, N.D.; Choo, J.H.; Hassan, C.R. Adsorbents for Removal of Cationic Dye: Nanocellulose Reinforced Biopolymer Composites. *J. Polym. Res.* **2020**, *27*, 373. [CrossRef]
61. Impert, O.; Katafias, A.; Kita, P.; Mills, A.; Pietkiewicz-Graczyk, A.; Wrzeszcz, G. Kinetics and Mechanism of a Fast Leuco-Methylene Blue Oxidation by Copper(Ii)–Halide Species in Acidic Aqueous Media. *Dalton Trans.* **2003**, *3*, 348–353. [CrossRef]
62. Salama, A.; Abouzeid, R.E.; Awwad, N.S.; Ibrahium, H.A. New Sustainable Ionic Polysaccharides Fibers Assist Calcium Phosphate Mineralization as Efficient Adsorbents. *Fibers Polym.* **2021**, *22*, 1526–1534. [CrossRef]
63. Thakur, S.; Pandey, S.; Arotiba, O.A. Development of a Sodium Alginate-Based Organic/Inorganic Superabsorbent Composite Hydrogel for Adsorption of Methylene Blue. *Carbohydr. Polym.* **2016**, *153*, 34–46. [CrossRef]
64. Singha, A.S.; Guleria, A. Chemical Modification of Cellulosic Biopolymer and Its Use in Removal of Heavy Metal Ions from Wastewater. *Int. J. Biol. Macromol.* **2014**, *67*, 409–417. [CrossRef] [PubMed]
65. Wang, W.; Zhao, Y.; Yi, H.; Chen, T.; Kang, S.; Zhang, T.; Rao, F.; Song, S. Pb(II) Removal from Water Using Porous Hydrogel of Chitosan-2D Montmorillonite. *Int. J. Biol. Macromol.* **2019**, *128*, 85–93. [CrossRef] [PubMed]
66. Zhang, N.; Zang, G.-L.; Shi, C.; Yu, H.-Q.; Sheng, G.-P. A Novel Adsorbent TEMPO-Mediated Oxidized Cellulose Nanofibrils Modified with PEI: Preparation, Characterization, and Application for Cu(II) Removal. *J. Hazard. Mater.* **2016**, *316*, 11–18. [CrossRef] [PubMed]
67. Haerifar, M.; Azizian, S. Fractal-Like Adsorption Kinetics at the Solid/Solution Interface. *J. Phys. Chem. C* **2012**, *116*, 13111–13119. [CrossRef]
68. Erdem, A.; Ngwabebhoh, F.A.; Yildiz, U. Novel MacroporousCryogels with Enhanced Adsorption Capability for the Removal of Cu(II) Ions from Aqueous Phase: Modelling, Kinetics and Recovery Studies. *J. Environ. Chem. Eng.* **2017**, *5*, 1269–1280. [CrossRef]
69. Muntean, S.G.; Nistor, M.A.; Ianoș, R.; Păcurariu, C.; Căpraru, A.; Surdu, V.-A. Combustion Synthesis of Fe_3O_4/Ag/C Nanocomposite and Application for Dyes Removal from Multicomponent Systems. *Appl. Surf. Sci.* **2019**, *481*, 825–837. [CrossRef]
70. Wang, Z.; Song, L.; Wang, Y.; Zhang, X.-F.; Yao, J. Construction of a Hybrid Graphene Oxide/Nanofibrillated Cellulose Aerogel Used for the Efficient Removal of Methylene Blue and Tetracycline. *J. Phys. Chem. Solids* **2021**, *150*, 109839. [CrossRef]
71. Wang, W.; Zhao, Y.; Bai, H.; Zhang, T.; Ibarra-Galvan, V.; Song, S. Methylene Blue Removal from Water Using the Hydrogel Beads of Poly(Vinyl Alcohol)-Sodium Alginate-Chitosan-Montmorillonite. *Carbohydr. Polym.* **2018**, *198*, 518–528. [CrossRef]
72. Shahnaz, T.; Padmanaban, V.C.; Narayanasamy, S. Surface Modification of Nanocellulose Using Polypyrrole for the Adsorptive Removal of Congo Red Dye and Chromium in Binary Mixture. *Int. J. Biol. Macromol.* **2020**, *151*, 322–332. [CrossRef]
73. Ngwabebhoh, F.A.; Erdem, A.; Yildiz, U. Synergistic Removal of Cu(II) and Nitrazine Yellow Dye Using an Eco-friendly Chitosan-montmorillonite Hydrogel: Optimization by Response Surface Methodology. *J. Appl. Polym. Sci.* **2016**, *133*, 43664. [CrossRef]
74. Zhao, H.; Ouyang, X.-K.; Yang, L.-Y. Adsorption of Lead Ions from Aqueous Solutions by Porous Cellulose Nanofiber–Sodium Alginate Hydrogel Beads. *J. Mol. Liq.* **2021**, *324*, 115122. [CrossRef]
75. Yang, Z.; Hou, J.; Miao, L.; Wu, J. Comparison of Adsorption Behavior Studies of Methylene Blue by Microalga Residue and Its Biochars Produced at Different Pyrolytic Temperatures. *Environ. Sci. Pollut. Res.* **2021**, *28*, 14028–14040. [CrossRef] [PubMed]
76. Zhang, H.; Omer, A.M.; Hu, Z.; Yang, L.-Y.; Ji, C.; Ouyang, X. Fabrication of Magnetic Bentonite/Carboxymethyl Chitosan/Sodium Alginate Hydrogel Beads for Cu (II) Adsorption. *Int. J. Biol. Macromol.* **2019**, *135*, 490–500. [CrossRef] [PubMed]
77. Mo, L.; Pang, H.; Tan, Y.; Zhang, S.; Li, J. 3D Multi-Wall Perforated Nanocellulose-Based Polyethylenimine Aerogels for Ultrahigh Efficient and Reversible Removal of Cu(II) Ions from Water. *Chem. Eng. J.* **2019**, *378*, 122157. [CrossRef]
78. Shariful, M.I.; Sepehr, T.; Mehrali, M.; Ang, B.C.; Amalina, M.A. Adsorption Capability of Heavy Metals by Chitosan/Poly(Ethylene Oxide)/Activated Carbon Electrospun Nanofibrous Membrane: Research Article. *J. Appl. Polym. Sci.* **2018**, *135*, 45851. [CrossRef]
79. Esmaeili, Z.; Izadyar, S.; Hamzeh, Y.; Abdulkhani, A. Preparation and Characterization of Highly Porous Cellulose Nanofibrils/Chitosan Aerogel for Acid Blue 93 Adsorption: Kinetics, Isotherms, and Thermodynamics Analysis. *J. Chem. Eng. Data* **2021**, *66*, 1068–1080. [CrossRef]
80. Wu, C.; McClements, D.J.; He, M.; Huang, Y.; Zhu, K.; Jiang, L.; Teng, F.; Li, Y. Okara Nanocellulose Fabricated Using Combined Chemical and Mechanical Treatments: Structure and Properties. *J. Mol. Liq.* **2021**, *335*, 116231. [CrossRef]

Disclaimer/Publisher's Note: The statements, opinions and data contained in all publications are solely those of the individual author(s) and contributor(s) and not of MDPI and/or the editor(s). MDPI and/or the editor(s) disclaim responsibility for any injury to people or property resulting from any ideas, methods, instructions or products referred to in the content.

Article

Evaluation of the Adsorption Efficiency of Graphene Oxide Hydrogels in Wastewater Dye Removal: Application of Principal Component Analysis

Omar Mouhtady, Emil Obeid, Mahmoud Abu-samha, Khaled Younes * and Nimer Murshid *

College of Engineering and Technology, American University of the Middle East, Kuwait; omar.mouhtady@aum.edu.kw (O.M.); emil.obeid@aum.edu.kw (E.O.); mahmoud.abusamha@aum.edu.kw (M.A.-s.)
* Correspondence: khaled.younes@aum.edu.kw (K.Y.); nimer.murshid@aum.edu.kw (N.M.)

Abstract: Industrial dye wastewater is one of the major water pollution problems. Adsorbent materials are promising strategies for the removal of water dye contaminants. Herein, we provide a statistical and artificial intelligence study to evaluate the adsorption efficiency of graphene oxide-based hydrogels in wastewater dye removal by applying Principal Component Analysis (PCA). This study aims to assess the adsorption quality of 35 different hydrogels. We adopted different approaches and showed the pros and cons of each one of them. PCA showed that alginate graphene oxide-based hydrogel (without polyvinyl alcohol) had better tolerance in a basic medium and provided higher adsorption capacity. Polyvinyl alcohol sulfonated graphene oxide-based hydrogels are suitable when higher adsorbent doses are required. In conclusion, PCA represents a robust way to delineate factors affecting hydrogel selection for pollutant removal from aqueous solutions.

Keywords: hydrogel; sustainability; wastewater treatment; principal component analysis; graphene oxide; adsorption; hydrogel composites; dye; machine learning; artificial intelligence

1. Introduction

Dyes are used primarily in the production of consumer products, including paints, textiles, printing inks, paper, and plastics. Each year, the discharged dyes reach 60,000 tons worldwide. Dyes consist of synthetic organic material with biological toxicity such as carcinogenicity and teratogenicity and are mutagenic [1]. The main source of synthetic and organic dyes is the textile dyeing process. Azo dyes are the largest group of artificial dyes, corresponding to 65% of the total production of dyes in the world [2]. Synthetic dyes are refractory to temperature [3] and very stable due to their complex molecular structure and, therefore, do not biodegrade easily [4]. Consequently, dye-contaminated water discharged by industrial activities, including dye production, is one of the major water pollution problems posing a serious risk to drinking-water supplies [5].

Enormous efforts and various physical, chemical, and biological remediation approaches have been developed to treat the aquatic environment. However, physical methods, including adsorption, have shown promising and sustainable efficiency for treating dye-contaminated water [6]. Adsorbent materials are yet considered one of the most promising strategies to remove contaminants [7]. By definition, adsorption is a phenomenon of surface in which a solute (atom, ion, or molecule) in a gas or liquid state) adheres to a solid sorbent. The advantages of adsorption processes are mainly their simple design, low cost, and their effectiveness towards a wide range of pollutants compared to other approaches (coagulation, filtration, precipitation, ion exchange, reverse osmosis, and oxidative processes) [8–10].

The tendency to favor better adsorption results is observed when the dye-contaminated waters, hydrophilic, and functional materials are taken into consideration. In recent years, studies have focused on using composite hydrogels for adsorption due to their promising properties compared to conventional hydrogels or other hydrophilic materials [6]. Hydrogels are three-dimensional networks of hydrophilic polymers that can absorb large amounts of water and swell while maintaining their structure due to the chemical or physical cross-linking of individual polymer chains [6]. These composites can be enriched with hydrophilic and functional groups, which enhance the adsorption of dyes and heavy metal ions from aqueous solutions.

Adsorbents and environmental applications of graphene-based composites have been reviewed for dye removal [8,11]. Activated carbon has been used intensively in dye manufacturing industries due to its sustainability and cost-effectiveness [12]. The new prospect of pollutant management is the combination of nanomaterials such as metal oxides, graphene, and carbon nanotubes. Graphene is massively used as a nano-adsorbent for environmental applications due to its high theoretical surface area ($\sim$2620 m^2g^{-1}) [13,14]. Graphene oxide (GO) is mainly produced from graphene by the Staudenmaier method [15]. GO has abundant oxygen-containing functional groups on its surface and can be processed into reduced graphene oxide (rGO) [16].

The surface functionality and electrostatic interactions of the adsorbate make GO a very promising material for environmental applications [17], such as the adsorption of charged species [18]. However, the efficiency of adsorption of GO depends on the charge on the dye [4].

To evaluate the adsorption efficiency of GO hydrogels in wastewater dye removal, Principal Component Analysis (PCA) with several parameters has been applied. In general, PCA is used to reduce the parameters of a dataset by producing linear combinations of the original parameters and, therefore, to identify the main parameters necessary to enhance and improve a given process [19].

Following the large number of parameters that affect the efficiency of GO for wastewater remediation, a PCA approach can be adopted to better seek intercorrelation in parameters related to adsorption efficiency. To the best of our knowledge, this work represents the first statistical and artificial intelligence study applied to evaluate the adsorption efficiency of GO hydrogels for dye removal.

2. Methodology

The aim of the study is to apply PCA based on the published study by Pereira et al. [6] (Table 1) to better understand the functional difference of multiple GO-based hydrogels depending on their adsorption properties. PCA is a method of revealing patterns among variables. These patterns were hidden from the bi-dimensional statistical approach. It presents an unsupervised machine-learning method since, once applied, no prior knowledge is assumed regarding the data or the investigated phenomena. The jth PC matrix (Fi) is expressed using a unit-weighting vector (Uj) and the original data matrix M with $m \times n$ dimensions. (m: number variables n: number of datasets) as follows [19–22]:

$$Fi = U_j^T M = \sum^{i=0} U_{ji} M_i \qquad (1)$$

where U is the loading coefficient and M is the data vector of size n. The variance matrix $M(Var(M))$, which is obtained by projecting M to U, should be maximized, following:

$$Var(M) = \frac{1}{n}(UM)(UM)^T = \frac{1}{n} UMM^T U \qquad (2)$$

$$MaxVar(M) = Max\left(\left(\frac{1}{n}\right) UMM^T U\right) \qquad (3)$$

Table 1. Adsorption data of different composite hydrogels containing graphene oxide (and derivatives) used for the removal of dyes from water (adapted with permission from Ref. [6]).

	Composite Hydrogel	C% [a]	D [b]	ET [c]	qm [d]	pH [e]	References
1	PMPTC/GO	0.3	-	150	13		Wang et al. [23]
2	PAAm/GO	50	0.2	20	293		Yang et al. [24]
3	CMC/Aam/GO	10	4	720	185	6	Varaprasad et al. [25]
4	Chitin/TA/GO	7		400	231	7	Liu et al. [26]
5	CTS/GO			4000	10		Zhao et al. [27]
6	CTS/amino-functionalized-GO	20		5	385	7	Omidi and Kakanejadifard [28]
7	PVP/Aac/GO	0.2	5	40	78	7	Atyaa et al. [29]
8	Double ALG/GO network		1	1200	2300	8	Zhuang et al. [30]
9	Single ALG/GO network		1	1200	1800	8	Zhuang et al. [30]
10	Double ALG/PVA/GO network	5	0.1	480	1437	6	Kong et al. [31]
11	Single ALG/PVA/GO network	5	0.1	480	1256	6	Kong et al. [31]
12	ALG/immobilized GO network	5	0.2	200	181	5.4	Li et al. [32]
13	ALG/GO		5	60	122	5.3	Balkız et al. [33]
14	CTA/PAAc/GO	0.5	1	2250	297	7	Chang et al. [34]
15	CTS/GO	50	0.13	70	390	6.5	Chen et al. [35]
16	CTS/GO			50	3.5		Zhao et al. [27]
17	PVA/sulfonated-GO	1	80	720	5.1	6.2	Li et al. [32]
18	Cellulose/GO	0.5	20	20	123	7	Soleimani et al. [36]
19	Cellulose/GO	10	2	70	46		Liu et al. [26]
20	CMC/PVA/GO	0.7	1.5	80	89	8	Dai et al. [37]
21	k-CARR/GO	30		6	658	5.3	Yang et al. [38]
22	PEGDMA-rGO	1	2.5	720	60	7.4	Halouane et al. [39]
23	PAMm/GO	5		75	26		Thompson et al. [40]
24	PEGD/thiolated-GO	17		75	6		Liu et al. [26]
25	PAAc-g-XG/GO	0.5	0.25			7	Hosseini et al. [41]
26	PEI/GO			240	334		Guo et al. [42]
27	PVA/sulfonated-GO	1	80		4.4	6.2	Li et al. [32]
28	ALG/PAAc/Graphite			60	629	7	Verma et al. [43]
29	XG-g-PAAc/rGO	5	0.5	30	1052	6	Makhado et al. [44]
30	PAMm/GO	50	0.025	20	288		Yang et al. [24]
31	CTS/GO			250	1.9		Zhao et al. [27]
32	PMPTC/GO	0.3		150	12		Wang et al. [23]
33	Cellulose/GO	0.5	20	40	62	7	Soleimani et al. [36]
34	PEI/GO			240	132		Guo et al. [42]
35	ALG-Fe^{3+}/rGO	50		360	18.4		Xiao et al. [45]

[a] C% = Content of graphene oxide (and derivatives) (wt-%) in the composite hydrogel. [b] D = Adsorbent dosage (g/L). [c] ET = time necessary to achieve the equilibrium condition (min). [d] qm = Adsorption capacity (mg/g). [e] pH = potential of hydrogen is a scale used to specify the acidity or basicity of an aqueous solution.

Since $\frac{1}{n} MM^T$ is the same as the covariance matrix of $M(cov(M))$, $Var(M)$ can be expressed, following:

$$Var(M) = U^T cov(M) U \qquad (4)$$

The Lagrangian function can be defined by performing the Lagrange multiplier method, following:

$$L = U^T$$
$$L = U^T cov(M) U - \delta(U^T U - 1) \qquad (5)$$

for (5), "$U^T U - 1$" is considered equal to zero since the weighting vector is a unit vector. Hence, the maximum value of $Var(M)$ can be calculated by equating the derivative of the Lagrangian function (L), with respect to U, following:

$$\frac{dL}{dU} = 0 \qquad (6)$$

$$cov(M)U - \delta U = (cov(M) - \delta I)U = 0 \quad (7)$$

where,

δ: eigenvalue of $cov(M)$
U: eigenvector of $cov(M)$

3. Results and Discussion

Figure 1 shows the PCA biplot for the published results on the adsorption data of different composite hydrogels containing GO (and derivatives) used for the removal of dyes from water [6]. The first two PCs accounted for 62.03% of the total variance (32.73% for PC1 and 29.30% for PC2). The factors: C%, D, and ET, exhibited the highest contribution to PC1, accounting for 26.43%, 34.12%, and 36.22%, respectively. As for PC2, qm and pH accounted for the highest contributions, yielding 45.91% and 35.66% of the total contribution of these factors, respectively. The difference in factors' contributions with respect to the investigated PCs indicates a high representation of the adsorption data of the investigated hydrogels. C% showed a negative influence on both PCs; however, it influenced PC2 to a lesser extent. For qm and pH, they presented certain proximity and were located on the top-right quarter of the biplot. More specifically, qm had a strong positive influence along PC2, with no influence along PC1. The factor pH had a slight positive influence along PC1 with a major positive effect along PC2. ET and D are located in the bottom-right corner of the biplot. More specifically, ET scored a strong positive influence along PC1, with no influence for PC2. For D, it scored a strong negative influence along both PCs.

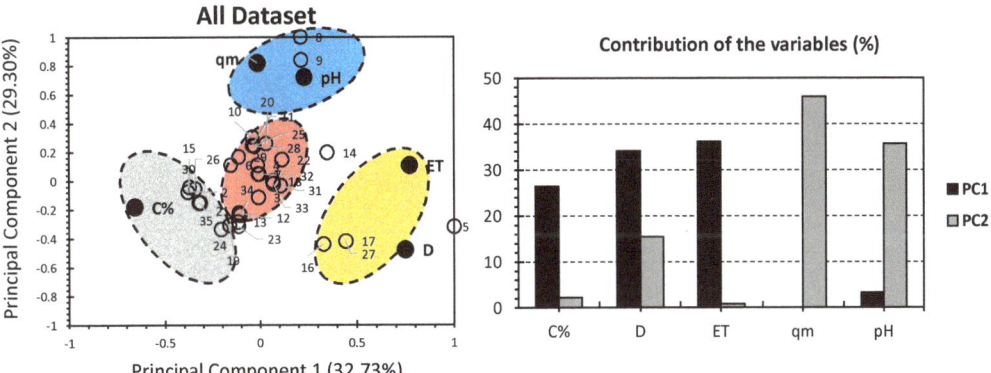

Figure 1. PCA for all datasets. Ref. [6] White bullets represent the 35 investigated graphene oxide hydrogels. Black bullets represent the adsorption properties involved. Different colors were used for clusters to make the interpretation of results easier.

PCA yielded four different distinguishable clusters of hydrogels: red, blue, yellow, and grey (Figure 1). It is quite interesting that the red cluster gathered most of the investigated hydrogels, indicating a poor to no influence of the studied factors on each hydrogel of this cluster. For the blue cluster, it gathered hydrogels 8 and 9 and showed a positive correlation along pH and qm. This indicates that alginate GO hydrogels (without polyvinyl alcohol) are more suitable for an elevated pH medium, and higher adsorption capacities are required. These findings are corroborated by Zhuang et al., where alginate GO hydrogels had the highest qm and the best tolerance for strong base [30]. For the yellow cluster, it gathered hydrogels 16, 17, and 27; and showed a positive correlation along ET and D. Since both 17 and 27 are the only sulfonated polyvinyl alcohol hydrogels, this could indicate that these hydrogels are better suitable for highly contaminated water. This is supported by Li and colleagues' results, where both 17 and 27 scored the highest sorbent dosage D [32]. For the time to achieve equilibrium conditions (ET), the findings in hand could not confirm or inform its relevance to these hydrogels, as a part of the data is missing (Figure 1). For the grey cluster, it gathered hydrogels 10, 15, 21, 24, 26, 30, and 35 and showed a positive correlation with C%. This could generally indicate the relevance of the content of GO in composite hydrogels. No other findings can be depicted since the included hydrogels show significantly different functional groups and are, therefore, not similar.

Even though most of the individuals have shown negligible influence by both PCs, all of the datasets for the PCA approach have shown quite interesting findings. Hence, ALG/GO hydrogels (without PVA) have shown more suitability for higher pH media and where higher adsorption capacities are required. PVA sulfonated hydrogels are estimated to be more likely applied where higher adsorbent doses (D) are required. To seek a better knowledge, the dataset will be split into: (a) high correlation individuals (having correlation factor, x > +0.2; Figure 2), (b) low correlation individuals (having correlation factor, −0.2 < x < +0.2; Figure 3).

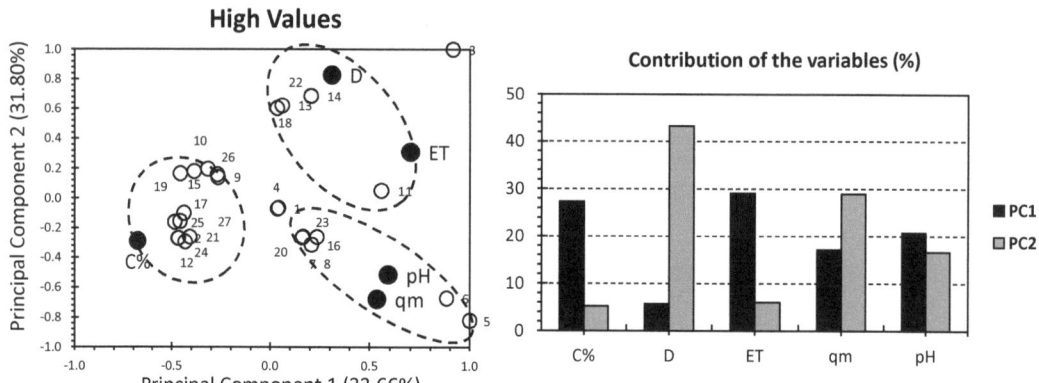

Figure 2. PCA for highly correlated values. Ref. [6] White bullets represent the 35 investigated graphene oxide hydrogels. Black bullets represent the adsorption properties involved.

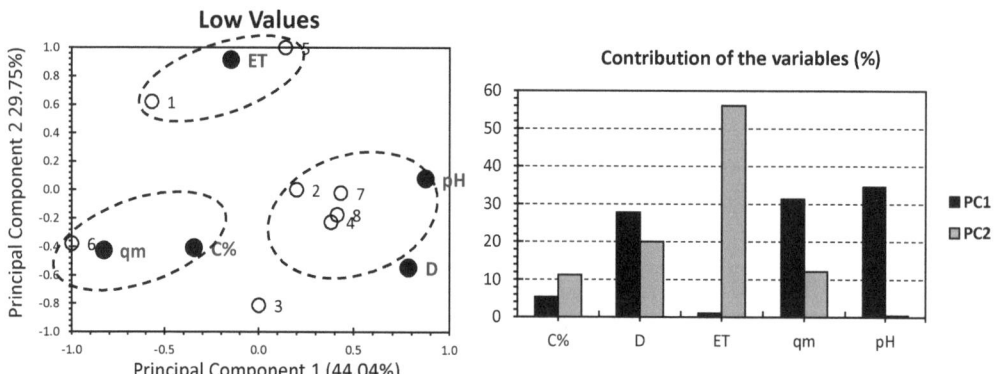

Figure 3. PCA for low correlated values. Ref. [6] White bullets represent the 35 investigated graphene oxide hydrogels. Black bullets represent the adsorption properties involved.

Figure 2 shows the PCA biplot for the highly correlated individuals of the investigated GO hydrogels. The first two PCs accounted for 65.46% of the total variance (33.66% for PC1 and 31.80% for PC2; Figure 2) The slightly higher variance, if compared to the all-dataset approach (Figure 2) indicates that the following findings are more reliable than the total dataset PCA. For the factors, C and ET exhibited the highest contribution of PC1, accounting for 27.32% and 29.12%, respectively. As for PC2, D and qm accounted for the highest contributions, yielding 43.22% and 28.91% of the total contribution of this factor. Interestingly, both qm and pH showed moderate contributions along both PCs. Similar to the case of all datasets, C% in the PCA analysis showed a negative influence on both PCs. For qm and pH, they presented certain proximity and were located in the bottom-right quarter of the PCA-biplot. Therefore, qm and pH scored a strong positive and negative influence along PC1 and PC2, respectively (Figure 2). ET and D are located on the top-right of the PCA biplot. More specifically, D scored a strong and moderate influence along PC2 and PC1, respectively (Figure 2). ET and D are located in the top-right of the PCA biplot. More specifically, D scored a strong and moderate influence along PC2 and PC1, respectively (Figure 2) For ET, it scored a moderate influence on both PCs. %C was individually located on the bottom-left corner of the PCA biplot and presented a moderate negative influence along with both PCs. Even though the factors showed different distributions on the PCA biplot than the all-dataset approach, it revealed the same grouping. Additionally, a better distribution of the individuals is clear (Figure 2). This reveals the efficiency of dividing the dataset into high and low-correlation individuals (Figures 2 and 3). In contrast, a high distribution of individuals makes seeking any relevant tendencies between hydrogels a rather tedious and time-consuming approach.

Figure 3 shows the PCA biplot for the low correlated individuals of the investigated GO hydrogels. The first two PCs accounted for 73.79% of the total variance (44.04% for PC1 and 29.75% for PC2; Figure 3). Once compared with the two previous approaches (Figures 1 and 2), the higher variance of the low correlation individuals indicates that the following strategy is the most reliable one, as it copes with the highest amount of the "truth" in the investigated dataset. Factors, D, qm, and pH exhibited the largest contribution of PC1, accounting for 27.77%, 31.28%, and 34.56%, respectively. As for PC2, ET accounted for the highest contribution, yielding 56.15% of the total contribution of this factor. It is worth mentioning that different groupings were yielded than the two previous approaches (Figures 1 and 2). Hence, ET is individually located on the upper part of the biplot, yielding a high positive influence following PC1 and a negligible one along PC2 (Figure 3). pH and D are located on the bottom-right quarter of the PCA biplot. More specifically, D scored strong positive and negative influences along PC1 and PC2, respectively. For pH, a strong positive influence along PC1, with a minor influence along PC, can be found. C% and qm are located in the bottom-left quarter of the PCA biplot. More specifically, qm scored a strong negative influence along with both PCs. For C%, it scored a moderate negative influence along with both PCs. For individuals, and similarly to the highly correlated individuals, it yielded multiple clusters containing hydrogels with very different matrices and functional groups, which prohibits any change of finding relevant findings between the hydrogels in hand.

4. Conclusions

This study aims to apply PCA to delineate interesting tendencies affecting the adsorption features of GO-based hydrogels. Different approaches were adopted, and each presented pros and cons. When PCA was run for the whole data set at once, ALG/GO hydrogels (without PVA) showed better tolerance in the basic medium and provided higher adsorption capacity to be implemented. PVA sulfonated hydrogels are considered preferably applied where higher adsorbent doses (D) are required.

Furthermore, we have attempted to develop a new strategy to reveal the outmost findings from the datasets. The adopted strategy involves splitting the individual hydrogels between high and low correlated ones. In our case, both groups of individual hydrogels showed a higher presentation of the total variance rather than having the total dataset analyzed all at once. Interestingly, the highest variance was yielded for the low correlated factors. This will allow a better seeking out of the tendencies between different hydrogels.

Even though no specific trends were yielded when the various hydrogels were separated, the highest variance makes this method better suited for the provided data-driven study.

Author Contributions: All authors, O.M., E.O., M.A.-s., K.Y. and N.M., contributed to writing, reviewing, and editing of this manuscript. All authors have read and agreed to the published version of the manuscript.

Funding: This research received no external funding.

Conflicts of Interest: The authors declare no conflict of interest.

References

1. Liu, Q. Pollution and Treatment of Dye Waste-Water. *IOP Conf. Ser. Earth Environ. Sci.* **2020**, *514*, 052001. [CrossRef]
2. Guaratini, C.C.I.; Zanoni, M.V.B. Corantes têxteis. *Quím. Nova* **2000**, *23*, 71–78. [CrossRef]
3. Barka, N.; Abdennouri, M.; Makhfouk, M.E. Removal of Methylene Blue and Eriochrome Black T from Aqueous Solutions by Biosorption on *Scolymus hispanicus*, L.: Kinetics, Equilibrium and Thermodynamics. *J. Taiwan Inst. Chem. Eng.* **2011**, *42*, 320–326. [CrossRef]
4. Khurana, I.; Saxena, A.; Bharti; Khurana, J.M.; Rai, P.K. Removal of Dyes Using Graphene-Based Composites: A Review. *Water Air Soil Pollut.* **2017**, *228*, 180. [CrossRef]
5. Carneiro, P.A.; Umbuzeiro, G.A.; Oliveira, D.P.; Zanoni, M.V.B. Assessment of Water Contamination Caused by a Mutagenic Textile Effluent/Dyehouse Effluent Bearing Disperse Dyes. *J. Hazard. Mater.* **2010**, *174*, 694–699. [CrossRef] [PubMed]
6. Pereira, A.G.B.; Rodrigues, F.H.A.; Paulino, A.T.; Martins, A.F.; Fajardo, A.R. Recent Advances on Composite Hydrogels Designed for the Remediation of Dye-Contaminated Water and Wastewater: A Review. *J. Clean. Prod.* **2021**, *284*, 124703. [CrossRef]
7. Malaviya, P.; Singh, A. Physicochemical Technologies for Remediation of Chromium-Containing Waters and Wastewaters. *Crit. Rev. Environ. Sci. Technol.* **2011**, *41*, 1111–1172. [CrossRef]
8. Bharathi, K.S.; Ramesh, S.T. Removal of Dyes Using Agricultural Waste as Low-Cost Adsorbents: A Review. *Appl. Water Sci.* **2013**, *3*, 773–790. [CrossRef]
9. Chen, Y.; Chen, L.; Bai, H.; Li, L. Graphene Oxide–Chitosan Composite Hydrogels as Broad-Spectrum Adsorbents for Water Purification. *J. Mater. Chem. A* **2013**, *1*, 1992–2001. [CrossRef]
10. Cui, W.; Ji, J.; Cai, Y.-F.; Li, H.; Ran, R. Robust, Anti-Fatigue, and Self-Healing Graphene Oxide/Hydrophobically Associated Composite Hydrogels and Their Use as Recyclable Adsorbents for Dye Wastewater Treatment. *J. Mater. Chem. A* **2015**, *3*, 17445–17458. [CrossRef]
11. Lü, K.; Zhao, G.; Wang, X. A Brief Review of Graphene-Based Material Synthesis and Its Application in Environmental Pollution Management. *Chin. Sci. Bull.* **2012**, *57*, 1223–1234. [CrossRef]
12. Carrott, P.J.M.; Carrott, M.M.L.R.; Roberts, R.A. Physical Adsorption of Gases by Microporous Carbons. *Colloids Surf.* **1991**, *58*, 385–400. [CrossRef]
13. Novoselov, K.S.; Geim, A.K.; Morozov, S.V.; Jiang, D.; Zhang, Y.; Dubonos, S.V.; Grigorieva, I.V.; Firsov, A.A. Electric Field Effect in Atomically Thin Carbon Films. *Sci. New Ser.* **2004**, *306*, 666–669. [CrossRef] [PubMed]
14. Liu, S.-Q.; Xiao, B.; Feng, L.-R.; Zhou, S.-S.; Chen, Z.-G.; Liu, C.-B.; Chen, F.; Wu, Z.-Y.; Xu, N.; Oh, W.-C.; et al. Graphene Oxide Enhances the Fenton-like Photocatalytic Activity of Nickel Ferrite for Degradation of Dyes under Visible Light Irradiation. *Carbon* **2013**, *64*, 197–206. [CrossRef]
15. Yang, Y.; Xie, Y.; Pang, L.; Li, M.; Song, X.; Wen, J.; Zhao, H. Preparation of Reduced Graphene Oxide/Poly(Acrylamide) Nanocomposite and its Adsorption of Pb(II) and Methylene Blue. *Langmuir* **2013**, *29*, 10727–10736. [CrossRef]
16. Haubner, K.; Murawski, J.; Olk, P.; Eng, L.M.; Ziegler, C.; Adolphi, B.; Jaehne, E. The Route to Functional Graphene Oxide. *Chem. Eur. J. Chem. Phys.* **2010**, *11*, 2131–2139. [CrossRef]
17. Kemp, K.C.; Seema, H.; Saleh, M.; Le, N.H.; Mahesh, K.; Chandra, V.; Kim, K.S. Environmental Applications Using Graphene Composites: Water Remediation and Gas Adsorption. *Nanoscale* **2013**, *5*, 3149. [CrossRef]
18. Ramesha, G.K.; Vijaya Kumara, A.; Muralidhara, H.B.; Sampath, S. Graphene and Graphene Oxide as Effective Adsorbents toward Anionic and Cationic Dyes. *J. Colloid Interface Sci.* **2011**, *361*, 270–277. [CrossRef]
19. Younes, K.; Grasset, L. The Application of DFRC Method for the Analysis of Carbohydrates in a Peat Bog: Validation and Comparison with Conventional Chemical and Thermochemical Degradation Techniques. *Chem. Geol.* **2020**, *545*, 119644. [CrossRef]
20. Younes, K.; Laduranty, J.; Descostes, M.; Grasset, L. Molecular Biomarkers Study of an Ombrotrophic Peatland Impacted by an Anthropogenic Clay Deposit. *Org. Geochem.* **2017**, *105*, 20–32. [CrossRef]
21. Younes, K.; Grasset, L. Analysis of Molecular Proxies of a Peat Core by Thermally Assisted Hydrolysis and Methylation-Gas Chromatography Combined with Multivariate Analysis. *J. Anal. Appl. Pyrolysis* **2017**, *124*, 726–732. [CrossRef]
22. Korichi, W.; Ibrahimi, M.; Loqman, S.; Ouhdouch, Y.; Younes, K.; Lemée, L. Assessment of Actinobacteria Use in the Elimination of Multidrug-Resistant Bacteria of Ibn Tofail Hospital Wastewater (Marrakesh, Morocco): A Chemometric Data Analysis Approach. *Environ. Sci. Pollut. Res.* **2021**, *28*, 26840–26848. [CrossRef] [PubMed]

23. Wang, H.; Li, Y.; Zhang, Y.; Chen, J.; Chu, G.; Shao, L. Preparation of CeO$_2$ Nano-Support in a Novel Rotor–Stator Reactor and Its Use in Au-Based Catalyst for CO Oxidation. *Powder Technol.* **2015**, *273*, 191–196. [CrossRef]
24. Yang, Y.; Song, S.; Zhao, Z. Graphene Oxide (GO)/Polyacrylamide (PAM) Composite Hydrogels as Efficient Cationic Dye Adsorbents. *Colloids Surf. A Physicochem. Eng. Asp.* **2017**, *513*, 315–324. [CrossRef]
25. Varaprasad, K.; Jayaramudu, T.; Sadiku, E.R. Removal of Dye by Carboxymethyl Cellulose, Acrylamide and Graphene Oxide via a Free Radical Polymerization Process. *Carbohydr. Polym.* **2017**, *164*, 186–194. [CrossRef]
26. Liu, J.; Zhu, K.; Jiao, T.; Xing, R.; Hong, W.; Zhang, L.; Zhang, Q.; Peng, Q. Preparation of Graphene Oxide-Polymer Composite Hydrogels via Thiol-Ene Photopolymerization as Efficient Dye Adsorbents for Wastewater Treatment. *Colloids Surf. A Physicochem. Eng. Asp.* **2017**, *529*, 668–676. [CrossRef]
27. Zhao, H.; Jiao, T.; Zhang, L.; Zhou, J.; Zhang, Q.; Peng, Q.; Yan, X. Preparation and Adsorption Capacity Evaluation of Graphene Oxide-Chitosan Composite Hydrogels. *Sci. China Mater.* **2015**, *58*, 811–818. [CrossRef]
28. Omidi, S.; Kakanejadifard, A. Eco-Friendly Synthesis of Graphene–Chitosan Composite Hydrogel as Efficient Adsorbent for Congo Red. *RSC Adv.* **2018**, *8*, 12179–12189. [CrossRef]
29. Atyaa, A.I.; Radhy, N.D.; Jasim, L.S. Synthesis and Characterization of Graphene Oxide/Hydrogel Composites and Their Applications to Adsorptive Removal Congo Red from Aqueous Solution. In Proceedings of the Journal of Physics: Conference Series; IOP Publishing: Bristol, UK, 2019; Volume 1234, p. 012095.
30. Zhuang, Y.; Yu, F.; Chen, J.; Ma, J. Batch and Column Adsorption of Methylene Blue by Graphene/Alginate Nanocomposite: Comparison of Single-Network and Double-Network Hydrogels. *J. Environ. Chem. Eng.* **2016**, *4*, 147–156. [CrossRef]
31. Kong, Y.; Zhuang, Y.; Han, Z.; Yu, J.; Shi, B.; Han, K.; Hao, H. Dye Removal by Eco-Friendly Physically Cross-Linked Double Network Polymer Hydrogel Beads and Their Functionalized Composites. *J. Environ. Sci.* **2019**, *78*, 81–91. [CrossRef]
32. Li, H.; Fan, J.; Shi, Z.; Lian, M.; Tian, M.; Yin, J. Preparation and Characterization of Sulfonated Graphene-Enhanced Poly (Vinyl Alcohol) Composite Hydrogel and Its Application as Dye Absorbent. *Polymer* **2015**, *60*, 96–106. [CrossRef]
33. Balkız, G.; Pingo, E.; Kahya, N.; Kaygusuz, H.; Bedia Erim, F. Graphene Oxide/Alginate Quasi-Cryogels for Removal of Methylene Blue. *Water Air Soil Pollut.* **2018**, *229*, 1–9. [CrossRef]
34. Chang, Z.; Chen, Y.; Tang, S.; Yang, J.; Chen, Y.; Chen, S.; Li, P.; Yang, Z. Construction of Chitosan/Polyacrylate/Graphene Oxide Composite Physical Hydrogel by Semi-Dissolution/Acidification/Sol-Gel Transition Method and Its Simultaneous Cationic and Anionic Dye Adsorption Properties. *Carbohydr. Polym.* **2020**, *229*, 115431. [CrossRef] [PubMed]
35. Chen, H.; Huang, M.; Liu, Y.; Meng, L.; Ma, M. Functionalized Electrospun Nanofiber Membranes for Water Treatment: A Review. *Sci. Total Environ.* **2020**, *739*, 139944. [CrossRef] [PubMed]
36. Soleimani, K.; Tehrani, A.D.; Adeli, M. Bioconjugated Graphene Oxide Hydrogel as an Effective Adsorbent for Cationic Dyes Removal. *Ecotoxicol. Environ. Saf.* **2018**, *147*, 34–42. [CrossRef]
37. Dai, H.; Huang, Y.; Huang, H. Eco-Friendly Polyvinyl Alcohol/Carboxymethyl Cellulose Hydrogels Reinforced with Graphene Oxide and Bentonite for Enhanced Adsorption of Methylene Blue. *Carbohydr. Polym.* **2018**, *185*, 1–11. [CrossRef]
38. Yang, M.; Liu, X.; Qi, Y.; Sun, W.; Men, Y. Preparation of κ-Carrageenan/Graphene Oxide Gel Beads and Their Efficient Adsorption for Methylene Blue. *J. Colloid Interface Sci.* **2017**, *506*, 669–677. [CrossRef]
39. Halouane, F.; Oz, Y.; Meziane, D.; Barras, A.; Juraszek, J.; Singh, S.K.; Kurungot, S.; Shaw, P.K.; Sanyal, R.; Boukherroub, R. Magnetic Reduced Graphene Oxide Loaded Hydrogels: Highly Versatile and Efficient Adsorbents for Dyes and Selective Cr (VI) Ions Removal. *J. Colloid Interface Sci.* **2017**, *507*, 360–369. [CrossRef]
40. Thompson, L.; Fu, L.; Wang, J.; Yu, A. Impact of Graphene Oxide on Dye Absorption in Composite Hydrogels. *Fuller. Nanotub. Carbon Nanostructures* **2018**, *26*, 649–653. [CrossRef]
41. Hosseini, S.M.; Shahrousvand, M.; Shojaei, S.; Khonakdar, H.A.; Asefnejad, A.; Goodarzi, V. Preparation of Superabsorbent Eco-Friendly Semi-Interpenetrating Network Based on Cross-Linked Poly Acrylic Acid/Xanthan Gum/Graphene Oxide (PAA/XG/GO): Characterization and Dye Removal Ability. *Int. J. Biol. Macromol.* **2020**, *152*, 884–893. [CrossRef]
42. Guo, H.; Jiao, T.; Zhang, Q.; Guo, W.; Peng, Q.; Yan, X. Preparation of Graphene Oxide-Based Hydrogels as Efficient Dye Adsorbents for Wastewater Treatment. *Nanoscale Res. Lett.* **2015**, *10*, 1–10. [CrossRef] [PubMed]
43. Verma, A.; Thakur, S.; Mamba, G.; Gupta, R.K.; Thakur, P.; Thakur, V.K. Graphite Modified Sodium Alginate Hydrogel Composite for Efficient Removal of Malachite Green Dye. *Int. J. Biol. Macromol.* **2020**, *148*, 1130–1139. [CrossRef] [PubMed]
44. Makhado, E.; Pandey, S.; Ramontja, J. Microwave Assisted Synthesis of Xanthan Gum-Cl-Poly (Acrylic Acid) Based-Reduced Graphene Oxide Hydrogel Composite for Adsorption of Methylene Blue and Methyl Violet from Aqueous Solution. *Int. J. Biol. Macromol.* **2018**, *119*, 255–269. [CrossRef] [PubMed]
45. Xiao, D.; He, M.; Liu, Y.; Xiong, L.; Zhang, Q.; Wei, L.; Li, L.; Yu, X. Strong Alginate/Reduced Graphene Oxide Composite Hydrogels with Enhanced Dye Adsorption Performance. *Polym. Bull.* **2020**, *77*, 6609–6623. [CrossRef]

Article

Metal Oxide Hydrogel Composites for Remediation of Dye-Contaminated Wastewater: Principal Component Analysis

Nimer Murshid [1], Omar Mouhtady [1], Mahmoud Abu-samha [1], Emil Obeid [1], Yahya Kharboutly [1], Hamdi Chaouk [1], Jalal Halwani [2] and Khaled Younes [1,*]

1. College of Engineering and Technology, American University of the Middle East, Kuwait
2. Water and Environment Sciences Lab, Lebanese University, Tripoli, Lebanon
* Correspondence: khaled.younes@aum.edu.kw

Citation: Murshid, N.; Mouhtady, O.; Abu-samha, M.; Obeid, E.; Kharboutly, Y.; Chaouk, H.; Halwani, J.; Younes, K. Metal Oxide Hydrogel Composites for Remediation of Dye-Contaminated Wastewater: Principal Component Analysis. *Gels* **2022**, *8*, 702. https://doi.org/10.3390/gels8110702

Academic Editors: Daxin Liang, Ting Dong, Yudong Li and Caichao Wan

Received: 18 October 2022
Accepted: 28 October 2022
Published: 30 October 2022

Publisher's Note: MDPI stays neutral with regard to jurisdictional claims in published maps and institutional affiliations.

Copyright: © 2022 by the authors. Licensee MDPI, Basel, Switzerland. This article is an open access article distributed under the terms and conditions of the Creative Commons Attribution (CC BY) license (https://creativecommons.org/licenses/by/4.0/).

Abstract: Water pollution is caused by multiple factors, such as industrial dye wastewater. Dye-contaminated water can be treated using hydrogels as adsorbent materials. Recently, composite hydrogels containing metal oxide nanoparticles (MONPs) have been used extensively in wastewater remediation. In this study, we use a statistical and artificial intelligence method, based on principal component analysis (PCA) with different applied parameters, to evaluate the adsorption efficiency of 27 different MONP composite hydrogels for wastewater dye treatment. PCA showed that the hydrogel composites CTS@Fe_3O_4, PAAm/TiO_2, and PEGDMA-rGO/Fe_3O_4@cellulose should be used in situations involving high pH, time to reach equilibrium, and adsorption capacity. However, as the composites PAAm-co-AAc/TiO_2, PVPA/Fe_3O_4@SiO_2, PMOA/ATP/Fe_3O_4, and PVPA/Fe_3O_4@SiO_2, are preferred when all physical and chemical properties investigated have low magnitudes. To conclude, PCA is a strong method for highlighting the essential factors affecting hydrogel composite selection for dye-contaminated water treatment.

Keywords: dye removal; hydrogel; hydrogel composites; machine learning; metal oxide nanoparticles; principal component analysis; wastewater

1. Introduction

Dyes are made of synthetic organic material. They are mutagenic and exhibit biological toxicity, such as teratogenicity and carcinogenicity [1,2]. Dyes are primarily used in the production of some consumer goods, including textiles, plastics, paints, paper, and printing inks. According to recent studies [3], approximately 60,000 tons of dyes are discharged annually worldwide. Synthetic and organic dyes are mainly produced through the textile dyeing process. Azo dyes, which correspond to more than half of the total global production of dyes, represent a major part of artificial dyes [4,5]. Due to their complex molecular structure, synthetic dyes are known to be refractory to temperature [6], and very stable; hence, they are not easily biodegradable [7]. Subsequently, dye-contaminated water discharged by industries is one of the major water pollution issues threatening drinking water supplies [8].

Huge efforts and numerous physical, chemical, and biological remediation methods have been devoted to the treatment of the aquatic environment [9]. In particular, physical processes, including adsorption, show promising and long-term sustainable efficacy in treating dye-contaminated water [10]. Indeed, adsorbent materials are very capable of eliminating contaminants [11]. By definition, adsorption is a surface phenomenon in which a solute adheres to a solid sorbent. The solute can be an atom, ion, or molecule in a gas or liquid state. Adsorption processes have several advantages over other methods, such as filtration, precipitation, coagulation, reverse osmosis, ion exchange, and oxidative processes. In addition, adsorption processes are effective against a wide range of pollutants while keeping a simple design and a low cost [12–14]. When dye-contaminated waters, hydrophilic materials, and functional materials are considered, there is a tendency to favor better improved adsorption results. Recently, the use of composite hydrogels for adsorption has been the focus, thanks to their promising properties in comparison to conventional hydrogels or some other hydrophilic materials [10]. Due to their three-dimensional network structure and polymeric hydrophilicity, hydrogels are able to adsorb large quantities of water and to swell while preserving their structures. This is due to individual polymer chains that are chemically or physically cross-linked [10]. These composites can also be enriched with a variety of functional groups to further improve the adsorption of dyes and heavy metal ions from aqueous media.

Recently, composite hydrogels containing metal oxide nanoparticles (MONPs) have been extensively prepared as they have been used in different areas, including environmental remediation [15–17]. The use of these composites in the treatment of dye-contaminated water has received particular attention [10,18,19]. MONPs have numerous characteristics, such as specific adsorption properties [20], magnetic features, and redox capabilities. Therefore, in addition to their ability to improve hydrogels' electrical, mechanical, and thermal properties, MONPs have been used to enhance adsorption selectivity and catalytic activity in pollutant species degradation [21,22]. By adjusting the external magnetic field, they can also allow for remote control of swelling and analyte adsorption/desorption. In fact, composite hydrogels containing magnetic MONPs can reversibly change shape and volume in response to external magnetic fields [23–25]. Because an external magnetic field imposes attractive/repulsive forces, the movements of the embedded magnetic nanoparticles direct the polymeric chains' contraction and distention [26]. As a result, liquid diffusion throughout the hydrogel matrix can be tailored, influencing the adsorption/desorption of the concerned solutes, such as dye molecules. The ability to easily recover from the treated media using magnets represents one more advantage for composite hydrogels containing magnetic compounds when compared to the use of more arduous processes, such as filtration, sedimentation, or centrifugation [27,28].

Principal component analysis (PCA) with several parameters was used to assess the adsorption efficiency of composite hydrogels containing MONPs in wastewater dye removal. PCA is generally used to reduce the parameters of a dataset by generating linear combinations of the original parameters, and thus to identify the main parameters required to enhance and improve a given process [29–31]. Following the huge number of parameters affecting the effectiveness of composite hydrogels containing MONPs for wastewater treatment, a PCA study can be implemented to pursue intercorrelation in parameters associated with adsorption efficiency. In this work, we used the same methodology as our previously published work on dye removal using graphene oxide hydrogels [29]. Herein, we conduct our analysis on 27 different MONPs hydrogel composites, and we examine the intercorrelation between five parameters, namely pH, adsorbent dosage (D), time to reach equilibrium (ET), adsorption surface (qm), and the content of MONPs in the hydrogel (MONP%). To the best of our knowledge, this is the first statistical and artificial intelligence study that has been used to assess the adsorption efficiency of MONPs containing hydrogels for dye removal.

2. Results and Discussion

PCA analysis was conducted on previously published data (Table 1) from the study of Pereira et al. [10]. Figure 1 presents the PCA bi-plot for previously published data on the physical and chemical properties of various composite hydrogels containing MONPs used for dye removal from water [10]. The first two PCs were responsible for 61.89% of the total variance (37.15% for PC1 and 24.74% for PC2) (Figure 1). When the physical and chemical properties of composite hydrogels containing MONPs (and derivatives) were considered, they yielded similar results to those of PCA [29]. This indicates that the PCA approach is equally efficient for both dataset approaches. ET provided the highest contribution to PC1 for the factor MONP% and accounted for 76% of its total contribution. The high contribution of MONP% was surprising, given that fewer data for this factor were provided following the various investigated samples (Table 1). In terms of PC2, pH was the most significant factor, accounting for 70% of its total contribution. The large disparity in factor contribution following the first two PCs indicates the representability of the investigated physical and chemical properties for the various hydrogels under consideration. MONP% and ET had a strong positive influence along PC1, with no to minor positive influence along PC2. This could probably indicate a high correlation between the necessary time to reach equilibrium, from one side, and the carbon content, from the other side. Nonetheless, this could not be confirmed or infirmed following the shortage in data with regard to the carbon content. For pH, it showed a high positive influence along PC2, with no influence along PC1. As for qm, it showed an average negative influence, and a positive influence along PC1, and PC2. Interestingly, D showed nearly no influence on either PC.

Table 1. Physical and chemical properties data of different composite hydrogels containing MONPs used for the removal of dyes from water (adapted with permission from Pereira et al. Ref [10]).

MONPs Composite Hydrogel	Composite #	MONP%	D	ET	qm	pH	Ref
CTS@Fe_3O_4	1	-	1	400	142	7	[32]
ALG@Yttrium	2	-	2	30	1087	6	[33]
Collagen-g-PAAc-co-NVP/Fe_3O_4@SiO_2	3	-	0.05	150	199	7	[34]
PAAm-co-AAc/TiO_2	4	20	1	-	2.2	-	[35]
PAAm/TiO_2	5	0.5	-	600	132	6.5	[36]
St-g-PAAc/ZnSe	6	-	1	30	189	6	[37]
PAAc/Co_3O_4	7	-	0.5	30	837	-	[38]
PEGDMA-rGO/Fe_3O_4@cellulose	8	30	2.5	720	112	7.4	[39]
CTS/Fe_3O_4@κ-CARR	9	-	2	30	123	5.5	[40]
CTS/MMT/γFe_2O_3	10	-	100	180	82	-	[41]
Collagen-g-PAAc-co-NVP/Fe_3O_4@SiO_2	11	-	0.05	125	202	7	[34]
PVPA/Fe_3O_4@SiO_2	12	0	1.4	-	14	-	[42]
AMPS/NIPAAm/Fe_3O_4	13	0	1	10	833	7	[43]
AMPS/NIPAAm/Cu_2O	14	0	1	35	341	7	[43]
AMPS/NIPAM/$Fe_3O_4 \cdot Cu_2O$	15	0	1	5	746	7	[43]
Cellulose/κ-CARR/TiO_2	16	0.7			115	7	[44]
ALG/AgNPs	17		1	120	214	-	[45]
CMSt/PVA/Fe_3O_4	18		10	600	24	7	[46]
PAAm/CTS/Fe_3O_4	19		0.1	125	1603	7	[47]
Cellulose/Fe_3O_4-diatomite	20		0.7	30	102	10	[48]
HPG@Fe_3O_4	21		4	30	459	8	[49]
PAAc-co-AAm/$Co_3O_4 \cdot Cu_2O$	22		0.5	40	238	7	[50]
PAAc-g-ALG/TiO_2	23		0.6		1157	7	[51]
HPG@Fe_3O_4	24		4	30	400	7	[49]
PMOA/ATP/Fe_3O_4	25	3		400	1.7	4.6	[52]
PAAc-g-salep/AgNPs	26		1	20	93	2	[53]
PVPA/Fe_3O_4@SiO_2	27		1.4		16	-	[42]

MONP% = Content of metal oxide (and derivatives) (wt-%) in the composite hydrogel. D = Adsorbent dosage (g/L). ET = time necessary to achieve the equilibrium condition (min). qm = Adsorption capacity (mg/g).

Individuals can be clustered in three ways (blue, red, and yellow) based on the different trends found in the samples (Figure 1). Surprisingly, the red cluster contained the vast majority of the samples examined. This cluster, along with D, qm, and pH, was positively correlated, indicating that these properties had the greatest influence on the investigated hydrogels. It put together samples 4, 12, 26, and 27 for the yellow cluster. All investigated factors had a negative to low correlation with these samples. This suggests that these hydrogels could be used in situations where low pH, adsorbent dosage (D), time to reach equilibrium (ET), and adsorption surface (qm) are required. Only samples 8 and 18 were collected for the blue cluster because they were positively correlated with ET and MONP%. Interestingly, both hydrogels included ferric oxide in their composite structure, despite the fact that this feature is not unique to them. In summary, when the entire dataset was considered, the PCA presentation demonstrated an acceptable presentation of the truth (around 60% of the total variance; Figure 1). However, one shortcoming may arise from the fact that MONP% was missing for the majority of the investigated hydrogels. This will almost certainly create a bias in the differences. As a result, overcoming this problem is as simple as ignoring the MONP% portion.

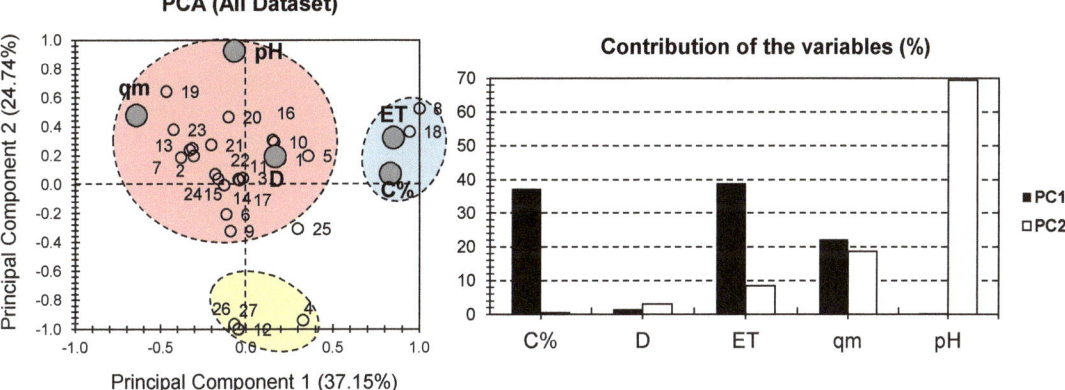

Figure 1. PCA for all datasets. Small empty bullets represent the 27 investigated hydrogels containing MONPs. Large gray bullets represent different physical and chemical properties.

The PCA bi-plot for the physical and chemical properties of the investigated hydrogels, excluding MONP%, is shown in Figure 2. The first two PCs were responsible for 67.87% of the total variance (40.18% for PC1 and 27.69% for PC2; Figure 2). The higher variance score, in comparison to the PCA bi-plot in Figure 1, indicates that the strategy used was effective. ET and qm were the factors that contributed the most to PC1, accounting for 82.85% of the total contributions. In terms of PC2, D contributed the most (67.62%), with pH having a moderate influence (30% of the PC2 contribution; Figure 2). Similar to the case in Figure 1, a high discrepancy in the factors' contributions is scored. Figure 2 shows a higher distribution of the factors, which is interesting. On one side of PC1, qm had a strong positive influence, while ET had a strong negative influence. D had a significant positive influence on PC2. Both positive influences on pH were observed in both PCs.

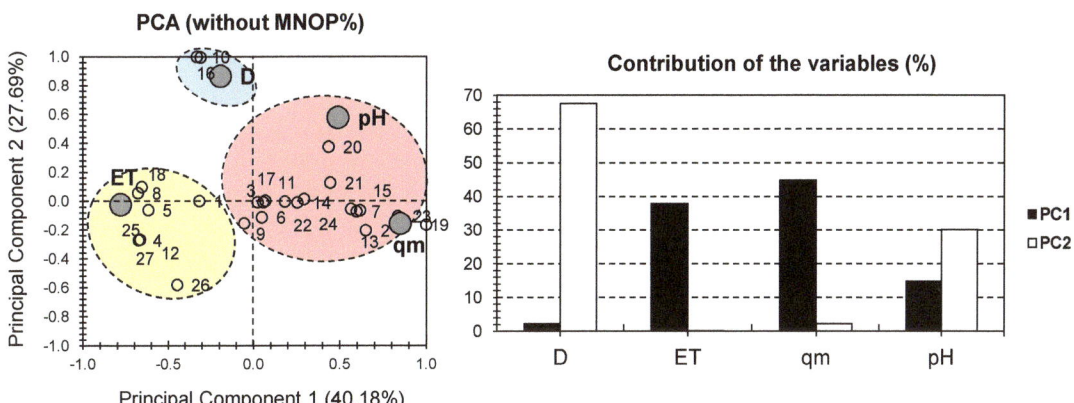

Figure 2. PCA for all datasets. Small empty bullets represent the 27 investigated hydrogels containing MONPs. Large gray bullets represent different physical and chemical properties, with the exclusion of MONP%.

Individually, and similarly to the "all dataset" case, three distinct clusters can be identified when MONP% is considered (Figure 2). The majority of the samples were found in the red cluster, which is positively correlated with both pH and qm. The yellow cluster contained a smaller number of samples than the red cluster. It included samples 1, 4, 5, 8, 12, 18, 25, and 27, all of which showed a strong positive correlation with ET. Interestingly, more samples were more likely to be influenced by the time to reach equilibrium when MONP% was excluded. Only hydrogel samples 10 and 16 were found in the blue cluster, which was positively correlated with D. Nonetheless, the lack of data input for these two samples makes a non-speculative conclusion about the origin of this proximity impossible. In summary, when the MONP% was excluded, the dataset's representativeness increased. This is demonstrated by the greater contribution of total variance in Figure 2 than in Figure 1. A "separation of individuals" approach was used to improve the presentation of the dataset. The goal was to perform a PCA on each cluster to gain a better understanding of the similarities and differences between the hydrogel samples under consideration.

Figure 3 presents the PCA bi-plot for the physical and chemical properties of the samples of the red cluster in Figure 2. The first two PCs accounted for 62.58% of the total variance (35.48% for PC1, and 27.19% for PC2; Figure 3). For the factors, the highest contribution was scored for D and ET, along PC1 (46.38% and 47.57%, respectively). For PC2, the highest contribution was scored for qm and pH (52.47% and 41.74%, respectively). Interestingly, a high distribution of the factors can be noticed, as in the case of Figure 2. ET had a strong positive influence on one side of PC1, while D had a strong negative influence on the other. Both pH and qm had a significant positive influence along PC2, with a minor positive influence along PC1.

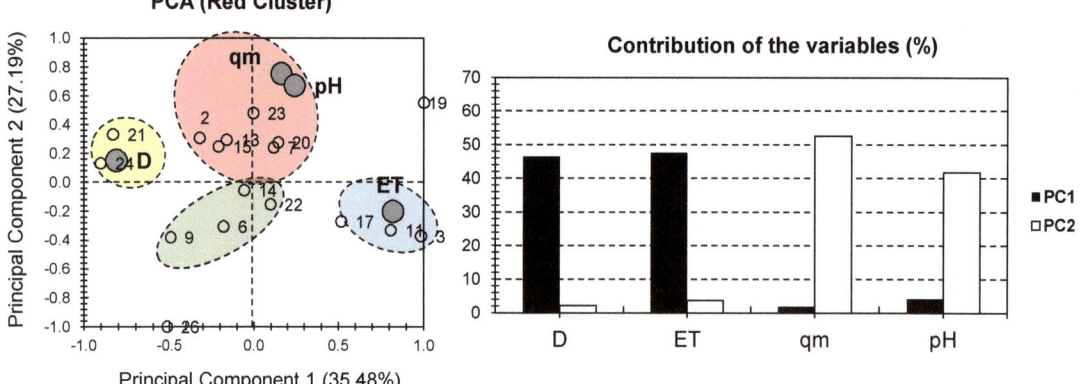

Figure 3. PCA for all datasets. Small bullets represent the 15 investigated hydrogels containing MONPs (red cluster components of Figure 2). Large gray bullets represent different physical and chemical properties, with the exclusion of metallic oxide nanoparticles (MONP%).

For individuals, four different clusters were distinguished. The red cluster contained samples 2, 7, 13, 15, 20, and 23 and showed a positive correlation along qm and pH factors. The blue cluster contained samples 3, 11, and 17 and showed a positive correlation along the ET. The yellow cluster only contained samples 21 and 24 and showed a positive correlation with factor D. As for the green cluster, it contained samples 6, 9, 14, and 22 and showed a negative correlation along all of the investigated factors. Even though the red cluster PCA presented a lower variance than the "all dataset" approach, it similarly showed a higher distribution of the factors along the first two PCs, and it distinctively showed a high distribution of the individuals (four clusters in Figure 3, rather than three clusters in Figure 2). This allows for a better distinction between the different features and conditions of the different investigated hydrogel composites.

Figure 4 depicts the PCA bi-plot for the physical and chemical properties of the samples in the yellow cluster of Figure 2. When only these samples were considered, a higher presentation of the total variance was observed, with a variance of 91.89% (64.11% for PC1 and 27.78% for PC2; Figure 4). This demonstrates the effectiveness of the strategy used, as more focus on "similar" individuals' results in a greater ability to compare them. In the case of PC1, the factors with the highest contributions were pH and ET (36.8% and 30.56%, respectively). In terms of PC2, D had the highest contribution (57.72 percent of the total contribution of PC1), while qm had a moderate contribution of 40% (Figure 4). When compared to the original PCA in Figure 2, a lower distribution of the factors were seen along the bi-plot of the yellow cluster. As a result, all of the factors were located on the positive side of PC1, with qm on the positive side of PC2 and D on the negative side. It had no effect on pH or ET along PC2.

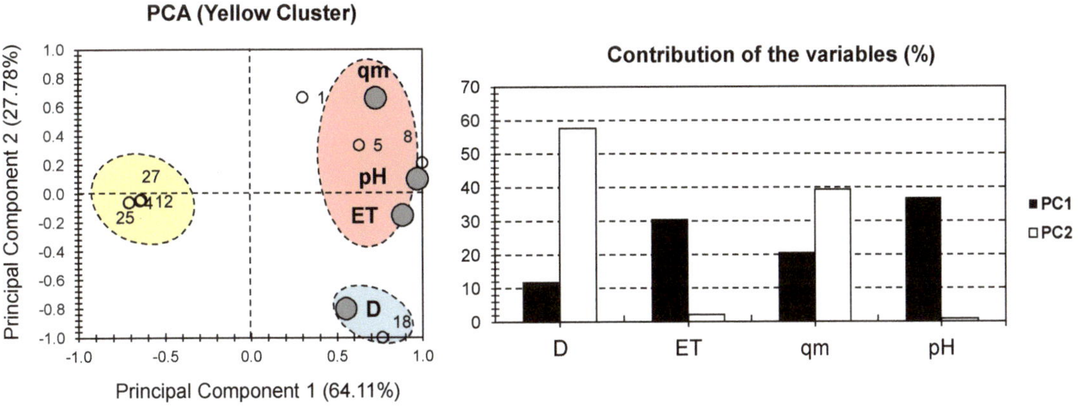

Figure 4. PCA for all datasets. Small bullets represent the 15 investigated hydrogels containing MONPs (yellow cluster components of Figure 2). Large gray bullets represent different physical and chemical properties, with the exclusion of metallic oxide nanoparticles (MONP%).

Individuals were divided into three clusters; the red cluster contained samples 1, 5, and 8, and had a high positive correlation with qm, pH, and ET. The blue cluster contained only hydrogel sample 18 and demonstrated a strong positive correlation with D. The yellow cluster contained samples 4, 12, 25, and 27, and it was located on the opposite side of the different factors (along the negative side of PC2). Given the high variance, it is safe to assume that the composite hydrogel CMSt/PVA/Fe_3O_4 should be used with high adsorbent doses. For CTS@Fe_3O_4, PAAm/TiO_2, and PEGDMA-rGO/Fe_3O_4@cellulose, it should be used where high pH, time to reach equilibrium, and adsorption capacity were implemented. For PAAm-co-AAc/TiO_2, PVPA/Fe_3O_4@SiO_2, PMOA/ATP/Fe_3O_4, and PVPA/Fe_3O_4@SiO_2, it should be used where all of the investigated factors are low.

3. Conclusions

In this study, we performed principal component analysis (PCA) for a better understanding of the correlation between several chemical and physical properties. The properties in-hand are: (a) Time to reach equilibrium (ET), (b) water acidity (pH), (c) Adsorbent dosage (D), and (d) adsorption capacity (qm). In order to seek a higher presentation of the dataset, a "separation of individuals" approach was acquired. The aim was to perform a PCA on each of the clusters to better seek the similarities and dissimilarities of the investigated hydrogel composites. Interestingly, a higher presentation of the total variance was shown in one of the cases, making the PCA-biplot reliable for seeking solid conclusions. The PCA (Figure 4) showed different potential applications for some of the investigated hydrogels. In fact, CTS@Fe_3O_4, PAAm/TiO_2, and PEGDMA-rGO/Fe_3O_4@cellulose should be used where high pH, time to reach equilibrium, and adsorption capacity are encountered. For PAAm-co-AAc/TiO$_2$, PVPA/Fe$_3$O$_4$@SiO$_2$, PMOA/ATP/Fe$_3$O$_4$, and PVPA/Fe$_3$O$_4$@SiO$_2$, it should be used where all of the investigated physical and chemical properties are at low magnitudes. A shortcoming arising from this study resides in the neglecting of structural differences between azo dyes compounds. In fact, we aimed to focus on the physical and chemical features of the adsorbent. Therefore, we have assumed that all azo dyes compounds have similar adsorption properties. Hence, it could be interesting to investigate the influence of adsorbent molecular discrepancies in further investigations.

4. Methodology

The purpose of this study is to apply PCA to a previously published study by Pereira et al. [10] (Table 1) in order to better understand the differences in the functioning of multiple metal oxide nanoparticle (MONP)-based hydrogels based on their adsorption properties. PCA is regarded as a technique for identifying patterns among variables. The bi-dimensional statistical approach failed to reveal these patterns. It presents an unsupervised machine-learning method because, once applied, no prior knowledge of the data or the investigated phenomena is assumed. A unit-weighting vector (W_j) and the original data matrix M with m × n dimensions (m: number of variables, n: number of datasets) are used to express the jth PC matrix (Pj) [31,54,55].

$$Pj = WM = \sum^{i=0} W_{ji} M_i \tag{1}$$

where W is the loading coefficient and M is the n-dimensional data vector. $M(Var(M))$, which is obtained by projecting M to W, should be maximized as follows:

$$Var(M) = \frac{1}{n}\left(W^T M\right)(WM)^T = \frac{1}{n} W^T M M^T W \tag{2}$$

$$MaxVar(M) = Max\left(\left(\frac{1}{n}\right) W^T M M^T W\right) \tag{3}$$

Since $\frac{1}{n} MM^T$ is the same as the covariance matrix of $M(cov(M))$, $Var(M)$ can be expressed as follows:

$$Var(M) = W^T cov(M) W \tag{4}$$

The Lagrangian function can be defined using the Lagrange multiplier method, which is as follows:

$$L = W^T \tag{5}$$

$$L = W^T cov(M) W - \delta\left(W^T W - 1\right) \tag{6}$$

Because the weighting vector is a unit vector, "$W^T W - 1$" is assumed to be equal to zero in Equation (6). As a result, the maximum value of $Var(M)$ can be calculated by equating the derivative of the Lagrangian function (L) with respect to W, as follows:

$$\frac{dL}{dW} = 0 \tag{7}$$

$$cov(M)W - \delta W = (cov(M) - \delta I)W = 0 \tag{8}$$

where, δ: eigenvalue of $cov(M)$, W: eigenvector of $cov(M)$.

Author Contributions: All authors, N.M., O.M., E.O., M.A.-s., Y.K., H.C., J.H. and K.Y., contributed to writing, reviewing, and editing of this manuscript. All authors have read and agreed to the published version of the manuscript.

Funding: This research received no external funding.

Institutional Review Board Statement: Not applicable.

Informed Consent Statement: Not applicable.

Data Availability Statement: The manuscript has no associated data.

Conflicts of Interest: The authors declare no conflict of interest.

References

1. Ismail, M.; Akhtar, K.; Khan, M.I.; Kamal, T.; Khan, M.A.; Asiri, A.M.; Seo, J.; Khan, S.B. Pollution, Toxicity and Carcinogenicity of Organic Dyes and Their Catalytic Bio-Remediation. *Curr. Pharm. Des.* **2019**, *25*, 3645–3663. [CrossRef] [PubMed]
2. Alderete, B.L.; da Silva, J.; Godoi, R.; da Silva, F.R.; Taffarel, S.R.; da Silva, L.P.; Garcia, A.L.H.; Júnior, H.M.; de Amorim, H.L.N.; Picada, J.N. Evaluation of Toxicity and Mutagenicity of a Synthetic Effluent Containing Azo Dye after Advanced Oxidation Process Treatment. *Chemosphere* **2021**, *263*, 128291. [CrossRef] [PubMed]
3. Liu, Q. Pollution and Treatment of Dye Waste-Water. *IOP Conf. Ser. Earth Environ. Sci.* **2020**, *514*, 052001. [CrossRef]
4. Mota, I.G.C.; Neves, R.A.M.D.; Nascimento, S.S.D.C.; Maciel, B.L.L.; Morais, A.H.D.A.; Passos, T.S. Artificial Dyes: Health Risks and the Need for Revision of International Regulations. *Food Rev. Int.* **2021**, 1–16. [CrossRef]
5. Rodríguez-López, M.I.; Pellicer, J.A.; Gómez-Morte, T.; Auñón, D.; Gómez-López, V.M.; Yáñez-Gascón, M.J.; Gil-Izquierdo, Á.; Cerón-Carrasco, J.P.; Crini, G.; Núñez-Delicado, E.; et al. Removal of an Azo Dye from Wastewater through the Use of Two Technologies: Magnetic Cyclodextrin Polymers and Pulsed Light. *Int. J. Mol. Sci.* **2022**, *23*, 8406. [CrossRef]
6. Barka, N.; Abdennouri, M.; Makhfouk, M.E. Removal of Methylene Blue and Eriochrome Black T from Aqueous Solutions by Biosorption on *Scolymus Hispanicus* L.: Kinetics, Equilibrium and Thermodynamics. *J. Taiwan Inst. Chem. Eng.* **2011**, *42*, 320–326. [CrossRef]
7. Khurana, I.; Saxena, A.; Bharti; Khurana, J.M.; Rai, P.K. Removal of Dyes Using Graphene-Based Composites: A Review. *Water Air Soil Pollut.* **2017**, *228*, 180. [CrossRef]
8. Carneiro, P.A.; Umbuzeiro, G.A.; Oliveira, D.P.; Zanoni, M.V.B. Assessment of Water Contamination Caused by a Mutagenic Textile Effluent/Dyehouse Effluent Bearing Disperse Dyes. *J. Hazard. Mater.* **2010**, *174*, 694–699. [CrossRef]
9. Katheresan, V.; Kansedo, J.; Lau, S.Y. Efficiency of Various Recent Wastewater Dye Removal Methods: A Review. *J. Environ. Chem. Eng.* **2018**, *6*, 4676–4697. [CrossRef]
10. Pereira, A.G.B.; Rodrigues, F.H.A.; Paulino, A.T.; Martins, A.F.; Fajardo, A.R. Recent Advances on Composite Hydrogels Designed for the Remediation of Dye-Contaminated Water and Wastewater: A Review. *J. Clean. Prod.* **2021**, *284*, 124703. [CrossRef]
11. Malaviya, P.; Singh, A. Physicochemical Technologies for Remediation of Chromium-Containing Waters and Wastewaters. *Crit. Rev. Environ. Sci. Technol.* **2011**, *41*, 1111–1172. [CrossRef]
12. Bharathi, K.S.; Ramesh, S.T. Removal of Dyes Using Agricultural Waste as Low-Cost Adsorbents: A Review. *Appl. Water Sci.* **2013**, *3*, 773–790. [CrossRef]
13. Cui, W.; Ji, J.; Cai, Y.-F.; Li, H.; Ran, R. Robust, Anti-Fatigue, and Self-Healing Graphene Oxide/Hydrophobically Associated Composite Hydrogels and Their Use as Recyclable Adsorbents for Dye Wastewater Treatment. *J. Mater. Chem. A* **2015**, *3*, 17445–17458. [CrossRef]
14. Chen, Y.; Chen, L.; Bai, H.; Li, L. Graphene Oxide–Chitosan Composite Hydrogels as Broad-Spectrum Adsorbents for Water Purification. *J. Mater. Chem. A* **2013**, *1*, 1992–2001. [CrossRef]
15. Wahid, F.; Zhong, C.; Wang, H.-S.; Hu, X.-H.; Chu, L.-Q. Recent Advances in Antimicrobial Hydrogels Containing Metal Ions and Metals/Metal Oxide Nanoparticles. *Polymers* **2017**, *9*, 636. [CrossRef] [PubMed]
16. Ong, W.-J.; Tan, L.-L.; Ng, Y.H.; Yong, S.-T.; Chai, S.-P. Graphitic Carbon Nitride (g-C_3N_4)-Based Photocatalysts for Artificial Photosynthesis and Environmental Remediation: Are We a Step Closer To Achieving Sustainability? *Chem. Rev.* **2016**, *116*, 7159–7329. [CrossRef] [PubMed]
17. Dannert, C.; Stokke, B.T.; Dias, R.S. Nanoparticle-Hydrogel Composites: From Molecular Interactions to Macroscopic Behavior. *Polymers* **2019**, *11*, 275. [CrossRef] [PubMed]
18. Thakur, S.; Chaudhary, J.; Kumar, V.; Thakur, V.K. Progress in Pectin Based Hydrogels for Water Purification: Trends and Challenges. *J. Environ. Manag.* **2019**, *238*, 210–223. [CrossRef]
19. Nangia, S.; Warkar, S.; Katyal, D. A Review on Environmental Applications of Chitosan Biopolymeric Hydrogel Based Composites. *J. Macromol. Sci. Part A* **2018**, *55*, 747–763. [CrossRef]
20. Cathcart, N.; Murshid, N.; Campbell, P.; Kitaev, V. Selective Plasmonic Sensing and Highly Ordered Metallodielectrics via Encapsulation of Plasmonic Metal Nanoparticles with Metal Oxides. *ACS Appl. Nano Mater.* **2018**, *1*, 6514–6524. [CrossRef]
21. Ningthoujam, R.; Singh, Y.D.; Babu, P.J.; Tirkey, A.; Pradhan, S.; Sarma, M. Nanocatalyst in Remediating Environmental Pollutants. *Chem. Phys. Impact* **2022**, *4*, 100064. [CrossRef]
22. Lu, F.; Astruc, D. Nanocatalysts and Other Nanomaterials for Water Remediation from Organic Pollutants. *Coord. Chem. Rev.* **2020**, *408*, 213180. [CrossRef]
23. Yan, E.; Cao, M.; Ren, X.; Jiang, J.; An, Q.; Zhang, Z.; Gao, J.; Yang, X.; Zhang, D. Synthesis of Fe_3O_4 Nanoparticles Functionalized Polyvinyl Alcohol/Chitosan Magnetic Composite Hydrogel as an Efficient Adsorbent for Chromium (VI) Removal. *J. Phys. Chem. Solids* **2018**, *121*, 102–109. [CrossRef]

24. Huang, J.; Liang, Y.; Jia, Z.; Chen, J.; Duan, L.; Liu, W.; Zhu, F.; Liang, Q.; Zhu, W.; You, W.; et al. Development of Magnetic Nanocomposite Hydrogel with Potential Cartilage Tissue Engineering. *ACS Omega* **2018**, *3*, 6182–6189. [CrossRef] [PubMed]
25. Peng, X.; Wang, H. Shape Changing Hydrogels and Their Applications as Soft Actuators. *J. Polym. Sci. Part B Polym. Phys.* **2018**, *56*, 1314–1324. [CrossRef]
26. Shankar, A.; Safronov, A.P.; Mikhnevich, E.A.; Beketov, I.V.; Kurlyandskaya, G.V. Ferrogels Based on Entrapped Metallic Iron Nanoparticles in a Polyacrylamide Network: Extended Derjaguin–Landau–Verwey–Overbeek Consideration, Interfacial Interactions and Magnetodeformation. *Soft Matter* **2017**, *13*, 3359–3372. [CrossRef]
27. Zhang, J.; Huang, Q.; Du, J. Recent Advances in Magnetic Hydrogels. *Polym. Int.* **2016**, *65*, 1365–1372. [CrossRef]
28. Zhao, W.; Huang, X.; Wang, Y.; Sun, S.; Zhao, C. A Recyclable and Regenerable Magnetic Chitosan Absorbent for Dye Uptake. *Carbohydr. Polym.* **2016**, *150*, 201–208. [CrossRef]
29. Mouhtady, O.; Obeid, E.; Abu-samha, M.; Younes, K.; Murshid, N. Evaluation of the Adsorption Efficiency of Graphene Oxide Hydrogels in Wastewater Dye Removal: Application of Principal Component Analysis. *Gels* **2022**, *8*, 447. [CrossRef]
30. Younes, K.; Moghrabi, A.; Moghnie, S.; Mouhtady, O.; Murshid, N.; Grasset, L. Assessment of the Efficiency of Chemical and Thermochemical Depolymerization Methods for Lignin Valorization: Principal Component Analysis (PCA) Approach. *Polymers* **2022**, *14*, 194. [CrossRef]
31. Younes, K.; Grasset, L. The Application of DFRC Method for the Analysis of Carbohydrates in a Peat Bog: Validation and Comparison with Conventional Chemical and Thermochemical Degradation Techniques. *Chem. Geol.* **2020**, *545*, 119644. [CrossRef]
32. Xu, P.; Zheng, M.; Chen, N.; Wu, Z.; Xu, N.; Tang, J.; Teng, Z. Uniform Magnetic Chitosan Microspheres with Radially Oriented Channels by Electrostatic Droplets Method for Efficient Removal of Acid Blue. *J. Taiwan Inst. Chem. Eng.* **2019**, *104*, 210–218. [CrossRef]
33. Liu, C.; Liu, H.; Tang, K.; Zhang, K.; Zou, Z.; Gao, X. High-Strength Chitin Based Hydrogels Reinforced by Tannic Acid Functionalized Graphene for Congo Red Adsorption. *J. Polym. Environ.* **2020**, *28*, 984–994. [CrossRef]
34. Nakhjiri, M.T.; Bagheri Marandi, G.; Kurdtabar, M. Adsorption of Methylene Blue, Brilliant Green and Rhodamine B from Aqueous Solution Using Collagen-g-p(AA-Co-NVP)/Fe$_3$O$_4$@SiO$_2$ Nanocomposite Hydrogel. *J. Polym. Environ.* **2019**, *27*, 581–599. [CrossRef]
35. Kangwansupamonkon, W.; Klaikaew, N.; Kiatkamjornwong, S. Green Synthesis of Titanium Dioxide/Acrylamide-Based Hydrogel Composite, Self Degradation and Environmental Applications. *Eur. Polym. J.* **2018**, *107*, 118–131. [CrossRef]
36. Raj, A.; Bethi, B.; Sonawane, S.H. Investigation of Removal of Crystal Violet Dye Using Novel Hybrid Technique Involving Hydrodynamic Cavitation and Hydrogel. *J. Environ. Chem. Eng.* **2018**, *6*, 5311–5319. [CrossRef]
37. Abdolahi, G.; Dargahi, M.; Ghasemzadeh, H. Synthesis of Starch-g-Poly (Acrylic Acid)/ZnSe Quantum Dot Nanocomposite Hydrogel, for Effective Dye Adsorption and Photocatalytic Degradation: Thermodynamic and Kinetic Studies. *Cellulose* **2020**, *27*, 6467–6483. [CrossRef]
38. Ansari, T.M.; Ajmal, M.; Saeed, S.; Naeem, H.; Ahmad, H.B.; Mahmood, K.; Farooqi, Z.H. Synthesis and Characterization of Magnetic Poly(Acrylic Acid) Hydrogel Fabricated with Cobalt Nanoparticles for Adsorption and Catalytic Applications. *J. Iran. Chem. Soc.* **2019**, *16*, 2765–2776. [CrossRef]
39. Halouane, F.; Oz, Y.; Meziane, D.; Barras, A.; Juraszek, J.; Singh, S.K.; Kurungot, S.; Shaw, P.K.; Sanyal, R.; Boukherroub, R.; et al. Magnetic Reduced Graphene Oxide Loaded Hydrogels: Highly Versatile and Efficient Adsorbents for Dyes and Selective Cr(VI) Ions Removal. *J. Colloid Interface Sci.* **2017**, *507*, 360–369. [CrossRef]
40. Mahdavinia, G.R.; Mosallanezhad, A. Facile and Green Rout to Prepare Magnetic and Chitosan-Crosslinked κ-Carrageenan Bionanocomposites for Removal of Methylene Blue. *J. Water Process Eng.* **2016**, *10*, 143–155. [CrossRef]
41. Bée, A.; Obeid, L.; Mbolantenaina, R.; Welschbillig, M.; Talbot, D. Magnetic Chitosan/Clay Beads: A Magsorbent for the Removal of Cationic Dye from Water. *J. Magn. Magn. Mater.* **2017**, *421*, 59–64. [CrossRef]
42. Sengel, S.B.; Sahiner, N. Poly(Vinyl Phosphonic Acid) Nanogels with Tailored Properties and Their Use for Biomedical and Environmental Applications. *Eur. Polym. J.* **2016**, *75*, 264–275. [CrossRef]
43. Atta, A.M.; Al-Hussain, S.A.; Al-Lohedan, H.A.; Ezzat, A.O.; Tawfeek, A.M.; Al-Otabi, T. In Situ Preparation of Magnetite/Cuprous Oxide/Poly(AMPS/NIPAm) for Removal of Methylene Blue from Waste Water. *Polym. Int.* **2018**, *67*, 471–480. [CrossRef]
44. Jo, S.; Oh, Y.; Park, S.; Kan, E.; Lee, S.H. Cellulose/Carrageenan/TiO$_2$ Nanocomposite for Adsorption and Photodegradation of Cationic Dye. *Biotechnol. Bioprocess Eng.* **2017**, *22*, 734–738. [CrossRef]
45. Karthiga Devi, G.; Senthil Kumar, P.; Sathish Kumar, K. Green Synthesis of Novel Silver Nanocomposite Hydrogel Based on Sodium Alginate as an Efficient Biosorbent for the Dye Wastewater Treatment: Prediction of Isotherm and Kinetic Parameters. *Desalination Water Treat.* **2016**, *57*, 27686–27699. [CrossRef]
46. Gong, G.; Zhang, F.; Cheng, Z.; Zhou, L. Facile Fabrication of Magnetic Carboxymethyl Starch/Poly(Vinyl Alcohol) Composite Gel for Methylene Blue Removal. *Int. J. Biol. Macromol.* **2015**, *81*, 205–211. [CrossRef] [PubMed]
47. Zhang, C.; Dai, Y.; Wu, Y.; Lu, G.; Cao, Z.; Cheng, J.; Wang, K.; Yang, H.; Xia, Y.; Wen, X.; et al. Facile Preparation of Polyacrylamide/Chitosan/Fe$_3$O$_4$ Composite Hydrogels for Effective Removal of Methylene Blue from Aqueous Solution. *Carbohydr. Polym.* **2020**, *234*, 115882. [CrossRef] [PubMed]

48. Dai, H.; Huang, Y.; Zhang, Y.; Zhang, H.; Huang, H. Green and Facile Fabrication of Pineapple Peel Cellulose/Magnetic Diatomite Hydrogels in Ionic Liquid for Methylene Blue Adsorption. *Cellulose* **2019**, *26*, 3825–3844. [CrossRef]
49. Song, Y.; Duan, Y.; Zhou, L. Multi-Carboxylic Magnetic Gel from Hyperbranched Polyglycerol Formed by Thiol-Ene Photopolymerization for Efficient and Selective Adsorption of Methylene Blue and Methyl Violet Dyes. *J. Colloid Interface Sci.* **2018**, *529*, 139–149. [CrossRef]
50. Naseer, F.; Ajmal, M.; Bibi, F.; Farooqi, Z.H.; Siddiq, M. Copper and Cobalt Nanoparticles Containing Poly(Acrylic Acid-Co-Acrylamide) Hydrogel Composites for Rapid Reduction of 4-Nitrophenol and Fast Removal of Malachite Green from Aqueous Medium. *Polym. Compos.* **2018**, *39*, 3187–3198. [CrossRef]
51. Thakur, S.; Arotiba, O. Synthesis, Characterization and Adsorption Studies of an Acrylic Acid-Grafted Sodium Alginate-Based TiO2 Hydrogel Nanocomposite. *Adsorpt. Sci. Technol.* **2018**, *36*, 458–477. [CrossRef]
52. Yuan, Z.; Wang, Y.; Han, X.; Chen, D. The Adsorption Behaviors of the Multiple Stimulus-Responsive Poly(Ethylene Glycol)-Based Hydrogels for Removal of RhB Dye. *J. Appl. Polym. Sci.* **2015**, *132*, 42244. [CrossRef]
53. Bardajee, G.R.; Azimi, S.; Sharifi, M.B.A.S. Ultrasonically Accelerated Synthesis of Silver Nanocomposite Hydrogel Based on Salep Biopolymer: Application in Rhodamine Dye Adsorption. *Iran. Polym. J.* **2016**, *25*, 1047–1063. [CrossRef]
54. Younes, K.; Grasset, L. Analysis of Molecular Proxies of a Peat Core by Thermally Assisted Hydrolysis and Methylation-Gas Chromatography Combined with Multivariate Analysis. *J. Anal. Appl. Pyrolysis* **2017**, *124*, 726–732. [CrossRef]
55. Korichi, W.; Ibrahimi, M.; Loqman, S.; Ouhdouch, Y.; Younes, K.; Lemée, L. Assessment of Actinobacteria Use in the Elimination of Multidrug-Resistant Bacteria of Ibn Tofail Hospital Wastewater (Marrakesh, Morocco): A Chemometric Data Analysis Approach. *Environ. Sci. Pollut. Res.* **2021**, *28*, 26840–26848. [CrossRef] [PubMed]

Article

Functional Aerogels Composed of Regenerated Cellulose and Tungsten Oxide for UV Detection and Seawater Desalination

Yanjin Tang [1,2], Yuhan Lai [1], Ruiqin Gao [3], Yuxuan Chen [1], Kexin Xiong [1], Juan Ye [1], Qi Zheng [1], Zhenxing Fang [1,*], Guangsheng Pang [4,*] and Hoo-Jeong Lee [2]

1. College of Science and Technology, Ningbo University, 521 Wenwei Road, Ningbo 315300, China
2. Department of Smart Fab. Technology, Sungkyunkwan University, Suwon 16419, Republic of Korea
3. School of Biological and Chemical Engineering, NingboTech University, No. 1 South Qianhu Road, Ningbo 315100, China
4. State Key Laboratory of Inorganic Synthesis and Preparative Chemistry College of Chemistry, Jilin University, Changchun 130012, China
* Correspondence: fangzhenxing128@163.com (Z.F.); panggs@jlu.edu.cn (G.P.)

Abstract: Functional aerogels composed of regenerated cellulose and tungsten oxide were fabricated by implanting tungsten-oxide nanodots into regenerated cellulose fiber. This superfast photochromic property benefitted from the small size and even distribution of tungsten oxide, which was caused by the confinement effect of the regenerated cellulose fiber. The composite was characterized using XRD and TEM to illustrate the successful loading of tungsten oxide. The composite turned from pale white to bright blue under ambient solar irradiation in five seconds. The evidence of solar absorption and electron paramagnetic resonance (EPR) demonstrated the fast photochromic nature of the composite and its mechanism. Furthermore, carbon fiber filled with preferential growth tungsten-oxide nanorods was obtained by annealing the photochromic composite in a N_2 atmosphere. This annealed product exhibited good absorption across the whole solar spectrum and revealed an excellent photothermal conversion performance. The water evaporation rate reached 1.75 kg m^{-2} h^{-1} under one sun illumination, which is 4.4 times higher than that of pure water. The photothermal conversion efficiency was 85%, which shows its potential application prospects in seawater desalination.

Keywords: functional aerogels; regenerated cellulose; color center; solar absorbent; desalination

Citation: Tang, Y.; Lai, Y.; Gao, R.; Chen, Y.; Xiong, K.; Ye, J.; Zheng, Q.; Fang, Z.; Pang, G.; Lee, H.-J. Functional Aerogels Composed of Regenerated Cellulose and Tungsten Oxide for UV Detection and Seawater Desalination. *Gels* **2023**, *9*, 10. https://doi.org/10.3390/gels9010010

Academic Editor: Annarosa Gugliuzza

Received: 18 November 2022
Revised: 20 December 2022
Accepted: 22 December 2022
Published: 25 December 2022

Copyright: © 2022 by the authors. Licensee MDPI, Basel, Switzerland. This article is an open access article distributed under the terms and conditions of the Creative Commons Attribution (CC BY) license (https:// creativecommons.org/licenses/by/ 4.0/).

1. Introduction

Developing advanced building materials with both excellent thermal insulating and tunable optical properties to replace common glass is highly desirable for improving humans' quality of life and reducing global energy consumption. Nanocellulose is a major composite of plants, whose resource is abundant on Earth and has been used in heat insulation by reflecting solar light [1]. Furthermore, nanocellulose is a promising dispersion material because of its outstanding advantages, such as being environmentally friendly, having a good film-forming property, and being a good dispersant for transition-metal oxide. Tungsten oxide has been widely used in the electrochromic and photochromic fields due to its unique crystal structure and electronic energy band structure [2,3]. The colored tungsten oxide can adsorb part of the visible light and near-infrared light, which has a potential application for cooling down the inside temperature, photothermal therapy, and catalysis [4–7]. The chromic nature of tungsten oxide is caused by the formation of a color center (reduced tungsten state W^{5+}), occurring in both the electrochromic and photochromic processes [8–10]. The electrochromic process has two parts, namely coloring and bleaching, which represent the macroscopic appearance of the cation intercalation and deintercalation processes [11]. The reduced tungsten state can also be formed by the self-trapping of the photon caused by exposure to UV light, i.e., the photochromic effect [12,13]. The reduced tungsten state can dramatically increase the light absorption of the visible range and even the near-infrared range, which has been applied in photothermal conversion and NIR shielding [14–17]. However, the slight color change, slow response, and poor irreversibility of WO_3 hinder its practical application. In general, photochromic efficiency has largely been limited by the separation and recombination of photon-induced electrons and holes. Thus, decreasing the particle size seems to be an effective way to reduce the migration path of the photon, which leads to an increase in photochromic efficiency. However, nanoparticles, even QDs, tend to aggregate due to the large surface energy, which hampers photochromic efficiency. Appling sub-nanoporous silica as a template has been reported to fabricate tungsten-oxide quantum dots (QDs) [18]. The spatial confinement effect of this porous silica prevents the WO_3 QDs from aggregating into large particles. Suzuko et al. increased the photochromic efficiency of tungsten oxide by using glyceric acid as a dispersant to accelerate electron diffusion. However, another disadvantage (suppressed bleaching process) emerged due to the low oxygen diffusion [9]. Lacking effective contact with oxygen prolonged the bleaching time of the colored tungsten oxide. In this research, nanocellulose was chosen as the dispersant because of its abundant OH and COOH functional groups, which have also been widely used as an anti-quenching agent by preventing the fluorescent nanoparticles from assembling [19]. The excellent dispersion of tungsten oxides on the cellulose surface provides a fast solar response and bleaching ability without solar irradiation. Furthermore, the obtained cellulose filled with tungsten-oxide quantum dots changes into a carbon fiber filled with preferential growth tungsten-oxide nanorods, which exhibits good solar absorption and excellent photothermal conversion performance. The annealed composites can be applied in many areas, such as photothermal evaporation membranes, NIR shielding, photothermal therapy, etc. [20,21]. This excellent photothermal conversion performance also benefits from the reduced tungsten oxide (W^{5+}) and the size, morphology, and orientation of the tungsten-oxide nanorods. Guo reported that nanorods may exhibit transverse and longitudinal surface plasmon resonances, which correspond to the electron oscillations perpendicular and parallel to the rod length direction, respectively.

2. Results and Discussion

Figure 1 shows the XRD patterns of the WO_{3-x}@-regenerated cellulose fiber composites. As known from the Debye–Scherrer equation, the broad diffraction peaks in the black line revealed the small nanoparticle size of the tungsten oxides. This was caused by the confinement of the regenerated cellulose fiber during the tungsten-oxide crystal growth process. As we can see in Figure 2, the size of the regenerated cellulose was hundreds of nanometers in diameter, which was composed of several nanocellulose fibers. When the alcoholysis intermediate product penetrated into the regenerated cellulose fiber, it was confined in the gaps between the cellulose fibers. Therefore, this confinement effect benefited from the gaps generated during the formation of the nanocellulose fiber in the regenerated cellulose fiber. This confinement guaranteed the small size of the tungsten-oxide crystal in the hydrolysis reaction of the alcoholysis intermediate product. This result is also evidenced by the partially enlarged TEM image shown in Figure 2b. As for the diffraction results of the annealed sample (the red line), there were only two obvious diffraction peaks, indexed to the (010) and (020) crystal face, which were well matched with JCPDS No. 71-2450. The disappearance of other diffraction peaks was caused by the oriented growth along with the <010> direction when it was annealed in an inert atmosphere. This was also seen in other reported works [22,23]. Tungsten oxide is a well-known semiconductor because of its varied crystal structure and affordability of oxygen deficiency. The WO_6 octahedron as the unit of the tungsten-oxide crystal structure presented a different arrangement under different temperatures. The crystal structure of tungsten oxide has a temperature-related nature. Furthermore, when being annealed under an inert atmosphere, oxygen deficiencies can be formed at the surface of the crystal structure, which has a large impact on the crystal structure. Thus, tungsten oxide always reveals preferential growth under an inert atmosphere at high temperatures. Therefore, the appearance of two diffraction peaks of parallel crystal faces showed that the tungsten oxides were loaded into the regenerated cellulose fiber. This is also evidenced by the enlarged TEM image shown in Figure 2d.

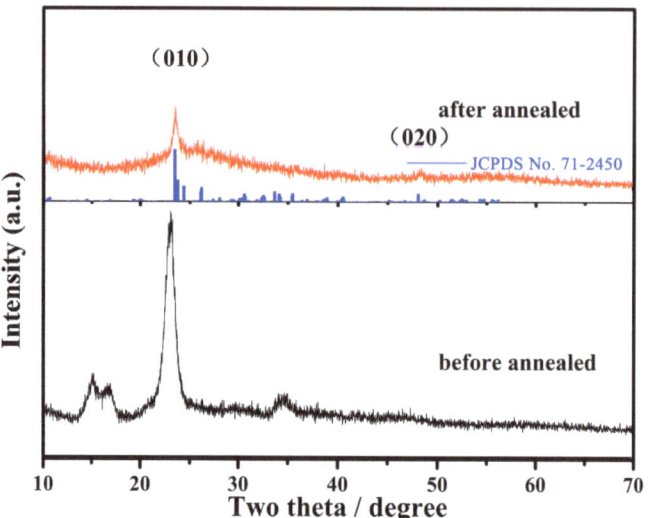

Figure 1. XRD patterns of the composite before and after heat treatment.

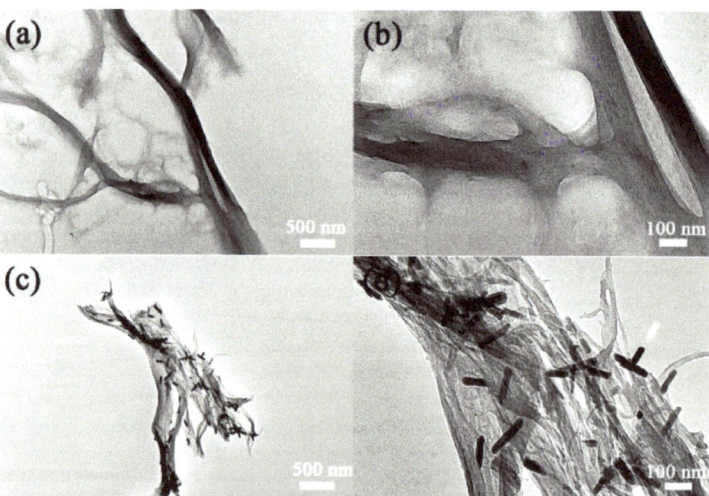

Figure 2. (a,b) TEM images of the composite before heat treatment. (c,d) TEM images of the composite after being annealed at 500 °C under N_2.

The morphology of the composite is shown in Figure 2. The morphology of the regenerated cellulose fiber with several um in length and a radius less than 100 nm can be seen in Figure 2a. This benefitted from the freezing and melting cycle process, which introduced ice crystals into the cellulose fibers to widen the distance between the fibers. This is a common way to decrease the hydrogen bonds between the cellulose fibers to induce a stable process of cellulose dispersion. As shown in the partially enlarged TEM image in Figure 2b, a large number of nanodots were anchored in the cellulose fiber. It is hard to see any small particles except in the gaps between the cellulose fibers. This result illustrates that the regenerated cellulose fiber acted as not only a growth substrate but also a confinement cage. These results were also consistent with the broad diffraction peaks of XRD diffraction shown with the black line. As shown in Figure 2c,d, these tungsten-oxide nanodots were transformed into nanorods after being annealed in a N_2 atmosphere. This correlated with the XRD result of the red line diffraction pattern. Tungsten oxide prioritized growth along the <010> direction during the heat treatment under an inert atmosphere. Furthermore, the regenerated cellulose fiber was carbonized into carbon fiber during the annealing process. The carbonization process inevitably causes shrinkage, which can be seen by comparing Figure 2a,c and Figure 2b,d (taken at the same magnification). The tungsten-oxide nanodots confined in the regenerated cellulose fiber were closer and contacted each other during the cellulose shrinkage, which enabled the nanodots' migration and the crystal's preferential growth. The orientational growth of the tungsten-oxide nanorods was also anchored in the carbon fibers. It is difficult to see any nanorods except in the gaps between the carbon fibers (Figure 2c,d). This is very important for its practical application, which effectively prevents the loss of active material.

As discussed above, the tungsten-oxide nanodots were exactly located in the regenerated cellulose fiber; thus, the composite should have a good photochromic performance. As shown in Figure 3, the black and red lines at the bottom are the absorption spectra of the regenerated cellulose fiber before and after solar irradiation, respectively. It should be noted that this solar absorption spectrum is collected on solid films, not solutions, because of its low dispersion in water. The absorption of cellulose had almost no change at all after solar irradiation throughout the UV–Vis range. However, the composite of the WO_{3-x}@-regenerated cellulose fiber exhibited a fast photochromic property. As the inset pictures reveal, the appearance of the composite turned from pale white to bright blue under ambient solar irradiation; the absorption intensity of the composite increased up to 0.3 from 0.01 during the long wavelength range from 600 to 800 nm. This fast photochromic performance can be attributed to the small particle size of tungsten oxide, just as Figure 2b shows. The appearance of the WO_{3-x}@-regenerated cellulose fiber composite quickly turned blue from pale in less than 10 s under solar irradiation (the photochromic process can be seen in the Supplemental Material Video S1). This photochromic nature was caused by a photon-induced electron self-trapping process, which introduced the formation of the color center W^{5+}. Furthermore, the formation of the color center was verified with the EPR experiment. The single electron in the 4f orbital of W^{5+} was detected by the fluctuating magnetic field. The signal peak in Figure 4 was exactly caused by the single-electron spinning resonance, which was evidenced by the value of the g factor (the same as that of the free radical). This result revealed that the composite could generate a large amount of W^{5+} under illumination by AM 1.5. This fast photochromic property came from the evenly distributed tungsten-oxide nanodots, which dramatically decreased the carrier migration distance so that the process of the carriers' recombination was depressed. Due to the surrounding regenerated cellulose fiber, the surface area of the tungsten-oxide dots exposed to contact with oxygen in the air was decreased. This is critical to the photochromic nature of this composite. However, the WO_{3-x}@-regenerated cellulose fiber composite turned back to pale white when in contact with the air for a period of time without solar irradiation, which demonstrated its good circularity. Thus, this product might have some potential applications in window films.

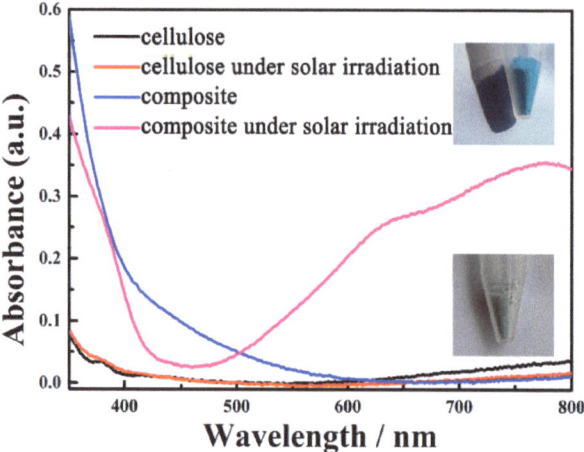

Figure 3. UV–Vis absorption spectra of the photochromic WO_{3-x}@-regenerated cellulose fiber composite.

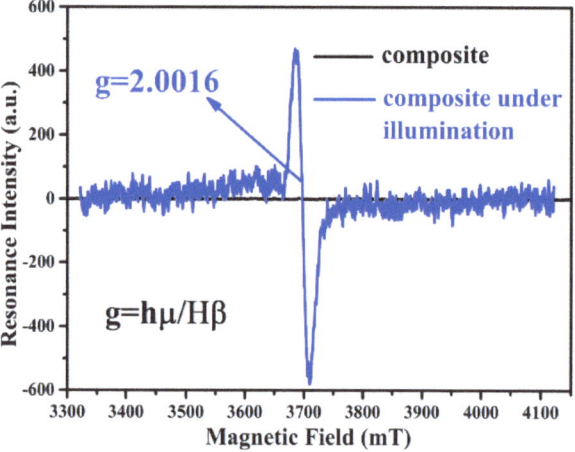

Figure 4. EPR spectra of the $WO_{2.72}$@-regenerated cellulose composite under natural conditions (black) and illuminated by AM 1.5 (blue).

As Figure 2c,d shows, the orientational growth of the nanorods was anchored in the carbonized fiber. The nanorod morphology depressed the light scattering to a large extent. The high concentration of free carriers in the semiconductor coupled with the carbonized cellulose fiber had excellent solar absorption, as shown in Figure 5a. As is known, the amorphous carbon material only absorbs visible light, not near-infrared light. Only some graphene-based carbon materials have a small near-infrared absorption capability due to their single electron on each carbon atom. Apparently, the carbon fiber obtained in this work could not be graphitized at the low carbonization temperature without catalysts. Therefore, the good absorption between 800 and 2500 nm of the annealed product (red line) mostly came from the oxygen deficiencies in $WO_{2.72}$. A new orbital energy level was introduced into the valance and conductive band when oxygen vacancies formed in the semiconductor. The collective oscillations of the surface free carriers (electron for n-type semiconductor $WO_{2.72}$) induced the surface plasmon resonance, which indicated a good photothermal conversion performance. As for the cellulose before being annealed, the weak absorption in the range from 1600 to 2500 nm was due to its surface functional groups such as COOH and OH, which is a common occurrence in biomass. Therefore, a water evaporation test was conducted under one sun illumination to evaluate the photothermal conversion performance. The results are shown in Figure 5b; the water evaporation rate of the WO$_{2.72}$@carbon fiber reached 1.75 kg m^{-2} h^{-1}, which was faster than that of carbonized fiber and 4.4 times higher than that of pure water (0.39 kg m^{-2} h^{-1}). The water evaporation efficiency of the carbonized cellulose fiber was 1.33 kg m^{-2} h^{-1}, which was lower than that of the graphene-based carbon materials [24]. This was determined by its electronic structure. The free-carrier concentration of the carbon fiber was very low, which could not effectively absorb the low-frequency photons to generate heat. According to the calculation equation of light to heat for water evaporation efficiency [25], the conversion efficiency of the WO$_{2.72}$@carbon fiber was 85%. This evaporation efficiency was higher than that of the reported 2D evaporators [26–28]. It should be mentioned that 2D evaporators aim to evaluate the photothermal conversion performance, and 3D evaporators aim to increase the water evaporation rate. Therefore, the water evaporation rate was limited by the finite evaporation surface. Here, water evaporation consumed a large amount of the heat generated by the photothermal conversion membrane to realize the phase transition; even with this, the surface temperature still reached 48 °C. Therefore, the composite of the WO$_{2.72}$@carbon fiber revealed an excellent photothermal conversion performance. As shown in Figure 5d, above the evaporation surface of the WO$_{2.72}$@carbon fiber membrane, the steam generation under one sun irradiation (simulated solar light irradiation, whose power was 100 mW/cm^2) was clear. The dynamic evaporation process recorded by a cell phone can be seen in Supplemental Material Video S2. Therefore, the photothermal evaporation membrane fabricated with the assistance of vacuum filtration might have a promising application in the solar desalination field.

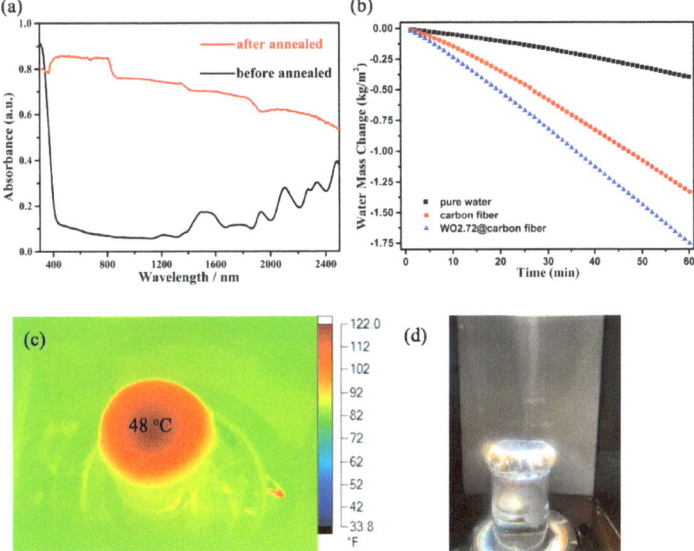

Figure 5. (**a**) Solar absorption spectrum of the cellulose fiber and the WO$_{2.72}$@carbonized cellulose fiber. (**b**) Water evaporation test of the WO$_{2.72}$@carbon fiber. (**c**) The photothermal image of the WO$_{2.72}$@carbon fiber under one sun irradiation. (**d**) Digital image of the evaporation for the WO$_{2.72}$@carbon fiber.

3. Conclusions

A superfast photochromic material was obtained by implanting tungsten-oxide nanodots into regenerated cellulose fiber. This superfast photochromic property benefitted from the small size and even distribution of the tungsten oxide, which was caused by the confinement effect of the regenerated cellulose fiber. This reversible photochromic nature shows promising applications for UV detectors, smart windows, NIR shielding, etc. The $WO_{2.72}$@carbon fiber composite had a broad and strong absorption across the whole solar spectrum. The anchoring effect of the substrate on the photothermal conversion material guarantees its long-term use. Its excellent photothermal conversion performance also makes it a promising application for solar desalination.

4. Materials and Methods

The materials described in this report were all purchased from Aladdin. There was no need to purify before use. All the materials were Analytical Reagents. The purity of NaOH was higher than 96%. The purity of urea and WCl_6 was 99%, and the purity of n-butanol and DMF was 99.5%.

4.1. Synthesis of the Regenerated Cellulose Fiber

The preparation solvent was as follows: First, an amount of 10 g NaOH was dissolved in 90 mL of deionized water to obtain a 10 wt% NaOH solution. Then, 0.9 g urea was added to decrease the polarity of the solution (prevent cellulose excessive dissociation). Next, 1 g of filter paper scraps was immersed into the mixture of NaOH and urea solution, followed by three repeated programs of freezing at −18 °C and melting at room temperature. The freezing process lasted at least 4 h to ensure the mixture was completely frozen. The regenerated cellulose fiber was obtained by centrifuging the mixture, followed by a vacuum-assisted drying process.

4.2. Synthesis of the WO_{3-x}@-Regenerated Cellulose Fiber

First, 100 mg WCl_6 and 1 g of the regenerated cellulose fiber were dispersed in 60 mL of n-butanol and DMF via sonication. This light-yellow dispersion was transferred into a 100 mL stainless autoclave and kept at 200 °C for 24 h. After cooling down to room temperature, the pale-white precipitate was centrifuged and washed with deionized water and ethanol three times. The WO_{3-x}@-regenerated cellulose fiber composite was dried in a vacuum oven at 60 °C overnight.

4.3. Synthesis of the $WO_{2.72}$@Carbon Fiber

The obtained WO_{3-x}@-regenerated cellulose fiber was annealed in a high-purity N_2 environment at 500 °C for 2 h with a 10 °C /min ramping rate. After cooling down to room temperature, dark black powders were obtained. The $WO_{2.72}$@carbon fiber was dispersed in DMF via sonication to induce stable dispersion. The photothermal evaporation membrane was fabricated with the assistance of vacuum filtration.

4.4. Characterization

The XRD patterns were recorded using PANalytical B.V. Empyrean X-ray powder diffraction with Cu Kα radiation over a range of 10–70° (2θ) with 0.02° per step. The transmission electron microscope (TEM) images were obtained with a Tecnai G2 FEI electron microscope. UV–vis adsorption spectra were recorded using a Lambda 950 spectroscope. Electron paramagnetic resonance (EPR) spectra were obtained with a JES-FA 200 EPR spectrometer. The EPR test of the colored composite was conducted under simulated light irradiation. The solar absorption spectrum was recorded with a Shimadzu UV-2450 spectroscope across the whole solar spectrum (200–2500 nm). The photothermal image was filmed using Fluke Ti 32s. The water-mass change was recorded with an electrical balance (FA 2004) under one sun irradiation (AM 1.5).

Supplementary Materials: The following supporting information can be downloaded at: https://www.mdpi.com/article/10.3390/gels9010010/s1, Video S1: photochromic; Video S2: evaporation.

Author Contributions: Y.T., Y.L., Y.C., K.X., and J.Y.: Conceptualization, Methodology, Formal analysis, and Investigation. R.G.: Methodology, Data Curation, and Characterization. Q.Z.: Characterization and Video Production. Z.F., G.P., and H.-J.L.: Conceptualization, Writing—Review and Editing, Supervision, and Funding acquisition. All authors have read and agreed to the published version of the manuscript.

Funding: This study was funded by the Natural Science Foundation of Ningbo (No. 202003N4157, 2021J159); the Start-up Fund from NingboTech University (20201110Z0134); the Open project of State Key Laboratory of Inorganic Synthesis and Preparative Chemistry, Jilin University (2022-25); and the K.C. Wong Magna Fund in Ningbo University.

Institutional Review Board Statement: Not applicable.

Informed Consent Statement: Not applicable.

Acknowledgments: Thanks are due to the Natural Science Foundation of Ningbo (No. 202003N4157, 2021J159); the Start-up Fund from NingboTech University (20201110Z0134); the Open project of State Key Laboratory of Inorganic Synthesis and Preparative Chemistry, Jilin University (2022-25); and the K.C. Wong Magna Fund in Ningbo University. The authors also thank Qi Zheng for the animation production.

Conflicts of Interest: The authors declare no conflict of interest.

References

1. Li, T.; Zhai, Y.; He, S.; Gan, W.; Wei, Z.; Heidarinejad, M.; Dalgo, D.; Mi, R.; Zhao, X.; Song, J.; et al. A radiative cooling structural material. *Science* **2019**, *364*, 760–763. [CrossRef] [PubMed]
2. Liu, B.J.; Zheng, J.; Wang, J.L.; Xu, J.; Li, H.H.; Yu, S.H. Ultrathin $W_{18}O_{49}$ nanowire assemblies for electrochromic devices. *Nano Lett.* **2013**, *13*, 3589–3593. [CrossRef] [PubMed]
3. Xi, G.; Ouyang, S.; Li, P.; Ye, J.; Ma, Q.; Su, N.; Bai, H.; Wang, C. Ultrathin $W_{18}O_{49}$ nanowires with diameters below 1 nm: Synthesis, near-infrared absorption, photoluminescence, and photochemical reduction of carbon dioxide. *Angew. Chem. Intternational Ed.* **2012**, *51*, 2395–2399. [CrossRef] [PubMed]
4. Guo, C.; Yin, S.; Yan, M.; Kobayashi, M.; Kakihana, M.; Sato, T. Morphology-controlled synthesis of $W_{18}O_{49}$ nanostructures and their near-infrared absorption properties. *Inorg. Chem.* **2012**, *51*, 4763–4771. [CrossRef] [PubMed]
5. Deng, K.; Hou, Z.; Deng, X.; Yang, P.; Li, C.; Lin, J. Enhanced Antitumor Efficacy by 808 nm Laser-Induced Synergistic Photothermal and Photodynamic Therapy Based on a Indocyanine-Green-Attached $W_{18}O_{49}$ Nanostructure. *Adv. Funct. Mater.* **2015**, *25*, 7280–7290. [CrossRef]
6. Tong, Y.; Guo, H.; Liu, D.; Yan, X.; Su, P.; Liang, J.; Zhou, S.; Liu, J.; Lu, G.Q.; Dou, S.X. Vacancy Engineering of Iron-Doped $W_{18}O_{49}$ Nanoreactors for Low-Barrier Electrochemical Nitrogen Reduction. *Angew. Chem. Int. Ed.* **2020**, *59*, 7356–7361. [CrossRef] [PubMed]
7. Zhang, M.; Cheng, G.; Wei, Y.; Wen, Z.; Chen, R.; Xiong, J.; Li, W.; Han, C.; Li, Z. Cuprous ion (Cu+) doping induced surface/interface engineering for enhancing the CO_2 photoreduction capability of $W_{18}O_{49}$ nanowires. *J. Colloid Interface Sci.* **2020**, *572*, 306–317. [CrossRef]
8. Fang, Z.; Jiao, S.; Kang, Y.; Pang, G.; Feng, S. Photothermal Conversion of $W_{18}O_{49}$ with a Tunable Oxidation State. *ChemistryOpen* **2017**, *6*, 261–265. [CrossRef]
9. Yamazaki, S.; Shimizu, D.; Tani, S.; Honda, K.; Sumimoto, M.; Komaguchi, K. Effect of Dispersants on Photochromic Behavior of Tungsten Oxide Nanoparticles in Methylcellulose. *ACS Appl. Mater. Interfaces* **2018**, *10*, 19889–19896. [CrossRef] [PubMed]
10. Yamazaki, S.; Ishida, H.; Shimizu, D.; Adachi, K. Photochromic Properties of Tungsten Oxide/Methylcellulose Composite Film Containing Dispersing Agents. *ACS Appl. Mater. Interfaces* **2015**, *7*, 26326–26332. [CrossRef]
11. Cong, S.; Tian, Y.; Li, Q.; Zhao, Z.; Geng, F. Single-Crystalline Tungsten Oxide Quantum Dots for Fast Pseudocapacitor and Electrochromic Applications. *Adv. Mater.* **2014**, *26*, 4260–4267. [CrossRef] [PubMed]
12. Tang, L.; Huang, S.; Wang, Y.; Liang, D.; Li, Y.; Li, J.; Wang, Y.; Xie, Y.; Wang, W. Highly Efficient, Stable, and Recyclable Hydrogen Manganese Oxide/Cellulose Film for the Extraction of Lithium from Seawater. *ACS Appl. Mater. Interfaces* **2020**, *12*, 9775–9781. [CrossRef] [PubMed]
13. Fang, Z.; Jiao, S.; Wang, B.; Yin, W.; Liu, S.; Gao, R.; Liu, Z.; Pang, G.; Feng, S. Synthesis of reduced cubic phase WO_{3-x} nanosheet by direct reduction of $H_2WO_4 \cdot H_2O$. *Mater. Today Energy* **2017**, *6*, 146–153. [CrossRef]
14. Quan, H.; Gao, Y.; Wang, W. Tungsten oxide-based visible light-driven photocatalysts: Crystal and electronic structures and strategies for photocatalytic efficiency enhancement. *Inorg. Chem. Front.* **2020**, *7*, 817–838. [CrossRef]

15. Yan, J.; Wang, T.; Wu, G.; Dai, W.; Guan, N.; Li, L.; Gong, J. Tungsten oxide single crystal nanosheets for enhanced multichannel solar light harvesting. *Adv. Mater.* **2015**, *27*, 1580–1586. [CrossRef]
16. Manthiram, K.; Alivisatos, A.P. Tunable localized surface plasmon resonances in tungsten oxide nanocrystals. *J. Am. Chem. Soc.* **2012**, *134*, 3995–3998. [CrossRef]
17. Cong, S.; Geng, F.; Zhao, Z. Tungsten Oxide Materials for Optoelectronic Applications. *Adv. Mater.* **2016**, *28*, 10518–10528. [CrossRef]
18. Watanabe, H.; Fujikata, K.; Oaki, Y.; Imai, H. Band-gap expansion of tungsten oxide quantum dots synthesized in sub-nano porous silica. *Chem. Commun.* **2013**, *49*, 8477–8479. [CrossRef]
19. Li, W.; Chen, Z.; Yu, H.; Li, J.; Liu, S. Wood-Derived Carbon Materials and Light-Emitting Materials. *Adv. Mater.* **2021**, *33*, e2000596. [CrossRef]
20. Wang, X.; Yang, Z.; Meng, Z.; Sun, S.-K. Transforming Commercial Copper Sulfide into Injectable Hydrogels for Local Photothermal Therapy. *Gels* **2022**, *8*, 319. [CrossRef]
21. Park, H.H.; Srisombat, L.O.; Jamison, A.C.; Liu, T.; Marquez, M.D.; Park, H.; Lee, S.; Lee, T.C.; Lee, T.R. Temperature-Responsive Hydrogel-Coated Gold Nanoshells. *Gels* **2018**, *4*, 28. [CrossRef] [PubMed]
22. Polleux, J.; Gurlo, A.; Barsan, N.; Weimar, U.; Antonietti, M.; Niederberger, M. Template-Free Synthesis and Assembly of Single-Crystalline Tungsten Oxide Nanowires and their Gas-Sensing Properties. *Angew. Chem.* **2006**, *118*, 267–271. [CrossRef]
23. Lee, K.; Seo, W.S.; Park, J.T. Synthesis and Optical Properties of Colloidal Tungsten Oxide Nanorods. *J. Am. Chem. Soc.* **2003**, *125*, 3408–3409. [CrossRef] [PubMed]
24. Li, G.; Wang, Q.; Wang, J.; Ye, J.; Zhou, W.; Xu, J.; Zhuo, S.; Chen, W.; Liu, Y. Carbon-supported nano tungsten bronze aerogels with synergistically enhanced photothermal conversion performance: Fabrication and application in solar evaporation. *Carbon* **2022**, *195*, 263–271. [CrossRef]
25. Ye, M.; Jia, J.; Wu, Z.; Qian, C.; Chen, R.; O'Brien, P.G.; Sun, W.; Dong, Y.; Ozin, G.A. Synthesis of Black TiO$_x$ Nanoparticles by Mg Reduction of TiO$_2$ Nanocrystals and their Application for Solar Water Evaporation. *Adv. Energy Mater.* **2017**, *7*, 1601811. [CrossRef]
26. Wang, C.; Wang, Y.; Song, X.; Huang, M.; Jiang, H. A Facile and General Strategy to Deposit Polypyrrole on Various Substrates for Efficient Solar-Driven Evaporation. *Adv. Sustain. Syst.* **2019**, *3*, 1800108. [CrossRef]
27. Shang, M.; Li, N.; Zhang, S.; Zhao, T.; Zhang, C.; Liu, C.; Li, H.; Wang, Z. Full-Spectrum Solar-to-Heat Conversion Membrane with Interfacial Plasmonic Heating Ability for High-Efficiency Desalination of Seawater. *ACS Appl. Energy Mater.* **2017**, *1*, 56–61. [CrossRef]
28. Zhang, L.; Tang, B.; Wu, J.; Li, R.; Wang, P. Hydrophobic Light-to-Heat Conversion Membranes with Self-Healing Ability for Interfacial Solar Heating. *Adv. Mater.* **2015**, *27*, 4889–4894. [CrossRef]

Disclaimer/Publisher's Note: The statements, opinions and data contained in all publications are solely those of the individual author(s) and contributor(s) and not of MDPI and/or the editor(s). MDPI and/or the editor(s) disclaim responsibility for any injury to people or property resulting from any ideas, methods, instructions or products referred to in the content.

Article

Nanocellulose-Linked MXene/Polyaniline Aerogel Films for Flexible Supercapacitors

Liying Xu [1,*], Wenxuan Wang [2], Yu Liu [2] and Daxin Liang [2,*]

1 School of Food Engineering, Harbin University, Harbin 150086, China
2 Key Laboratory of Bio-Based Material Science and Technology (Ministry of Education), Northeast Forestry University, Harbin 150040, China
* Correspondence: xuliying202206@126.com (L.X.); daxin.liang@nefu.edu.cn (D.L.)

Abstract: In the development of energy supply systems for smart wearable devices, supercapacitors stand out owing to their ability of quick and efficient energy supply. However, their application is limited due to their low energy density and poor mechanical energy. Herein, a strategy for the preparation of flexible supercapacitors is reported, which is based on the fabrication of aerogel films by simultaneously utilising cellulose nanofiber (CNFs) as an MXene intercalation material and polyaniline (PANI) as a template material. CNFs, which can form hydrogen-bonded networks, enhance the mechanical properties of MXene from 44.25 to 119.56 MPa, and the high electron transport properties of PANI endow MXene with a capacitance of 327 F g^{-1} and a resistance of 0.23 Ω. Furthermore, the combination of CNFs and PANI enables a 71.6% capacitance retention after 3000 charge/discharge and 500 folding cycles. This work provides a new platform for the development of flexible supercapacitors.

Keywords: supercapacitors; MXene; CNFs; polyaniline

Citation: Xu, L.; Wang, W.; Liu, Y.; Liang, D. Nanocellulose-Linked MXene/Polyaniline Aerogel Films for Flexible Supercapacitors. *Gels* **2022**, *8*, 798. https://doi.org/10.3390/gels8120798

Academic Editor: Pavel Gurikov

Received: 17 November 2022
Accepted: 1 December 2022
Published: 5 December 2022

Publisher's Note: MDPI stays neutral with regard to jurisdictional claims in published maps and institutional affiliations.

Copyright: © 2022 by the authors. Licensee MDPI, Basel, Switzerland. This article is an open access article distributed under the terms and conditions of the Creative Commons Attribution (CC BY) license (https://creativecommons.org/licenses/by/4.0/).

1. Introduction

As an alternative to fossil energy, electric power is a renewable and clean energy that is easy to store and transport [1]. For daily life applications, structural electrode materials with high flexibility and high electrochemical performance attracted have great attraction [2–4], because they are essential for manufacturing lightweight and flexible electronic products [5–7]. In particular, capacitors offer fast charging and discharging speed and good recyclability for practical applications [8]. To fabricate flexible capacitors, carbon-based materials, including carbon nanotubes and graphene, have been widely used as electrode materials with stable performance and structure [9–11]. Unfortunately, their electrochemical energy storage capacity is insufficient [12]. Although this problem can be significantly circumvented by adding metal oxides and conductive polymers, the reduction of the mechanical properties stemming from the lack of interaction between the two materials cannot be ignored [13,14].

MXenes (Ti_3AlC_x) are a family of two–dimensional (2D) transition metal carbides or nitrides having the general formula $M_{n+1}X_nTX$ (n = 1–3), where M represents an early transition metal such as Ti, Zr, V, Nb, Ta or Mo, X is the carbon or nitrogen element and T represents a surface functional group, generally =O, –OH or –F [12,15]. MXenes are prepared by selectively etching the A element layer from the corresponding MAX phase precursor ($M_{n+1}AX_n$) [16]. Since Gogotsi et al. synthesised the first MXene in 2011 [17,18], these materials have attracted a wide interest for application in supercapacitors. Their unique 2D layered structure and rich surface functional groups provide them with excellent conductivity and high surface hydrophilicity. However, their mechanical properties are still not ideal [19,20].

Cellulose nanofibrils (CNFs) have useful properties, such as strong mechanical strength, distinctive biological flexibility, great stability and outstanding capacity of liquid absorption [21–23]. Moreover, as a nanolinear structured material, CNFs can not only enhance the mechanical properties of MXene by dispersing between its interlayer gaps but also improve the electrochemical performance by inhibiting the stacking of MXene and facilitating the transport of electrolyte ions as an electrolyte reservoir [24–26]. However, the charge storage capacity of MXene in supercapacitors stems mainly from =O functional groups, whereas –F and –OH functional groups, which are abundant in CNFs, are detrimental to the electrical performance [12,27].

Polyaniline (PANI) is a typical conducting polymer that has been widely used to prepare supercapacitor electrodes due to its fast redox rate, high pseudocapacitance, fully reversible doping, low cost and easy synthesis [28–30]. However, the application of PANI is limited by its tendency to undergo agglomeration, which leads to the blockage of the conductive paths, increase in resistance and decrease in energy density decreases [31–33]. Interestingly, PANI can be uniformly dispersed using CNFs as templates and excess –OH as linkages, thereby leading to the improvement of their electrochemical performance.

Herein, MXene/CNF–PANI aerogel films with layered porous structures were prepared via in situ polymerisation and using MXene/CNF aerogel films. The films were obtained by self-assembling MXene and CNFs via vacuum suction filtration after preparing single-layer MXene suspensions by LiF/HCl etching of Ti$_3$AlC$_2$ (Scheme 1). CNFs served as an intercalation material to link MXene with PANI, which greatly enhanced the mechanical properties of the MXene films, and as a template material to enable PANI to disperse. Moreover, PANI provided a communicating pathway between MXene layers and improved the electrochemical properties of materials, demonstrating a new structural model for the preparation of flexible supercapacitors for daily life applications.

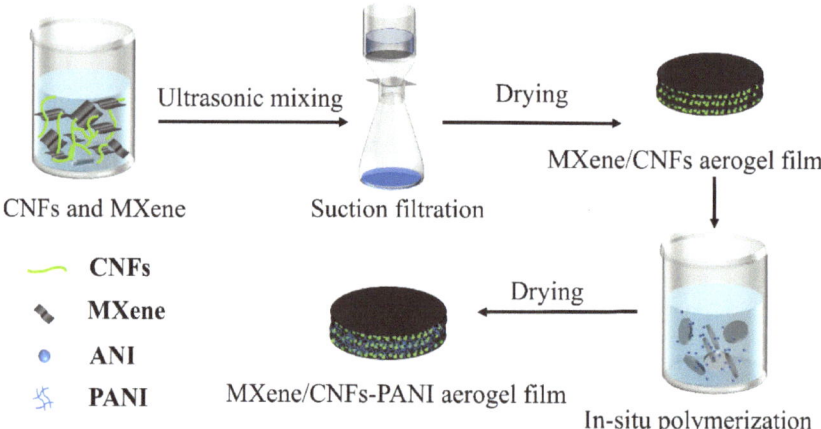

Scheme 1. Preparation of MXene/CNF aerogel films and MXene/CNF–PANI aerogel films.

2. Results and Discussion

2.1. Characterisation of MXene/CNF Aerogel Films and MXene/CNF–PANI Aerogel Films

2.1.1. Scanning Electron Microscopy (SEM) Analysis

The micro-morphologies of the MXene/CNF and MXene/CNF–PANI aerogel films are shown in Figure 1. Due to the insertion of CNFs between the MXene layers having a rough surface (Figure 1a), the MXene/CNF aerogel films were endowed with a layered and porous structure (Figure 1b), which is conducive to the infiltration of electrolytes. Moreover, this uniform dispersion can be expected to enhance the mechanical properties of MXenes. The polymerisation of PANI on the CNFs transformed the smooth surface (Figure 1c) to an overall layered porous structure (Figure 1d), illustrating the main roles of

PANI, which were to coat the CNFs and connect the MXenes that were disconnected after the incorporation of the CNFs for an improved electrochemical performance.

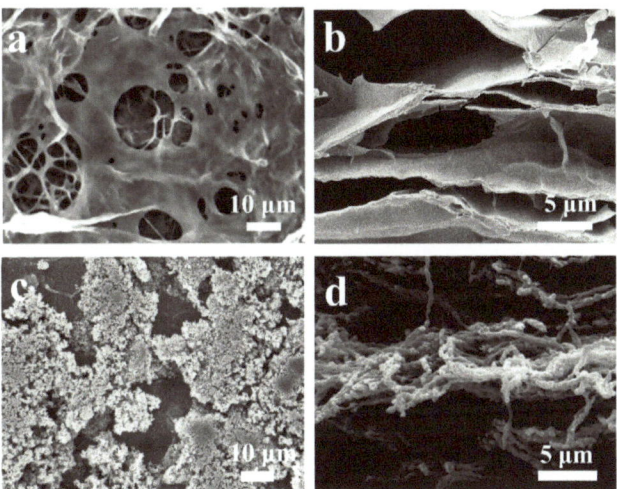

Figure 1. SEM images of (**a**) the surface; (**b**) cross section of MXene/CNF aerogel films; (**c**) the surface; (**d**) cross section of MXene/CNF–PANI aerogel films.

2.1.2. Fourier Transform Infrared (FTIR) Analysis

The FTIR spectra of CNFs, MXene, MXene/CNF aerogel films and MXene/CNF–PANI aerogel films are shown in Figure 2. In the spectrum of CNFs, the peaks at 3344, 2932–2868 and 1031 cm^{-1} can be assigned to the stretching vibration of –OH, the stretching vibrations of –CH and the C–O–C pyranose ring skeleton of CNFs, respectively. The spectrum of MXene shows a characteristic peak at 1100 cm^{-1} corresponding to the surface C–F end groups. [34]. When CNFs were added into MXene, the absorption peaks of C–F were masked and the intensity of the –OH peaks increased because a large amount of –OH groups was introduced, whereas the spectrum of the MXene/CNF–PANI aerogel films only exhibited stretching vibrations of N–H between 3350–3344 cm^{-1} and double-bond vibrations of benzene and quinone ring molecules of the PANI chains at 1486 and 1579 cm^{-1}, respectively [35]. This result demonstrates that CNFs serve only as a structural framework to construct the MXene/CNF–PANI aerogel films, in which PANI connects the MXenes by reducing the –OH on the surface of the CNFs, thereby improving the electrochemical performance of the MXene/CNF–PANI aerogel films.

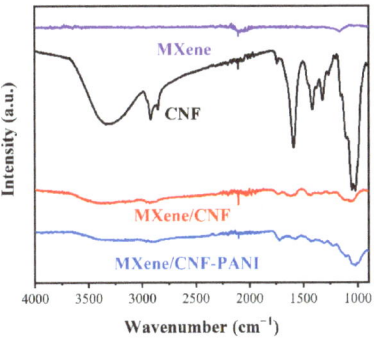

Figure 2. FTIR spectra of CNFs, MXene, MXene/CNF aerogel films and MXene/CNF–PANI aerogel films.

2.1.3. X-ray Diffraction (XRD) Analysis

Figure 3 shows the XRD patterns of CNFs, MXene, MXene/CNF aerogel films and MXene/CNF–PANI aerogel films. A distinct (002) diffraction peak was observed for MXene at 2θ = 7.1° [36], but it decreased with the addition of CNFs, indicating that the interlayer spacing of the MXene increased as a result of the successful penetration of cellulose between those layers. After the polymerisation of PANI, the diffraction peaks of the CNFs disappeared, which is in agreement with the SEM results, indicating that PANI coated the surface of cellulose.

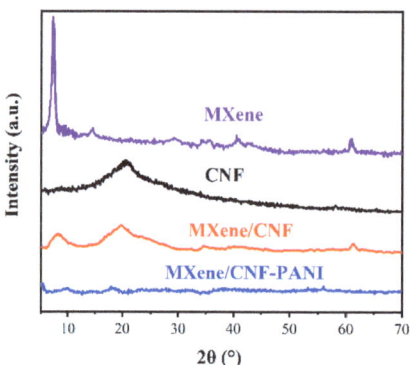

Figure 3. XRD patterns of CNFs, MXene, MXene/CNF aerogel films and MXene/CNF–PANI aerogel films.

2.1.4. Tensile Stress–Strain Analysis

Figure 4 shows the tensile stress–strain curves of CNFs, MXene, MXene/CNF aerogel films and MXene/CNF–PANI aerogel films. After the introduction of CNFs, which resulted in a uniform dispersion of stress in the MXene film, the stress strength of the MXene/CNF aerogel films increased from 44.25 to 132.84 MPa and their strain increased from 3.26% to 15.01% due to the hydrogen bonding interactions between the CNF networks. However, when PANI was introduced, it covered the surface of the CNFs and decreased the number of hydrogen bonds, reducing the stress strength and the strain of the MXene/CNF–PANI aerogel films, albeit only slightly, from 132.84 to 119.56 MPa and from 15.01% to 13.71%, respectively.

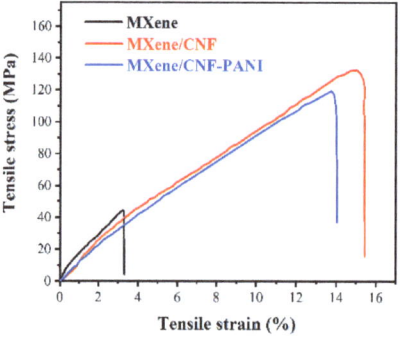

Figure 4. Tensile stress–strain curves of MXene, MXene/CNF aerogel films and MXene/CNF–PANI aerogel films.

2.2. Electrochemical Performance of MXene/CNF Aerogel Films and MXene/CNF–PANI Aerogel Films

As shown in Figure 5a, the cyclic voltammetry (CV) curve of the MXene films recorded at a scan rate of 2 mV s^{-1} exhibited a rectangular shape, which is indicative of their ideal capacitive behaviour. Upon addition of CNFs and PANI, an obvious increase in the peak current density in the CV curves at the same scan rate was observed. The results of a galvanostat charge/discharge (GCD) test were consistent with those of the CV measurements (Figure 5b). The GCD curves at the same discharge current density (3 mA cm^{-2}) further demonstrate the higher mass capacitance of the MXene/CNF aerogel films; the insertion of CNFs endowed the MXene with a more free interlayer space for charge storage while retaining the interlayer electron transport channels for high interlayer conductivity. Two pairs of broad redox peaks were clearly observed in the potential window of the MXene/CNF aerogel films after the addition of PANI, indicating that the capacitance is mainly a pseudocapacitance because the reversible intercalation/deintercalation of protons leads to a valence state change of the redox element Ti. The GCD curves exhibited a distorted triangle due to the redox reaction of MXene, which was consistent with the CV results. The mass capacitance results calculated from the GCD curves are shown in Figure 5c. After the introduction of CNFs, the capacitance of the MXene/CNF aerogel films reached 294 F g^{-1}, whereas pure MXene produced only a capacitance of 271 F g^{-1} at the same scan rate. When the scan rate was increased to 10 mV s^{-1}, the MXene/CNF aerogel films still maintained a capacitance of 227 F g^{-1} and a capacitance ratio of 77%, which was much higher than that of the MXene films with a capacitance of 132 F g^{-1} (48%). The results showed that the improvement of the capacitive properties of the MXene/CNF aerogel films was related to the electrode structure, and the CNF nanonetwork structure not only prevented the stacking of MXene but also accelerated the intercalation/extraction transport rate of ions and improved the capacitance of MXene [20]. After PANI loading, the capacitance of the resulting aerogel films reached 327 F g^{-1} with a ratio of 77% when the scan rate was increased to 10 mV s^{-1}. This large improvement was due to the fact that the introduction of PANI in the MXene/CNF aerogel films resulted in the formation of the loops inside, allowing a smooth electron transport.

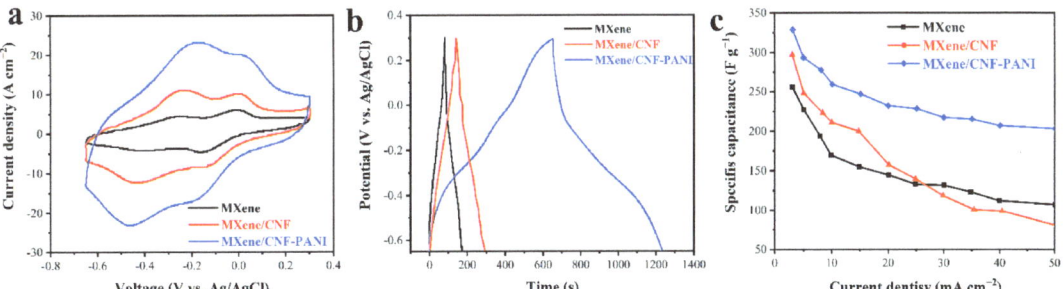

Figure 5. (**a**) CV curves, (**b**) GCD curves and (**c**) specific capacitance of MXene films, MXene/CNF aerogel films and MXene/CNF–PANI aerogel films.

To understand the kinetic process of the charge/discharge profiles of the films, electrochemical impedance spectroscopy (EIS) measurements were conducted. Figure 6 shows the Nyquist curves of CNFs, MXene, MXene/CNF aerogel films and MXene/CNF–PANI aerogel films, in which two regions can be observed, i.e., semicircular arcs at high frequencies and a straight line at low frequencies [34,37–39]. For the MXene/CNF and MXene/CNF–PANI aerogel films, the straight-line regions are almost parallel to the Z″ axis, indicating that ions could diffuse rapidly from the electrolyte solution to the film interface. MXene exhibited a relatively high resistance (2.5 Ω) due to the stacking of MXene nanosheets in a bi-continuous structure. In contrast, the addition of CNFs improved the stacking in the

MXene/CNF aerogel films, reducing the internal resistance to 2.15 Ω. Then, the internal pathway constructed through the action of PANI was beneficial for a smooth electron transport, and the resistance value was reduced to 0.23 Ω, which proved that the introduction of MXene/CNF with PANI was effective.

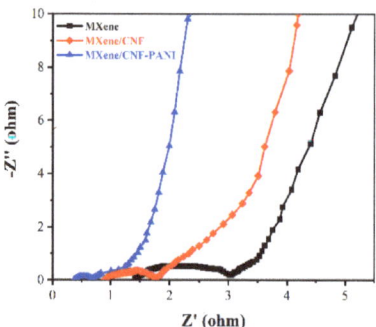

Figure 6. Nyquist curves of MXene films, MXene/CNF aerogel films and MXene/CNF–PANI aerogel films.

To demonstrate the effectiveness of this structural design for aerogel films, CV curves of the MXene/CNF–PANI aerogel films were recorded at varying scan rates from 1 to 20 mV s^{-1} (Figure 7a). At low scan rates, redox peaks appeared in the CV curves, which might be attributed to the special morphological features of the MXene/CNF–PANI aerogel films facilitating the protonation of the oxygen functional groups at positive and negative potential sites. As shown in Figure 7b, the charge/discharge profiles in a symmetric state reflect the nearly 100% Coulombic efficiency of the MXene/CNF–PANI aerogel films, indicating the occurrence of reversible redox reactions in the electrode. According to the results calculated from the GCD curves (Figure 5c), the MXene/CNF–PANI aerogel films showed high capacitive properties, i.e., 327 F g^{-1} at 3 mA cm^{-2} and 203 F g^{-1} at 50 mA cm^{-2}, indicating that the MXene/CNF–PANI aerogel films exhibited good ion transport ability and that the three-dimensional structure design of MXene was useful and efficient.

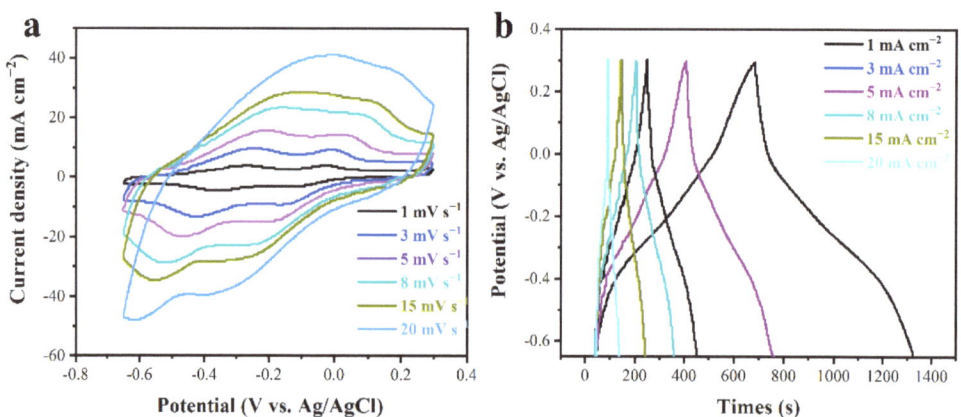

Figure 7. (**a**) CV curves and (**b**) GCD curves.

To evaluate the applicability of the MXene/CNF–PANI aerogel films, 3000 charge/discharge cycles were performed at a current density of 50 mA cm^{-2} (Figure 8). The combination of CNFs and PANI prevented the stacking of MXene and increased the film toughness, whereas the resistance was not increased, resulting in a capacitance retention of 84.1%

after 3000 charge/discharge cycles. Furthermore, a capacitance retention of 71.6% was obtained after subjecting the MXene/CNF–PANI aerogel films to 500 folding cycles at the same charge/discharge current density, indicating that the electrode material had excellent flexibility for practical applications [40]. The comprehensive performance of MXene/CNF–PANI aerogel films could be shown through Table 1, and the applicability of MXene/CNF–PANI aerogel films was shown through comparison.

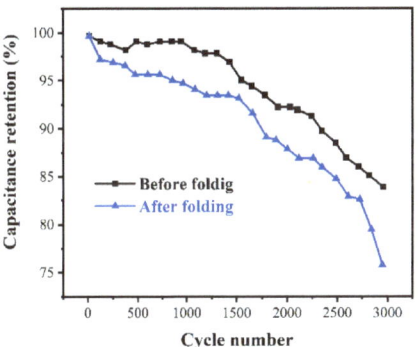

Figure 8. Capacitance of MXene/CNF–PANI aerogel films before and after 500 folding cycles and 3000 charge/discharge cycles at a constant charge/discharge current density.

Table 1. MXene/CNF–PANI aerogel films comparison of the performance of capacitors with other capacitors.

Composites	Capacitance	Mechanical Properties (MPa)	Resistance (Ω)	Ref.
TEMPO-oxidized-nanocellulose carbon nanotubes	65 F g^{-1}	0.065	26	[41]
N-doped porous carbon	193 F g^{-1}	—	0.97	[42]
Ti$_3$C$_2$T$_x$/CNF	298 F g^{-1}	0.004	0.003	[12]
MWNT/cellulose nanofibers	145 F g^{-1}	—	0.37	[43]
Brazilian-pine fruit coat	0.87 F cm^{-2}	6.1	13.5	[39]
MXene/CNF–PANI	327 F g^{-1}	119.56	0.23	This study

3. Conclusions

In this study, porous layered MXene/CNF–PANI aerogel films with a novel structure were prepared, which can be used as supercapacitors. Using PANI for cladding after intercalation of CNFs into the MXene layers boosted the performance of MXene in terms of the three following properties: (1) The MXene/CNF–PANI aerogel films possessed a capacitance of 327 F g^{-1} and a resistance of 0.23 Ω because the introduction of CNFs prevented the stacking of MXene, enabling more free interlayer space for charge storage. Meanwhile, the introduction of PANI provided communicating pathways for a smooth electron transport. (2) A high mechanical strength of 119.56 MPa originating from the introduction of CNFs resulted in the formation of abundant hydrogen bonds to enhance the overall MXene structure. (3) The MXene/CNF–PANI aerogel films showed a capacitance retention of 71.6% after 500 folding and 3000 charge/discharge cycles. In summary, the MXene/CNF–PANI aerogel films are promising as a supercapacitor material with enhanced electrochemical, mechanical and cycling properties.

4. Materials and Methods

4.1. Materials

All chemicals were used without further purification. Titanium aluminium carbide (Ti$_3$AlC$_2$, ~400 mesh, 98%) was purchased from Yiyi Technology Co., Ltd., Jilin, China, CNF suspension (0.55 wt%) from Hongqi Technology Co., Ltd., Guilin, China and sulphuric acid (H_2SO_4, 99.8%) from Sinopharm Chemical Reagent Co., Ltd., Shanghai, China). Lithium fluoride (LiF, 99.7%), ammonium persulfate ((NH_4)$_2S_2O_8$, 99.7%), aniline (C_6H_7N, 97.7%) and potassium hydroxide (KOH, 97.7%) were obtained from Sigma-Aldrich Co., Ltd., Shanghai China. Hydrochloric acid (HCl, 37%) and ethanol (C_2H_6O, 99%) were sourced from a local supplier.

4.2. Characterisation

The morphology and structural characteristics of the samples were observed via SEM using a TM3030 microscope (Hitachi, Tokyo, Japan). The crystal structures were analysed using a D/max-2200VPC X-ray diffractometer (Rigaku, Tokyo, Japan). FTIR spectroscopy (Perkin Elmer, Waltham, MA, USA) was used to characterise the functional groups in the samples. A universal testing machine (CMT5504, MTS, Eden Prairie, MN, USA) was used to study the mechanical properties. All electrochemical performance tests were conducted on an electrochemical workstation (51-XMX1004, Oxford, UK).

4.3. Preparation of MXene, MXene/CNF Aerogel Films and MXene/CNF-PANI Aerogel Films

To an aqueous solution of LiF (6 g) and 9 M HCl, 5 g Ti_3AlC_2 was slowly added under continuous stirring at 45 °C for 24 h. After centrifugation of the solution, the resulting precipitate was washed first with HCl (1 M) for 2–3 times and then with deionised water until pH 7. After sonication for 60 min at 0 °C in water and centrifugation at 3500 rpm min^{-1} for 10 min, a dark green supernatant was obtained, which was collected and subjected to freeze-drying to furnish layered MXene flakes.

CNF films and MXene films were obtained by suction filtration from CNF and MXene solutions, respectively. The MXene/CNF aerogel films were prepared via insertion and suction filtration by blending 3.6 mL CNF solution (0.55 wt%) with 6.6 mL MXene solution (3 mg mL^{-1}), followed by magnetic stirring at 500 rpm for 3 h, sonication for 30 min and then vacuum suction. The MXene/CNF aerogel films were immersed in 20 mL (NH_4)$_2S_2O_8$ solution (0.57 g mL^{-1}) containing 1 M HCl solution for 3 h and then removed and immersed in 20 mL aniline solution (0.19 g mL^{-1}) with 1 M HCl solution for 3 h for in situ polymerisation of aniline. MXene/CNF–PANI films were finally obtained by repeated rinsing with deionised water and drying.

4.4. Electrochemical Performance of MXene, MXene/CNFs Aerogel Films and MXene/CNFs-PANI Aerogel Films

The electrochemical performance of MXene, MXene/CNF aerogel films and MXene/CNF–PANI aerogel films was investigated by means of CV, GCD and EIS measurements using an electrochemical workstation. The composite films were cropped into 2×1.8 cm^2 electrode sheets and assembled into symmetric supercapacitors in the electrochemical workstation using a three electrodes system with Ag/AgCl as the reference electrode and 3 M H_2SO_4 as the electrolyte.

Author Contributions: Conceptualization, W.W.; Formal analysis, Y.L.; Funding acquisition, L.X.; Investigation, Y.L.; Methodology, L.X. and W.W.; Project administration, L.X.; Resources, Y.L.; Supervision, D.L.; Validation, W.W.; Visualization, D.L.; Writing—original draft, D.L.; Writing—review and editing, D.L. All authors have read and agreed to the published version of the manuscript.

Funding: This research was supported by the Fundamental Research Funds for the Central Universities (2572021BB02) and Undergraduate Training Programs for Innovations by NEFU (202210225635).

Institutional Review Board Statement: Not applicable.

Informed Consent Statement: Not applicable.

Data Availability Statement: Not applicable.

Conflicts of Interest: The authors declare no conflict of interest.

References

1. Tomain, J.P. A perspective on clean power and the future of US energy politics and policy. *Util. Policy* **2016**, *39*, 5–12. [CrossRef]
2. El-Mahdy, A.F.M.; Mohamed, M.G.; Mansoure, T.H.; Yu, H.H.; Chen, T.; Kuo, S.W. Ultrastable tetraphenyl-p-phenylenediaminebased covalent organic frameworks as platforms for high-performance electrochemical supercapacitors. *Chem. Commun.* **2019**, *55*, 14890–14893. [CrossRef]
3. Goonetilleke, D.; Sharma, N.; Pang, W.K.; Peterson, V.K.; Petibon, R.; Li, J.; Dahn, J.R. Structural Evolution and High-Voltage Structural Stability of Li(Ni$_x$Mn$_y$Co$_z$)O$_2$ Electrodes. *Chem. Mater.* **2019**, *31*, 376–386. [CrossRef]
4. Jiang, G.Y.; Wang, G.; Zhu, Y.; Cheng, W.K.; Cao, K.Y.; Xu, G.W.; Zhao, D.W.; Yu, H.P. A Scalable Bacterial Cellulose Ionogel for Multisensory Electronic Skin. *Research* **2022**, *2022*, 9814767. [CrossRef]
5. Ge, W.J.; Cao, S.; Yang, Y.; Rojas, O.J.; Wang, X.H. Nanocellulose/LiCl systems enable conductive and stretchable electrolyte hydrogels with tolerance to dehydration and extreme cold conditions. *Chem. Eng. J.* **2021**, *408*, 127306. [CrossRef]
6. Wan, C.C.; Jiao, Y.; Liang, D.X.; Wu, Y.Q.; Li, J. A Geologic Architecture System-Inspired Micro-/Nano-Heterostructure Design for High-Performance Energy Storage. *Adv. Energy Mater.* **2018**, *8*, 1802388. [CrossRef]
7. Javed, M.S.; Shah, S.S.A.; Najam, T.; Siyal, S.H.; Hussain, S.; Saleem, M.; Zhao, Z.; Mai, W. Achieving high-energy density and superior cyclic stability in flexible and lightweight pseudocapacitor through synergic effects of binder-free CoGa$_2$O$_4$ 2D-hexagonal nanoplates. *Nano Energy* **2020**, *77*, 105276. [CrossRef]
8. Doustkhah, E.; Hassandoost, R.; Khataee, A.; Luque, R.; Assadi, M.H.N. Hard-templated metal-organic frameworks for advanced applications. *Chem. Soc. Rev.* **2021**, *50*, 2927–2953. [CrossRef]
9. Rashidi, N.A.; Chai, Y.H.; Ismail, I.S.; Othman, M.F.H.; Yusup, S. Biomass as activated carbon precursor and potential in supercapacitor applications. *Biomass Convers. Biorefinery* **2022**, *45*, 649–657. [CrossRef]
10. Li, C.; Liu, Y.; Zhao, X.; Shen, Q.; Zhao, W.; Tan, S.; Zhang, N.; Li, P.; Jiao, L.; Qu, X. Sandwich-Like Heterostructures of MoS2/Graphene with Enlarged Interlayer Spacing and Enhanced Hydrophilicity as High-Performance Cathodes for Aqueous Zinc-Ion Batteries. *Adv. Mater.* **2021**, *33*, e2007480. [CrossRef]
11. Zheng, Z.; Zheng, Y.; Luo, Y.; Yi, Z.; Zhang, J.; Liu, Z.; Yang, W.; Yu, Y.; Wu, X.; Wu, P. A switchable terahertz device combining ultra-wideband absorption and ultra-wideband complete reflection. *Phys. Chem. Chem. Phys.* **2022**, *24*, 2527–2533. [CrossRef] [PubMed]
12. Tian, W.; VahidMohammadi, A.; Reid, M.S.; Wang, Z.; Ouyang, L.; Erlandsson, J.; Pettersson, T.; Wagberg, L.; Beidaghi, M.; Hamedi, M.M. Multifunctional Nanocomposites with High Strength and Capacitance Using 2D MXene and 1D Nanocellulose. *Adv. Mater.* **2019**, *31*, e1902977. [CrossRef]
13. Wen, Y.; Li, R.; Liu, J.; Wei, Z.; Li, S.; Du, L.; Zu, K.; Li, Z.; Pan, Y.; Hu, H. A temperature-dependent phosphorus doping on Ti3C2Tx MXene for enhanced supercapacitance. *J. Colloid Interface Sci.* **2021**, *604*, 239–248. [CrossRef] [PubMed]
14. Jalali, H.; Khoeini, F.; Peeters, F.M.; Neek-Amal, M. Hydration effects and negative dielectric constant of nano-confined water between cation intercalated MXenes. *Nanoscale* **2021**, *13*, 922–929. [CrossRef] [PubMed]
15. Javed, M.S.; Mateen, A.; Ali, S.; Zhang, X.; Hussain, I.; Imran, M.; Shah, S.S.A.; Han, W. The Emergence of 2D MXenes Based Zn-Ion Batteries: Recent Development and Prospects. *Small* **2022**, *18*, e2201989. [CrossRef]
16. Bai, Y.; Liu, C.; Chen, T.; Li, W.; Zheng, S.; Pi, Y.; Luo, Y.; Pang, H. MXene-Copper/Cobalt Hybrids via Lewis Acidic Molten Salts Etching for High Performance Symmetric Supercapacitors. *Angew. Chem. Int. Ed.* **2021**, *60*, 25318–25322. [CrossRef]
17. Anasori, B.; Lukatskaya, M.R.; Gogotsi, Y. 2D metal carbides and nitrides (MXenes) for energy storage. *Nat. Rev. Mater.* **2017**, *2*, 16098. [CrossRef]
18. Javed, M.S.; Shaheen, N.; Hussain, S.; Li, J.; Shah, S.S.A.; Abbas, Y.; Ahmad, M.A.; Raza, R.; Mai, W. An ultra-high energy density flexible asymmetric supercapacitor based on hierarchical fabric decorated with 2D bimetallic oxide nanosheets and MOF-derived porous carbon polyhedra. *J. Mater. Chem. A* **2019**, *7*, 946–957. [CrossRef]
19. Zhang, Y.; Chen, P.; Wang, Q.; Wang, Q.; Zhu, K.; Ye, K.; Wang, G.; Cao, D.; Yan, J.; Zhang, Q. High-Capacity and Kinetically Accelerated Lithium Storage in MoO$_3$ Enabled by Oxygen Vacancies and Heterostructure. *Adv. Energy Mater.* **2021**, *11*, 2101712. [CrossRef]
20. Javed, M.S.; Zhang, X.; Ali, S.; Mateen, A.; Idrees, M.; Sajjad, M.; Batool, S.; Ahmad, A.; Imran, M.; Najam, T.; et al. Heterostructured bimetallic–sulfide@layered Ti$_3$C$_2$T$_x$–MXene as a synergistic electrode to realize high-energy-density aqueous hybrid-supercapacitor. *Nano Energy* **2022**, *101*, 107624. [CrossRef]
21. Nallapureddy, R.R.; Pallavolu, M.R.; Joo, S.W. Construction of Functionalized Carbon Nanofiber-g-C$_3$N$_4$ and TiO$_2$ Spheres as a Nanostructured Hybrid Electrode for High-Performance Supercapacitors. *Energy Fuels* **2021**, *35*, 1796–1809. [CrossRef]
22. Bao, Q.; Bao, S.; Li, C.M.; Qi, X.; Pan, C.; Zang, J.; Lu, Z.; Li, Y.; Tang, D.Y.; Zhang, S.; et al. Supercapacitance of solid carbon nanofibers made from ethanol flames. *J. Phys. Chem. C* **2008**, *112*, 3612–3618. [CrossRef]
23. Tian, N.; Wu, S.H.; Han, G.T.; Zhang, Y.M.; Li, Q.; Dong, T. Biomass-derived oriented neurovascular network-like superhydrophobic aerogel as robust and recyclable oil droplets captor for versatile oil/water separation. *J. Hazard. Mater.* **2022**, *424*, 127393. [CrossRef]
24. Luo, S.; Xiang, T.; Dong, J.; Su, F.; Ji, Y.; Liu, C.; Feng, Y. A double crosslinking MXene/cellulose nanofiber layered film for improving mechanical properties and stable electromagnetic interference shielding performance. *J. Mater. Sci. Technol.* **2022**, *129*, 127–134. [CrossRef]

25. Wang, Y.; Ma, L.; Xu, F.; Ren, R.; Wang, J.; Hou, C. Ternary ZIF-67/MXene/CNF aerogels for enhanced photocatalytic TBBPA degradation via peroxymonosulfate activation. *Carbohydr. Polym.* **2022**, *298*, 120100. [CrossRef]
26. Huang, H.; Dong, Y.; Wan, S.; Shen, J.; Li, C.; Han, L.; Dou, G.; Sun, L. A transient dual-type sensor based on MXene/cellulose nanofibers composite for intelligent sedentary and sitting postures monitoring. *Carbon* **2022**, *200*, 327–336. [CrossRef]
27. Wu, N.; Yang, Y.; Wang, C.; Wu, Q.; Pan, F.; Zhang, R.; Liu, J.; Zeng, Z. Ultrathin Cellulose Nanofiber Assisted Ambient-Pressure-Dried, Ultralight, Mechanically Robust, Multifunctional MXene Aerogels. *Adv. Mater.* **2022**, e2207969. [CrossRef]
28. Fang, Y.-S.; He, P.; Cai, Y.-Z.; Cao, W.-Q.; Cao, M.-S. Bifunctional $Ti_3C_2T_x$-CNT/PANI composite with excellent electromagnetic shielding and supercapacitive performance. *Ceram. Int.* **2021**, *47*, 25531–25540. [CrossRef]
29. Lyu, W.; Li, J.; Zheng, L.; Liu, H.; Chen, J.; Zhang, W.; Liao, Y. Fabrication of 3D compressible polyaniline/cellulose nanofiber aerogel for highly efficient removal of organic pollutants and its environmental-friendly regeneration by peroxydisulfate process. *Chem. Eng. J.* **2021**, *414*, 128931. [CrossRef]
30. Mahmoud, Z.H.; Al-Bayati, R.A.; Khadom, A.A. Synthesis and supercapacitor performance of polyaniline-titanium dioxide-samarium oxide ($PANI/TiO_2$-Sm_2O_3) nanocomposite. *Chem. Pap.* **2022**, *76*, 1401–1412. [CrossRef]
31. Singh, G.; Kumar, Y.; Husain, S. Improved electrochemical performance of symmetric polyaniline/activated carbon hybrid for high supercapacitance: Comparison with indirect capacitance. *Polym. Adv. Technol.* **2021**, *32*, 4490–4501. [CrossRef]
32. Liu, J.; Ma, X.; Zi, Z. Proton Acid Doped Superior Capacitive Performances of Pseudocapacitance Electrodes for Energy Storage. *Chemelectrochem* **2022**, *9*, e202200082. [CrossRef]
33. Yang, Z.X.; Yang, D.; Zhao, X.Z.; Zhao, Q.Y.; Zhu, M.; Liu, Y.; Wang, Y.; Lu, W.H.; Qi, D.P. From liquid metal to stretchable electronics: Overcoming the surface tension. *Sci. China-Mater.* **2022**, *65*, 2072–2088. [CrossRef]
34. Xu, C.; Jiang, W.-Y.; Guo, L.; Shen, M.; Li, B.; Wang, J.-Q. High supercapacitance performance of nitrogen-doped $Ti_3C_2T_x$ prepared by molten salt thermal treatment. *Electrochim. Acta* **2022**, *403*, 139528. [CrossRef]
35. Jose, J.; Jose, S.P.; Prasankumar, T.; Shaji, S.; Pillai, S.; Sreeja, P.B. Emerging ternary nanocomposite of rGO draped palladium oxide/polypyrrole for high performance supercapacitors. *J. Alloy. Compd.* **2021**, *855*, 157481. [CrossRef]
36. Wu, N.; Zhao, W.; Zhou, B.; Wu, Y.; Hou, W.; Xu, W.; Du, J.; Zhong, W. 3D nitrogen-doped $Ti_3C_2T_x$/rGO foam with marco- and microporous structures for enhance supercapacitive performance. *Electrochim. Acta* **2022**, *404*, 139852. [CrossRef]
37. Tang, Y.; Zhu, J.; Yang, C.; Wang, F. Enhanced supercapacitive performance of manganese oxides doped two-dimensional titanium carbide nanocomposite in alkaline electrolyte. *J. Alloy. Compd.* **2016**, *685*, 194–201. [CrossRef]
38. Habib, I.; Ferrer, P.; Ray, S.C.; Ozoemena, K.I. Interrogating the impact of onion-like carbons on the supercapacitive properties of MXene (Ti_2CT_X). *J. Appl. Phys.* **2019**, *126*, 134301. [CrossRef]
39. Chen, W.; Li, Z.; Jiang, F.; Luo, M.; Yang, K.; Zhang, D.; Xu, W.; Liu, C.; Zhou, X. Water Evaporation Triggered Self–Assembly of MXene on Non–Carbonized Wood with Well–Aligned Channels as Size–Customizable Free–Standing Electrode for Supercapacitors. *Energy Environ. Mater.* **2022**, *126*, 134301. [CrossRef]
40. Na, Y.W.; Cheon, J.Y.; Kim, J.H.; Jung, Y.; Lee, K.; Park, J.S.; Park, J.Y.; Song, K.S.; Lee, S.B.; Kim, T.; et al. All-in-one flexible supercapacitor with ultrastable performance under extreme load. *Sci. Adv.* **2022**, *8*, eabl8631. [CrossRef]
41. Wu, Y.; Sun, S.; Geng, A.; Wang, L.; Song, C.; Xu, L.; Jia, C.; Shi, J.; Gan, L. Using TEMPO-oxidized-nanocellulose stabilized carbon nanotubes to make pigskin hydrogel conductive as flexible sensor and supercapacitor electrode: Inspired from a Chinese cuisine. *Compos. Sci. Technol.* **2020**, *196*, 108226. [CrossRef]
42. Chen, Z.; Peng, X.; Zhang, X.; Jing, S.; Zhong, L.; Sun, R. Facile synthesis of cellulose-based carbon with tunable N content for potential supercapacitor application. *Carbohydr. Polym.* **2017**, *170*, 107–116. [CrossRef] [PubMed]
43. Deng, L.; Young, R.J.; Kinloch, I.A.; Abdelkader, A.M.; Holmes, S.M.; De Haro-Del Rio, D.A.; Eichhorn, S.J. Supercapacitance from Cellulose and Carbon Nanotube Nanocomposite Fibers. *ACS Appl. Mater. Interfaces* **2013**, *5*, 9983–9990. [CrossRef]

Article

Cr(III) Ion-Imprinted Hydrogel Membrane for Chromium Speciation Analysis in Water Samples

Ivanka Dakova *, Penka Vasileva and Irina Karadjova

Faculty of Chemistry and Pharmacy, University of Sofia "St. Kliment Ohridski", 1, J. Bourchier Blvd., 1164 Sofia, Bulgaria
* Correspondence: i.dakova@chem.uni-sofia.bg

Abstract: Novel Cr(III)-imprinted poly(vinyl alcohol)/sodium alginate/AuNPs hydrogel membranes (Cr(III)-IIMs) were obtained and characterized and further applied as a sorbent for chromium speciation in waters. Cr(III)-IIMs were prepared via solution blending method using blends of poly(vinyl alcohol) and sodium alginate as film-forming materials, poly(ethylene glycol) as a porogen agent, sodium alginate stabilized gold nanoparticles (SA-AuNPs) as a crosslinking and mechanically stabilizing component, and Cr(III) ions as a template species. The physicochemical characteristics of pre-synthesized AuNPs and obtained hydrogel membranes Cr(III)-IIM were studied by UV-vis and FTIR spectroscopy, TEM and SEM observations, N_2 adsorption–desorption measurements, and XRD analysis. The mechanism of the adsorption process toward Cr(III) was best described by pseudo-first-order kinetic and Langmuir models. Experiments performed showed that quantitative retention of Cr(III) is attained in 20 h at pH 6 and temperature 40 °C. Under the same conditions, the adsorption of Cr(VI) is below 5%. A simple and sensitive analytical procedure was developed for the speciation of Cr in an aquatic environment using dispersive solid phase extraction of Cr(III) by Cr(III)-IIM prior to selective Cr(VI) measurement by ETAAS in the supernatants. The detection limits and reproducibility achieved for the Cr speciation analysis fulfill the requirements for their monitoring in waters under the demand of the Water Framework Directive.

Keywords: ion-imprinted hydrogel membrane; Cr(III); sodium alginate; polyvinyl alcohol; gold nanoparticles; chromium speciation; surface waters

1. Introduction

The importance of selective and sensitive determination of the two most common chemical forms of chromium, Cr (III) and Cr (VI), demanded by their very different toxic effects, is still an analytical problem. In contrast to the relatively non-toxicity of Cr(III), Cr(VI) is highly toxic to most living organisms, causing strong adverse effects and diseases [1]. Chromium exists mostly as Cr(III) in the aquatic environment; toxic Cr(VI) is also present, however, at much lower concentrations as a result of its ongoing industrial application. That is why developed speciation methods should allow direct, selective determination of low levels of toxic Cr(VI) in order to ensure reliable speciation results. Moreover, considering the high oxidizing power and chemical activity of Cr(VI), the proposed method should preserve the original concentrations of Cr(III) and Cr(VI) during the sample transportation to the analytical laboratory, meaning that the separation step should be performed during sampling. From such a point of view, the creation of conditions for the separation of the two chemical forms immediately after sample collection and during its transportation is both preferable and encourages the preparation of innovative materials for non-chromatographic, selective determination of toxic Cr(VI). Very recently, advanced techniques for the selective removal of Cr(VI) from aqueous samples have been presented in a review article [2]. Modern analytical strategies and efficient nanosized sorbents used for chromium speciation in various matrices have been summarized and critically discussed in several review articles [3–8].

The application of ion-imprinted polymers (IIPs) as sorbents for elemental speciation analysis attracts extensive research interest due to their advantages, such as selectivity, stability, ease of preparation, low cost, and reusability [9]. Several Cr(III)-IIPs have been studied and characterized as effective sorbents in non-chromatographic speciation analysis of chromium [10–15]. These polymer sorbents have been synthesized as micro- or nanoparticles so that the proposed analytical procedures could not avoid the filtration/centrifugation steps. This drawback might be overcome by using membranes instead of particles as sorbents for solid phase extraction (SPE). It is known that hydrogel membranes can be successfully used to adsorb pollutants from water samples [16]. Hydrogel membranes are crosslinked three dimensional (3D) networks composed of hydrophilic polymers (natural or synthetic). Their selectivity might be additionally improved by the introduction of the ion template species resulting in the high recognition ability of ion imprinted polymers [17,18]. Studies on the synthesis of IIMs and their application for selective adsorption of Cr(III) [19,20] or Cr(VI) [21,22] ions, mostly from water samples, have been reported, but no studies are known about the green synthesis of Cr(III)-IIMs and their application for chromium speciation analysis.

One of the strategies for the green synthesis of hydrogel membranes is based on the use of renewable or natural materials in the membrane formation process. Sodium alginate (SA) is a natural, non-toxic, biocompatible, and biodegradable anionic polysaccharide composed of 1,4-linked β-D-mannuronic acid and 1,4-linked α-L-guluronic acid residues, containing carboxyl and hydroxyl groups [23]. It is well known as an environmentally friendly polymer for membrane preparation. Poly(vinyl alcohol) (PVA) is characterized by properties such as non-toxicity, biocompatibility, high hydrophilicity, film-forming ability, and chemical and mechanical resistance [23]. Blending SA with PVA results in polymeric materials that possess the desired properties, such as improved physical characteristics and film-forming abilities. Since traditionally used crosslinking agents for SA and PVA, such as epichlorohydrin and glutaric dialdehyde, are toxic [24,25], it is recommended to use inorganic crosslinking agents [26]. For example, gold nanoparticles (AuNPs), known to be non-toxic [27], could be used in this case. A significant additional advantage of AuNPs as cross-linkers is their capacity to form multiple bonds (so-called multivalency) within the gel networks [28].

In the present work, Cr(III)-IIMs are synthesized using sodium alginate and poly(vinyl alcohol) as film-forming materials, poly(ethylene glycol) (PEG) as a porogen agent, SA-coated gold nanoparticles (SA-AuNPs) as a crosslinking and mechanically stabilizing component, and Cr(III) ions as template species. The literature survey showed that the preparation of Cr(III)-imprinted PVA/SA/AuNPs hydrogel membrane has not yet been published. The physicochemical characteristics of pre-synthesized SA-AuNPs and hydrogel Cr(III)-IIM were studied by UV-vis and FTIR spectroscopy, TEM and SEM observations, and XRD analysis. Cr(III) imprinting is confirmed by FTIR spectroscopy and sorption characteristics by N_2 adsorption–desorption measurements. Experiments performed showed that the mechanism of sorption of Cr(III) was best described by pseudo-first-order kinetic and Langmuir models. A novel analytical procedure for solid phase extraction, which combines the selectivity of ion imprinting with the practicality of hydrogel membrane application, is developed for direct selective determination of toxic Cr(VI) in water samples. The procedure proposed might be performed in one reaction vessel, avoiding desorption steps and any operations leading to contamination or loss of analyte. Moreover, the whole procedure for Cr(VI) quantification could be performed during sampling on a membrane previously transferred in a polypropylene vessel and brought to the sampling site.

2. Results and Discussion

2.1. Cr(III) Ion-Imprinted Hydrogel Membrane Synthesis

Cr(III)-IIMs were synthesized by an approach based on "crosslinking of linear chain polymers" [29]. In the preparation process of Cr(III)-IIM adsorbent, two kinds of polymer materials, SA and PVA, were used as the functional hydrogel matrix. Preparation of Cr(III)–IIM included several steps, shown schematically in Figure 1. Initially, the solution of SA was blended with PVA, and then a solution of Cr(III) ions (template) was added. As a result, the carboxylate ions of SA formed complexes with Cr(III) cations, while the hydroxyl groups of SA could form coordinate bonds with Cr(III), confirmed by FTIR spectra. Based on these two kinds of chemical bonds, many stable structures like "egg box" were formed by SA chains [23]. In the next step, PVA chains were physically crosslinked with SA-AuNPs due to the coordination interaction between sodium alginate-capped AuNPs and hydroxyl groups of PVA [30]. Then the hydrogel matrix dispersion prepared was cast on the bottom of glass beakers and allowed the solvent to evaporate and dry until the formation of the membranes. The obtained self-standing hydrogel membrane can be considered a novel double crosslinking interpenetrating polymer network [31]. In the final step, Cr(III) ions were removed from the membrane prepared, resulting in the formation of a cavity with geometry and functional groups oriented specifically to the complex formation with template specie.

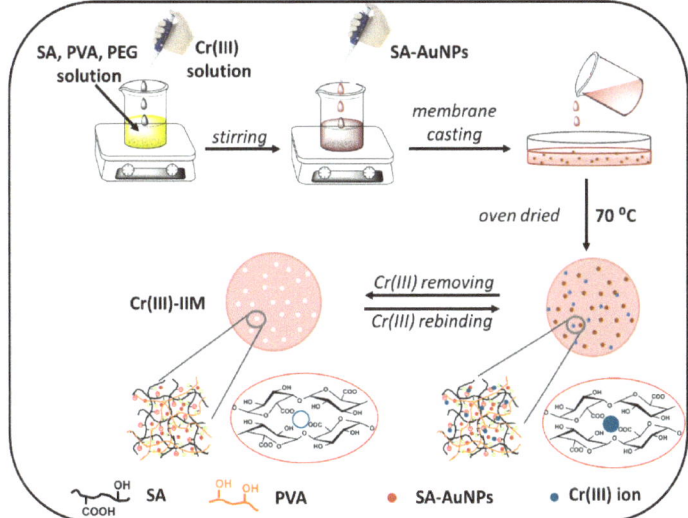

Figure 1. Schematic representation of the hydrogel Cr(III)-IIMs preparation.

2.2. Characterization of SA-AuNPs and Cr(III)-IIM

The optical, morphological and structural properties of SA-AuNPs before and after their incorporation in the hydrogel polymer matrix of membranes are studied and compared.

The UV-vis absorption spectrum of SA-AuNPs, recorded right after their preparation by sodium borohydride reduction of $AuCl_4^-$, is shown in Figure 2. The pink-red SA-AuNPs dispersion shows a surface plasmon resonance (SPR) band at 508 nm, and no aggregation was observed for at least up to six months. The narrow absorption band suggests the preparation of small gold nanoparticles with a narrow size distribution, as further confirmed by the TEM and XRD analysis. Figure 2 also displays the effect of SA-AuNPs incorporation in PVA/PEG/SA hydrogel polymer matrix solution before casting the membranes on the optical properties of gold nanoparticles. A red shift of the absorption band of gold nanoparticles (from 508 nm to 515 nm) was observed after the incorporation of AuNPs in the hydrogel polymer matrix, probably due to the partial sintering. No aggregation was observed, as further confirmed by TEM observation.

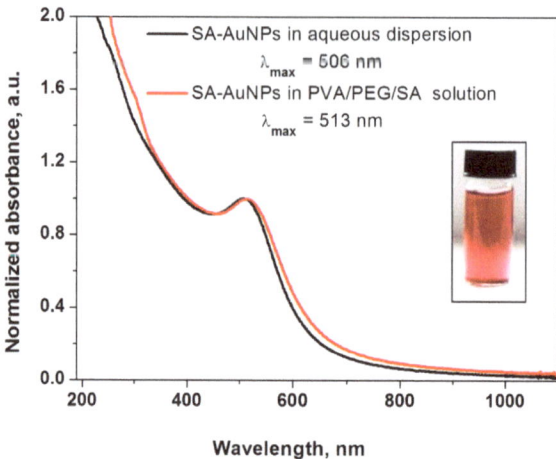

Figure 2. UV-vis absorption spectra of SA-AuNPs in: aqueous dispersion (red line) and PVA/PEG/SA hydrogel matrix solution (black line); inset: optical photo of SA-AuNPs aqueous dispersion.

Figure 3 shows TEM images at different magnifications of gold nanoparticles prepared by chemical reduction of $AuCl_4^-$ and stabilized by sodium alginate.

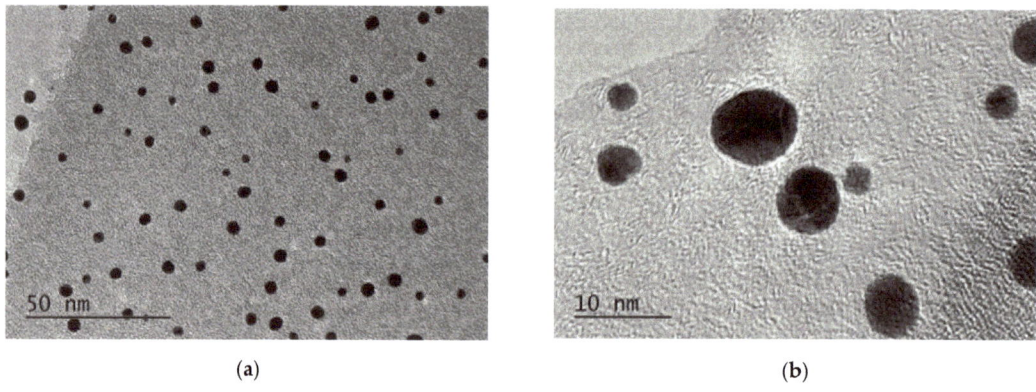

(a)   (b)

Figure 3. (a) TEM and (b) HRTEM micrographs of SA-AuNPs.

Most gold nanoparticles have a nearly spherical morphology, while a small part of them is polyhedral. It can be clearly seen that the nanoparticles in the aqueous dispersion are well separated from each other due to the protection by SA and have a small particle size—the average particle diameter is 4.9 ± 0.6 nm. An insignificant number of very small gold nanoparticles are also seen in TEM micrographs, which confirms the effective stabilization of nanoparticles with SA in aqueous dispersion.

TEM micrographs of hydrogel Cr(III)-IIM (Figure 4) display uniformly distributed gold nanoparticles with efficient stabilization by SA throughout the entire Cr(III)-imprinted PVA/SA hydrogel network. Local congregations of gold nanoparticles in the Cr(III)-IIM are observed in the TEM micrograph at higher magnification, probably due to the role of Cr(III) ions as a linker between nanoparticles, a natural result of which is a reduction of the distances between them in the ion-imprinted membrane. This observation is in excellent agreement with the previously commented red shift of the SPR band of AuNPs after their incorporation into the PVA/PEG/SA polymer hydrogel matrix.

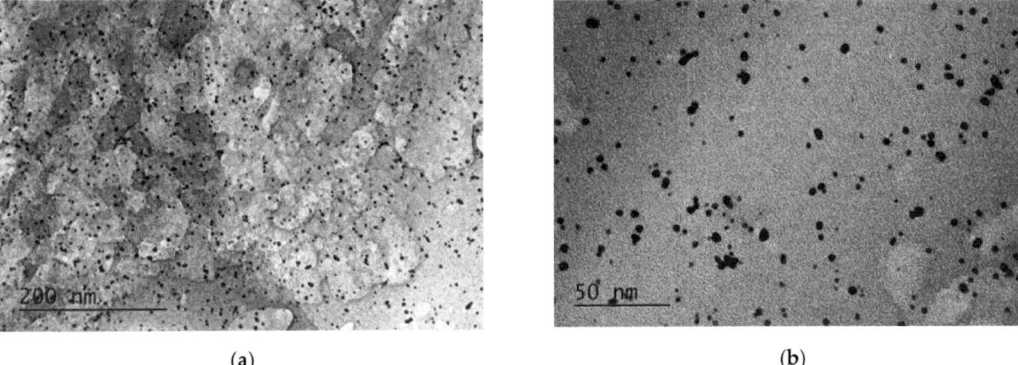

Figure 4. TEM micrographs at different magnifications (**a**,**b**) of Cr(III)-IIM.

An energy dispersive X-ray (EDX) elemental analysis was conducted for further investigation of the Cr(III)-IIM surface characteristics. The results are shown in Figure S1a,b. The EDX mapping confirmed the homogeneous dispersion of both Cr and Au elements in the polymer hydrogel matrix of the membrane. From the EDX spectrum and the inset table of Figure S1b, giving the elemental composition of Cr(III)-IIM, the presence of Cr and gold nanoparticles is confirmed.

The surface morphology characteristics of a non-imprinted membrane (NIIM) and Cr(III)-IIM were compared using SEM. As shown in Figure 5, the Cr(III)-IIM and NIIM display considerable differences in surface morphology. The SEM images of NIIM (Figure 5a,b) represent non-uniformity and some conglomeration of SA-AuNPs on the membrane surface. In contrast, the SEM images at different magnifications of Cr(III)-IIM (Figure 5c,d) clearly indicate a more uniform distribution of SA-AuNPs. Surface pores of Cr(III)-IIM can be distinguished with average sizes around 0.2–0.3 µm, while for NIIM, there are no pores on the membrane surface. The formation of a double crosslinking interpenetrating polymer network in Cr(III)-IIM can help generate a regularly distributed surface morphology, which does not exist in NIIM since Cr(III) ions are absent.

FTIR spectroscopy was used to elucidate the structure of hydrogel Cr(III)-IIM (see Figure S2). The chelate complex formation between alginic acid and metal ions is thoroughly studied, and the structure of complexes formed is confirmed by FTIR spectroscopy in the published literature [19,32]. As expected, a comparison of FTIR spectra of NIIM and Cr(III)-IIM shows that bands of the asymmetric (v_{as}) and symmetric (v_s) stretching vibrations of alginic acid −OCO− group are shifted from 1654 cm^{-1} and 1419 cm^{-1} for NIIM to lower frequencies of 1601 cm^{-1} and 1409 cm^{-1} for Cr(III)-IIM, respectively. These results mean that carboxylic functional groups take part in chelate formation. The shift of the broad v_{OH} band at around 3400 cm^{-1} to lower frequencies indicates that the OH groups are also involved in the chelation. FTIR spectra proved the coordination process between Cr(III) and alginic acid confirming the successful imprinting of Cr(III) in hydrogel Cr(III)-IIM. A schematic presentation of the interactions between SA and Cr(III) ions is shown in Figure S3.

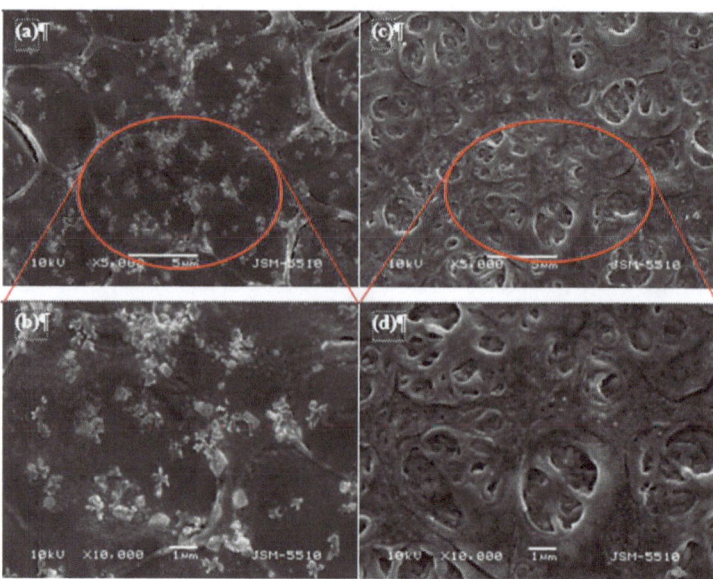

Figure 5. SEM images at different magnifications of (**a,b**) NIIM and (**c,d**) Cr(III)-IIM.

Nitrogen adsorption–desorption isotherm studies performed for hydrogel Cr(III)-IIM membrane indicated that the specific surface area is 5 m^2/g with a total pore volume of 0.04 cm^3/g. Similar results have already been reported for hydrogel membranes based on SA/PVA blend and different inorganic constituents [33,34].

The XRD pattern of the PVA/PEG/SA polymer membrane shows a significant semicrystalline peak at 2θ value of 19.65°, which is connected to the PVA structure, generated from strong intra- and intermolecular hydrogen bonding [35,36] (see Figure S4a). In cases of NIIM and Cr(III)-IIM, this semicrystalline peak appears at the same 2θ value along with other broad diffraction peaks of low intensity centered at 2θ values of 38.8°, 44.4°, 64.7°, 77.5°, which can be indexed to the (111), (200), (220), and (311) crystal planes corresponding to the face-centered crystal (fcc) structure of gold [37] (see Figure S4b).

2.3. Adsorption Behavior of Cr-IIM toward Cr(III) and Cr(VI)—Optimization Studies

In order to evaluate the suitability of hydrogel Cr(III)-IIM as a sorbent for the selective separation of Cr(III) ions, chemical conditions for quantitative retention of Cr(III) were optimized. Taking into account the kinetic inertness of $Cr(H_2O)_6^{3+}$ complexes, three important parameters were optimized—pH, temperature, and time for adsorption. As a first step, the progress of Cr(III) retention on the Cr(III)-IIM at different times was studied. Experimental data for the degree of Cr(III) sorption, D_s, were obtained at initial concentration 5 mg/L, pH 6, and temperature 40 °C. The kinetic adsorption curve is shown in Figure 6, where the duration of the sorption process varied from 1 to 24 h. It can be seen that as the contact time increases, the degree of sorption D_s also increases. According to this curve, quantitative sorption > 95% for Cr(III) in the Cr(III)-IIM adsorption system was achieved within 20 h. The retention time considered optimal was set to 20 h. A similar relatively slow process (equilibrium time of 18 h) has already been reported for quantitative Pd(II) sorption using palladium imprinted membrane based on a chitosan matrix with azo-derivative ligand [38]. It is reasonable to assume that such a slow reaching of the adsorption equilibrium is due to the large diffusion barrier in the thin ion-imprinted membrane. The greater diffusion resistance leads both to the difficult entry of Cr(III) ions into the membrane cavities and to their limited association with the recognition centers.

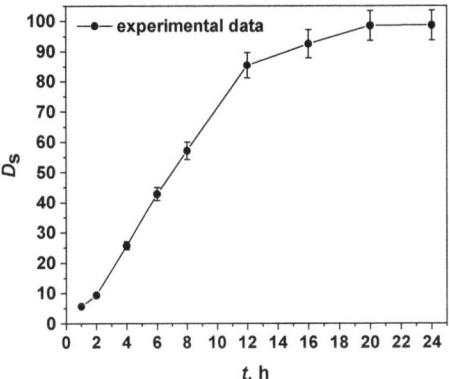

Figure 6. Effect of contact time on the degree of sorption D_s of Cr(III) onto Cr(III)-IIM at initial concentration 5 mg/L, pH 6, temperature 40 °C, and adsorbent dose (one membrane) 0.140 g.

The acidity of the solution is an important parameter determining the effectiveness of the SPE procedure because the pH value affects both the binding sites on the surface of the sorbent and the metal chemistry in aqueous solutions. In order to preserve the original concentrations of Cr(III) and Cr(VI) during the Cr(III) sorption onto Cr(III)-imprinted PVA/SA/AuNPs membrane, it is very important to take into account the possibility of reduction of Cr (VI) by the carboxyl groups of SA—a process that also depends on pH. The influence of pH on the reduction of Cr(VI) to Cr(III) with alginic acid is well established—at pH 1–3, alginic acid slowly reduces Cr(VI); at pH 6.0, the redox reaction of Cr(VI) with alginic acid proceeds very slowly, with negligible reduction of Cr(VI) [39].

The effect of pH (in the range 4–9) and temperature (25, 40, 50, and 60 °C) on the degree of Cr(III) sorption onto Cr(III)-IIM is illustrated in Figure 7. The hydrogel membranes prepared contain carboxylic groups (in SA) and hydroxylic groups (in SA and PVA) in the polymer matrix, suggesting that at lower pH (pH < pKa = 3.6 for alginic acid), the functional groups are protonated, and in this way, the Cr(III) adsorption onto Cr(III)-IIM is restricted. Hence, the values of D_s are very small (these results are not presented in Figure 7). It is seen from Figure 7 that the degree of Cr(III) sorption is enhanced with an increasing pH of 6 for all studied temperatures. At pH values in the range 4–6, the fraction of deprotonated carboxyl groups in SA grows and (–COO$^-$) becomes available for binding and adsorption of Cr(III) cations. In addition, the positively charged Cr^{3+} and $CrOH^{2+}$ ions (species existing at pH < 6 [40]) can be bound to the negatively charged groups of the membrane by electrostatic attraction, leading to an increased degree of sorption. However, at pH values higher than 6, a decrease in the degree of Cr(III) sorption is noticed (Figure 7), which may be attributed to the precipitation of the metal ions as $Cr(OH)_3$ [40]. Furthermore, the temperature dependence of the degree of Cr(III) sorption on Cr(III)-IIM is clearly visible from the results in Figure 7.

Quantitative Cr(III) sorption (D_s > 95%) is achieved at temperatures in the range of 40–50 °C, while the degree of Cr(III) sorption is lower at temperatures outside this range (91.0% and 93.7% at temperatures of 25 °C and 60 °C, respectively). These results can be explained by the kinetic stability of the $Cr(H_2O)_6^{3+}$ complex, for which ligand exchange in the inner coordination sphere requires elevated temperatures. Finally, quantitative retention of Cr(III) on the Cr(III)-IIM was achieved at optimal pH 6 for 20 h at a temperature of 40 °C. Under the established optimal conditions for quantitative sorption of Cr(III), the degree of Cr(VI) sorption is found to be less than 5%. These results unambiguously confirm that the hydrogel Cr(III)-IIM can be used for the quantitative separation of Cr species in order to perform successful speciation analysis. Results from parallel adsorption experiments carried out with NIIM membrane showed similar sorption behavior (not presented in Figure 7) toward Cr(III), however, with about a 30% lower value of D_s. Under defined optimal conditions, the sorption capacity of the Cr(III)-IIM and NIIM were evaluated after saturation of the membranes with Cr(III) ions. The effect of the initial concentration of Cr(III) ions (5–35 mg/L) on the sorption capacity of Cr(III)-IIM and NIIM is displayed in Figure 8.

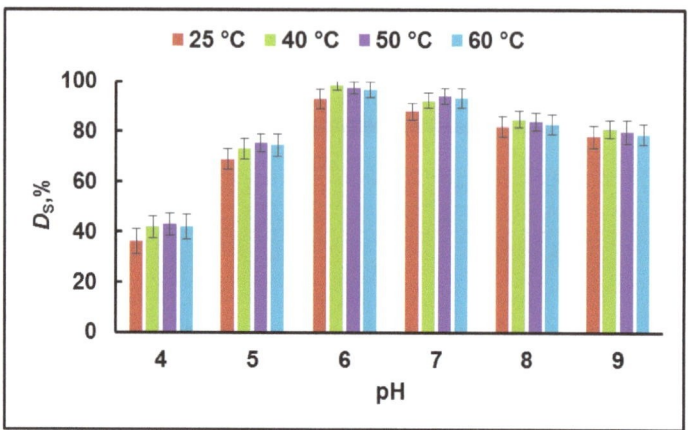

Figure 7. Dependence of the degree of sorption (D_S, %) of Cr(III) ions onto Cr(III)-IIM on pH and temperature.

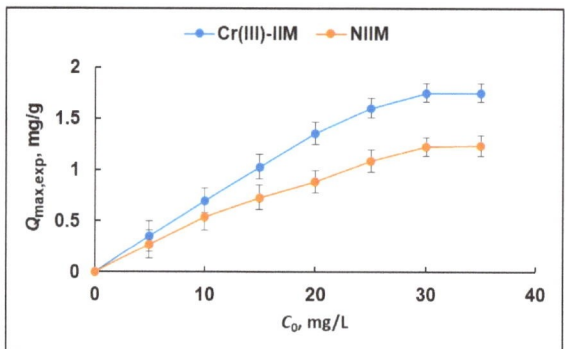

Figure 8. Effect of the initial concentration of Cr(III) on the adsorption capacity of Cr(III)-IIM and NIIM (pH 6; contact time 20 h; temperature 40 °C).

Adsorption isotherms (Figure 8) clearly show that the amount of adsorbed Cr(III) per unit mass of the membrane increases with growing Cr(III) concentration and reaches a plateau determining the maximum adsorption capacity ($Q_{max,exp}$)—1.75 mg/g for Cr(III)-IIM and 1.23 mg/g for NIIM. As expected, the adsorption capacity of Cr(III)-IIM exceeds the NIIM's capacity, indicating that the binding sites created after the removal of template ions ensure higher affinity of the hydrogel Cr(III)-IIM toward Cr(III) in this way proving the advantages of ion-imprinting approach for the preparation of sorbent materials with higher adsorption capacity.

2.4. Elution Studies

The elution step should ensure quantitative desorption of sorbed Cr(III) in this way, ensuring further use of synthesized Cr(III)-IIM. Eluent solutions containing HCl or NH_4-EDTA were tested for Cr quantitative extraction from loaded Cr(III)-IIMs. The results obtained are presented in Table 1. It can be concluded that hydrochloric acid at any concentration level is not suitable for the elution of Cr(III)—the elution is not quantitative, and AuNPs in the membranes are dissolved at the higher acid concentration (1 mol/L). The most suitable eluent is NH_4-EDTA solution (0.1 mol/L), which provides complete elution of Cr(III) (>99%) from the membranes, and at the same time, the membrane composition and stability are unaffected. The effect of desorption agent volume was also studied (Table 1). A 10 mL NH_4-EDTA solution was found to be the optimum volume to provide quantitative Cr(III) elution from the membranes. The kinetics of the Cr(III) desorption process studied according to the procedure described in Section 4.5 for 1–5 h showed that quantitative desorption was reached for 2 h. Optimal conditions defined for quantitative elution of Cr(III) include 10 mL 0.1 mol/L NH_4-EDTA for 2 h desorption time.

Table 1. Degree of elution D_E (%) of Cr(III) from Cr(III)-IIM using different eluents.

Eluent	c, mol/L	D_E, %
HCl (V = 10 mL)	0.1	67.3 ± 3
	0.5	80.6 ± 3
	1.0	AuNPs dissolution
NH_4-EDTA (V = 10 mL)	0.05	68.7 ± 4
	0.1	>99
	0.2	>99
NH_4-EDTA (V = 5 mL)	0.1	75.6 ± 3
NH_4-EDTA (V = 10 mL)	0.1	>99
NH_4-EDTA (V = 20 mL)	0.1	>99

2.5. Investigations on the Mechanism of Cr(III) Adsorption onto Cr(III)-IIM

2.5.1. Adsorption Isotherm Models

The adsorption data for Cr(III)-IIM and NIIM as a function of the initial Cr(III) concentrations were analyzed using the Freundlich and Langmuir adsorption isotherm models. The applicability of the isotherm models was evaluated by comparing the calculated values for the R^2 coefficient.

The Freundlich isotherm model can be applied in the case of multilayer adsorption of the adsorbate on a heterogeneous surface [41]. Equation (1) presents the Freundlich isotherm in the linear form:

$$ln\ Q_e = ln\ k_F + n^{-1}.ln\ C_e \quad (1)$$

where C_e (mg/L) and Q_e (mg/g) are Cr(III) equilibrium concentration in the solution and equilibrium capacity of the membranes, respectively; k_F is the Freundlich isotherm constant; n is the adsorption intensity. The value of n gives information about the adsorbent–adsorbate interaction. The adsorption process is favorable, when $0 < 1/n < 1$; unfavorable—$1/n > 1$; and irreversible—$1/n = 1$ [42].

The Langmuir isotherm model describes a sorption process occurring in a surface monolayer of homogeneous sites [41]. The linear form of Langmuir isotherm is presented by (Equation (2)):

$$\frac{C_e}{Q_e} = \frac{C_e}{Q_{max}} + \frac{1}{b.Q_{max}} \quad (2)$$

where Q_{max} (mg/g) is the calculated maximum adsorption capacity, b (L/mg) is the Langmuir constant.

To predict the favorability of a given adsorption system, it is recommended to use the dimensionless factor R_L (Equation (3)). The isotherm is irreversible, favorable, linear, or unfavorable if $R_L = 0$, $0 < R_L < 1$, $R_L = 1$, or $R_L > 1$, respectively [41].

$$R_L = \frac{1}{1 + b.C_0} \quad (3)$$

The final calculation results of Langmuir and Freundlich isotherm parameters are exhibited in Table 2, and the graphical visualization is in Figure S5.

Table 2. Experimental adsorption capacity values and Langmuir and Freundlich isotherm parameters obtained by linear fitting for the Cr(III)-IIM and NIIM at temperature of 40 °C.

Polymer Hydrogel Membrane	$Q_{max,exp}$ mg/g	Langmuir Isotherm Model				Freundlich Isotherm Model		
		$Q_{max,calc}$ mg/g	b L/mg	R^2	R_L	k_F	n	R^2
Cr(III)-IIM	1.75	1.74	3.52	0.9997	0.01–0.05	1.15	3.47	0.8956
NIIM	1.23	1.25	0.32	0.9993	0.08–0.38	11.47	2.38	0.9592

From Table 2, it can be concluded that the values of coefficient of determination R^2 obtained for the Langmuir model (0.9997 and 0.9993 for Cr(III)-IIM and NIIM, respectively) are higher than values obtained when using Freundlich isotherm (0.8956 and 0.9592 for Cr(III)-IIM and NIIM, respectively). The calculated values of adsorption capacity $Q_{max,calc}$ are in good agreement with experimentally obtained values (Table 2). These results confirm the correctness of the assumption that the adsorption process occurs in a surface monolayer of homogeneous sites.

The calculated values of Langmuir dimensionless factor R_L are in the range $0 < R_L < 1$ (Table 2), indicating that the adsorption of Cr(III) ions onto Cr(III)-IIM and NIIM is favorable. This conclusion for favorable adsorption is also confirmed by the values of the Freundlich coefficient n related to the adsorption intensity that satisfies the condition $0 < 1/n < 1$ ($1/n$ is 0.28 and 0.46 for Cr(III)-IIM and NIIM, respectively).

2.5.2. Modeling of Cr(III) Sorption Kinetics

In order to understand the behavior of Cr(III) ions adsorbed by the novel hydrogel Cr(III)-IIM and to determine the controlling mechanism of the adsorption process, several kinetic models, which contain two undetermined parameters, have been used to fit the experimental data [43]:

$$\text{pseudo-first-order model}: q_t = q_e \left(1 - e^{-k_1 \cdot t}\right) \quad (4)$$

where q_t and q_e (mg/g) are the adsorbed amounts at different times t (h) and at an equilibrium, respectively, and k_1 (1/h) is the rate constant. The pseudo-first-order kinetic model better describes an adsorption process controlled by diffusion and is mainly used to simulate a simple single reaction.

$$\text{pseudo-second-order model}: q_t = k_2 \cdot q_e^2 \frac{t}{1 + k_2 \cdot q_e \cdot t} \quad (5)$$

where k_2 (g/(mg·h)) is the rate constant. The pseudo-second-order model assumes that the chemisorption is a rate-limiting step.

$$\text{Elovich equation}: q_t = \frac{1}{\beta} \ln (\alpha \cdot \beta) + \frac{1}{\beta} \ln t \quad (6)$$

where α (mg/(g·h)) is the initial rate of the adsorption process and β (g/mg) is the desorption constant of this process related to the extent of surface coverage and activation energy of chemisorption. Elovich equation is useful in describing the chemical sorption on highly heterogeneous surfaces [44].

In order to find other important correlations of experimental kinetic data in this study, the Weber and Morris equation was tested for evaluation of adsorption kinetics of Cr(III) ions onto Cr(III)-IIM:

$$\text{intra-particle diffusion model}: q_t = k_i \cdot t^{0.5} + C_i \quad (7)$$

where k_i (mg/(g·h$^{0.5}$)) is the equilibrium rate constant of intra-particle diffusion, and C_i (mg/g) is the intercept associated with the thickness of the boundary layer. The intra-particle diffusion model describes the kinetics of the diffusion process inside a particle; it is not suitable for describing the kinetics of the diffusion process on the surface of a particle [45].

Kinetic parameters of pseudo-first-order, pseudo-second-order, and Elovich kinetic models estimated by regression analysis are summarized in Table 3, and the fitted curves are plotted in Figure S6a–c. To choose the superior model, both coefficient of determination (R^2) and the equilibrium adsorption capacity predicted by the model ($q_{e,calc}$) should be considered [46].

Table 3. Fitted kinetic parameters of pseudo-first-order, pseudo-second-order, Elovich, and intra-particle diffusion models for adsorption of Cr(III) ions onto the Cr(III)-IIM at concentration 5 mg/L, pH 6, temperature 40 °C, and adsorbent dose (one membrane) 0.140 g.

Model	Parameters	Values
Pseudo-first-order model	k_1 (1/h) $q_{e,calc}$ * (mg/g) R^2	0.07529 0.4496 0.9694
Pseudo-second-order model	k_2 (g/(mg·h)) $q_{e,calc}$ * (mg/g) R^2	0.07234 0.6974 0.9626
Elovich equation	α (mg/(g·h)) β (g/mg) R^2	0.09449 8.1739 0.9331
Intra-particle diffusion model Region 1	k_i (mg/(g·h$^{0.5}$)) C_i (mg/g) R^2	0.1034 −0.09989 0.9558
Intra-particle diffusion model Region 2	k_i (mg/(g·h$^{0.5}$)) C_i (mg/g) R^2	0.03459 0.1892 0.8710

* $q_{e,exp}$ = 0.3521 mg/g.

The low value of the determination coefficient (0.9331, Table 3) shows that the Elovich model is unsuitable to represent the adsorption of Cr(III) ions onto the Cr(III)-IIM and also indicates that the adsorption process is not controlled by chemisorption [13]. Curve fitting results (Table 3) implied that the pseudo-first order kinetic model (R^2 = 0.9694) is more suitable to describe the adsorption behavior than the pseudo-second order model (R^2 = 0.9626), and the values of $q_{e,exp}$ and $q_{e,calc}$ are closer to each other under pseudo-first-order kinetic model than that of pseudo-second-order model, indicating that the adsorption is mainly controlled by diffusion.

The rate constant of intra-particle diffusion k_i could be obtained from the slope of the plot presented in Figure S6d. It is seen that the plot does not pass through the origin and is nonlinear. It can be concluded that the adsorption of Cr(III) ions onto Cr(III)-IIM is a complex process [47]. Two straight lines simulating the experimental results and the values of kinetic parameters are presented in Table 3. The slope of the line for the first region (responsible for external diffusion; $k_{i,1}$ = 0.10342) is higher than the slope of the line for the second region (corresponding to intra-particle diffusion; $k_{i,2}$ = 0.03459), which confirms the conclusion that the active sorption sites for Cr(III) ions are distributed onto the outer sorbent surface and penetration into the inside of the membrane is insignificant [48]. A negative C_i

value in Equation (7) (see Table 3, Region 1) could be explained by the combined effects of surface reaction control and film diffusion processes [49].

2.6. Analytical Applications

The experimental results obtained showed that an analytical procedure for Cr speciation might be developed based on the sorption of Cr(III) on the hydrogel membrane and selective determination of Cr(VI) in the supernatant (see Section 4.7). Model experiments were performed with various waters such as river, sea, and mineral water aiming to assess the selective recovery of Cr(VI) independent of the water matrix. As a first step, interference studies according to the procedure described in Section 4.6 were performed in order to confirm that even in the presence of different levels of matrix cations and anions, quantitative separation of both Cr(III) and Cr(VI) is still achieved (see Table S1). Results obtained undoubtedly showed that independently of the sample matrix degree of sorption of Cr(III) is in the range between 95–98%, and for Cr(VI), in all cases degree of sorption is below 5%. As a next step, the separation of both species was studied at different ratios more relevant to the environmental conditions, e.g., relatively low concentrations of Cr(VI) in the presence of high amounts of Cr(III) and for different types of surface waters, using added/found method. River water, groundwater, and seawater, filtered through a cellulose membrane filter (0.45 μm), were spiked with different concentration ratios of Cr(VI) to Cr(III) and passed through the procedure described in Section 4.7. The results obtained are presented in Table 4.

Table 4. Determination of Cr(III) and Cr(VI) in different types of waters (three parallel determinations).

Sample	Cr(III), μg/L Mean ± SD	Cr(VI), μg/L Mean ± SD	Recovery for Cr(VI), %
River water	2.3 ± 0.2	<DL	
River water + 0.5 μg/L Cr(VI)	2.2 ± 0.2	0.49 ± 0.02	94 ± 2
Seawater	0.52 ± 0.04	<DL	
Seawater + 0.2 μg/L Cr(VI)	0.54 ± 0.04	0.21 ± 0.02	95 ± 4
Groundwater	1.3 ± 0.1	0.25± 0.02	
Groundwater + 0.4 μg/L Cr(VI)	1.2 ± 0.1	0.63 ± 0.03	93 ± 4

Evidently, for all studied ratios and for all types of waters, recoveries for toxic Cr(VI) are between 93–95%, confirming the applicability and reliability of the developed analytical procedure.

In addition, the results for Cr (VI) content in waters obtained by the proposed analytical method were compared at a bit higher concentration level to the results obtained using a standard procedure based on the spectrophotometric method with 1,5-diphenylcarbazide (ISO 11083:1994). Natural ground waters from polluted aquifers in north Bulgaria were used for this comparison. The very good agreement observed between parallel results for more than 10 samples verifies the accuracy and versatility of the proposed approach for Cr(VI) quantification using hydrogel Cr(III)-IIM.

The experiments performed showed that the hydrogel membrane might be used for four adsorption/desorption cycles using 0.1 mol/L NH_4-EDTA for elution (extraction efficiency above 95%). The extraction efficiency toward Cr(III) achieved by using hydrogel membranes from different batches showed very good repeatability, most probably due to the sustainability and robustness of the developed synthesis procedure.

2.7. Analytical Figures of Merit

Analytical figures of merit were defined after Cr(III) and Cr(VI) determination in five parallel samples. Detection and determination limits were calculated based on 3σ and 10σ criteria taking into account standard deviations of a blank sample (Cr measurement in 10 mL high-purity water passed through the whole developed analytical procedure. The results obtained are depicted in Table 5.

Table 5. Analytical figures of merit determined after five parallel determinations.

Species	Detection Limit, µg/L	Determination Limit, µg/L	RSD, % for the Range 0.05–50 µg/L
Cr(III)	0.001	0.003	7–11
Cr(VI)	0.01	0.03	4–6

As seen from the results in Table 5, the analytical procedure developed is characterized by low determination limits and very good reproducibility. The most serious advantage is the possibility for direct determination of Cr(VI), avoiding any parallel determination and additional calculations.

A comparison of analytical figures of merit reported in the literature for Cr speciation procedures using different sorbent materials is presented in Table S2 [10–15,50]. As can be seen, the proposed in this work analytical method for the selective determination of Cr(III) and Cr(VI) ensures the lowest detection limits and allows the determination of environmentally relevant concentrations of Cr in surface/ground waters, even at background levels in unpolluted sites.

3. Conclusions

In this study, a novel hydrogel membrane, Cr(III)-imprinted poly(vinyl alcohol)/sodium alginate/AuNPs, was prepared by green synthesis and tested for Cr(III)/Cr(VI) separation. The formation of a double crosslinking interpenetrating polymer network and obtained good dispersion of gold nanoparticles in a polymer hydrogel matrix restricts the chain movements and thereby supports a mechanical strength of membrane produced and easy operation in sorption experiments. Optimization studies performed showed quantitative retention of Cr(III) at pH 6 and temperature 40 °C, while sorption of Cr(VI) is below 5%. The adsorption equilibrium for Cr(III) was attained within 20 h. The kinetics adsorption data for Cr(III) were well-fitted with a pseudo-first-order kinetic model, and the equilibrium data were best described by the Langmuir isotherm model. The maximum adsorption capacity of the Cr(III)-IIM for Cr(III) ions under the optimal condition was 1.75 mg/g. The successive adsorption–desorption experiment indicated that 0.1 mol/L NH_4-EDTA solution could be effectively applied for Cr(III) elution from the Cr(III)-IIM, and the membrane can be used for additional three adsorption/desorption cycles.

A simple and sensitive analytical procedure was developed for the speciation of Cr in an aquatic environment using dispersive solid phase extraction of Cr(III) by Cr(III)-IIM membranes prior to selective Cr(VI) determination in the supernatants. The determination limit achieved for toxic species Cr(VI) fulfills the requirements for their monitoring in surface water bodies under the demand of the Water Frame Directive. The developed procedure avoids any additional calculations or parallel determinations for Cr(VI) quantification. In addition, if necessary, Cr(III) might be determined in the same sample with an even lower determination limit.

4. Materials and Methods

4.1. Materials, Reagents, and Instruments

High-purity water (Millipore Corp., Milford, MA, USA) was used to prepare all aqueous solutions. The working standard solutions were prepared daily by appropriately diluting the stock solutions of Cr(III) (Spex Certiprep 1000 mg/L in 2% HNO_3) and Cr(VI) (Spex Certiprep 1000 mg/L in H_2O).

Tetrachloroauric(III) acid ($HAuCl_4 \cdot 3H_2O$, 99%, Panreac, Poland) and sodium tetrahydridoborate ($NaBH_4$, GR for analysis, Merck, Germany) were used for AuNPs preparation. Sodium alginate (SA, low viscosity, Alfa Aesar, MA, USA), poly(vinyl alcohol) (PVA, relative molecular mass 72000, Sigma-Aldrich, St. Louis, MO, USA), and poly(ethylene glycol) (PEG, relative molecular mass 400, Sigma-Aldrich, St. Louis, MO, USA) were used to prepare the hydrogel Cr(III)-IIM and NIIM. Hydrochloric acid (Fisher Chemical™, Waltham,

MA, USA) and ethylenediamine tetraacetic acid (EDTA, Sigma-Aldrich, St. Louis, MO, USA) were used for Cr desorption in the optimization experiments. After the dissolution of EDTA in NH_3 solution (25%, Merck, Darmstadt, Germany), EDTA diammonium salt (NH_4-EDTA) was prepared. The pH value of water samples was adjusted with NH_3 solution or HNO_3.

Absorption spectra of gold nanoparticles were recorded on a Thermo Scientific Evolution 300 UV–V spectrometer in the range 190–1100 nm, using quartz cuvette with 1 cm optical path. Quartz cuvette containing high-purity water served as a reference sample for background absorption. The morphology and sizes of gold nanoparticles were examined by a transmission electron microscope (TEM, JEOL JEM-2100 operating at 200 kV). The surface morphology of membranes was observed by scanning electron microscope (SEM, JEOL JSM-5510 operating at 10 kV). X-ray diffraction (XRD) patterns were registered on an X-ray powder diffractometer Siemens D500 equipped with the CuKα radiation (λ = 1.54 Å) in 2θ ranging from 10° to 90°.

The texture parameters were determined by nitrogen adsorption at temperature 77.4 K in NOVA 1200e (Quantachrome, Boynton Beach, FL, USA) instrument. The BET equation and the Gurvich rule (at a relative pressure close to 0.99) were used to calculate the specific surface area (S_{BET}) T and the total pore volume (V_t), respectively.

ATR-FTIR spectra were recorded by using Nicolet iS50 (Thermo Scientific, Waltham, MA, USA) Fourier Transform Infrared (FTIR) spectrophotometer with Attenuated Total Reflectance Attachment. In general, 32 scans and 4 cm^{-1} resolution were applied. The spectral data were processed with OMNIC Software (version 9.12.1002., (Thermo Fisher Scientific Inc., Waltham, MA, USA).

The concentrations of Cr were measured by Electrothermal atomic absorption spectrometry (Perkin Elmer Model AAnalyst 400, equipped with HGA 900 and AS 800 autosampler). Samples of effluate and eluate solutions (10–20 μL) were injected into pyrolytically coated graphite tubes using AS-800. Optimized temperature program consists of drying step at 120 °C, pretreatment step at 1100 °C, and atomization step at 2500 °C. Integrated absorbance signals (three replicates) were used for Cr quantification against external calibration.

The solutions' pH was measured with a pH meter (Mettler Toledo; Seven Compact S220-K, Greifensee, Switzerland).

4.2. Synthesis of SA-AuNPs

The aqueous dispersions of sodium alginate stabilized gold nanoparticles were prepared by chemical reduction method based on the reduction of Au(III) (8 mL 0.001 mol/L $HAuCl_4$) using strong reductant sodium tetrahydridoborate (24 mL 0.002 mol/L $NaBH_4$) and alginate ions (1.5 mL 1% SA) as a non-toxic capping agent. The reduction was carried out in ice bath under magnetic stirring, and at the end of reaction sodium alginate solution was added for steric stabilization of gold nanoparticles by different functional groups, such as −COOH and −OH. The synthesis process is schematized in Figure S7. The noble metal nanoparticle dispersion was stored in dark bottles at room temperature. The wine-red dispersion of SA-AuNPs was stable for several months under storage conditions.

4.3. Preparation of Cr(III)-IIM and NIIM

The preparation of hydrogel Cr(III)-IIM includes using blends of poly(vinyl alcohol) and sodium alginate as film-forming materials, poly(ethylene glycol) as porogen agent, gold nanoparticles (SA-AuNPs) as cross-linking and mechanically stabilizing component, and Cr(III) ions as template species. In the typical procedure, aqueous solutions of SA (1% w/v) and PVA (2% w/v) were prepared in high-purity water with stirring at 85–90 °C for 90 min; then the hot solutions were filtered. Prepared mixture from PVA solution (30 mL) and PEG (115 mg) was poured into SA solution (30 mL) and stirred well for 30 min. This was followed by the addition drop by drop of Cr(III) solution (3 mL, 1000 mg/L), pH adjusting up to 5–6 by NaOH (2 mol/L). The resulting mixture was stirred vigorously for 60 min. In the next step, the pre-synthesized SA-AuNPs aqueous dispersion (80 mL) was added into the above polymer hydrogel matrix solution and stirred vigorously for 60 min. Then the solution was cast on plastic Petri dishes in portions of 7.5 mL and dried in hot air oven at 70 °C for 12 h. In order to remove the porogen PEG, the dried hydrogel membranes were immersed in high-purity water for one day. Then, chromium was extracted from the produced membranes by elution with 0.2 mol/L NH_4-EDTA solution until the Cr concentration in the eluate solution was below the LOQ as measured by ETAAS. Similarly, in the absence of matrix ions, non-imprinted membranes (called NIIMs) were prepared. The whole imprinting process is schematized in Figure 1.

4.4. Static Adsorption/Desorption Experiments

The model solutions for static adsorption experiments were prepared by addition of 50 µg Cr(III) to 10 mL high-purity water. a The pH value between 4–9 was adjusted by HNO_3 or NH_3 solution. Cr(III)-IIM or NIIM was immersed in this solution and stirred with an electric shaker for 20 h at temperature of 40 °C. The membrane was removed and remaining solution (effluate) was analyzed by ETAAS. The membrane was treated twice with high-purity water, and Cr(III) was eluted with 0.1 mol/L NH_4-EDTA solution. Chromium content was measured in the eluate by ETAAS.

The degree of sorption (D_S, %) and degree of elution (D_E, %) of Cr(III) ions were calculated by the following equations:

$$D_S = \frac{A_i - A_{eff}}{A_i} \times 100 \tag{8}$$

$$D_E = \frac{A_{el}}{A_i - A_{eff}} \times 100 \tag{9}$$

where A_i (µg) is the initial amount of Cr(III) in contact with the membrane; A_{eff} (µg) is the amount of Cr(III) in the effluate solution after Cr(III)-IIM extraction; A_{el} (µg) is the amount of Cr(III) in the eluate.

4.5. Isotherm and Kinetic Studies

The following procedure was used for determination of the adsorption capacities of the hydrogel Cr(III)-IIM and NIIM: 10 mL solutions (pH 6) with various concentrations of Cr(III) ions (from 5 to 35 mg/L) were added to the tested membrane and shaken for 20 h at temperature 40 ± 1 °C. The Cr concentrations were measured in the effluate solutions by ETAAS under optimized instrumental parameters. All the experiments were performed in triplicate, and the average value was used to calculate the maximum adsorption capacity of Cr(III)-IIM and NIIM ($Q_{max,exp}$) using the following expression:

$$Q_{max,exp} = \frac{(C_0 - C_e) \cdot V}{m} \tag{10}$$

where $Q_{max,exp}$ (mg/g) is the mass of Cr(III) ions adsorbed per unit mass of the membrane; V (L)—solution volume; m (g)—mass of the membrane; C_0 and C_e (mg/L)—initial and equilibrium concentrations of Cr(III) ions in the solution, respectively.

The sorption kinetics of Cr(III) was investigated using one Cr(III)-IIM in contact with 10 mL 5 mg/L Cr(III) standard solution at pH 6, placed in 15 mL centrifuge tubes on an electrical shaker at 150 rpm at 40 ± 1 °C. The sorption time was varied in the range of 1–24 h, and the residual Cr content in the effluate solutions was determined by ETAAS. Each experiment was repeated in triplicate. The amount of Cr(III) adsorbed at time t, q_t (mg/g), was calculated from Equation (11) by the difference between the initial chromium concentration in the solution (C_i, mg/L) at t = 0 and the residual chromium concentration at t adsorption time (C_t, mg/L):

$$q_t = \frac{(C_i - C_t) \cdot V}{m} \tag{11}$$

4.6. Interference Studies on the Selective Separation of Cr(III) and Cr(VI)

A standard solution containing 50 µg Cr(III) or Cr (VI) was added separately to each one of the 10 mL model solutions, containing 5% NaCl, 400 mg/L $SO_4{}^{2-}$, 400 mg/L $PO_4{}^{3-}$, 100 mg/L Fe(III), Cu(II), or Zn(II) at pH 6. The hydrogel Cr(III)-IIMs were immersed in these solutions and stirred with an electric shaker for 20 h at temperature of 40 ± 1 °C. The membrane is removed, and remaining solution is analyzed by ETAAS. Chromium(III) content was quantified by ETAAS after membrane elution with 0.1 mol/L NH_4-EDTA.

4.7. Analytical Procedure

A sample of surface water 20 mL was filtered through 45 µm membrane filter, and Cr(III)-IIM was immersed in this solution and stirred with an electric shaker for 20 h at temperature of 40 ± 1 °C. The supernatant solution is removed, and Cr(VI) is measured in this solution by ETAAS. In the case of very low concentrations of Cr (III), it might be eluted and also determined by ETAAS. The whole procedure could be performed during sampling—filtered sample is added to polypropylene vessel with inserted membrane. Supernatant after sorption is analyzed for Cr(VI) later in the laboratory.

Supplementary Materials: The following supporting information can be downloaded at: https://www.mdpi.com/article/10.3390/gels8110757/s1, Figure S1: (a) EDX mapping images and (b) EDX spectrum of Cr(III)-IIM after adsorption of Cr(III); Figure S2: FTIR spectra of Cr(III)-IIM and NIIM; Figure S3: Schematic presentation of the interactions between SA and Cr(III) ions; Figure S4: XRD patterns of (a) PVA/PEG/SA polymer membrane, NIIM, Cr(III)-IIM, SA-AuNPs (layer on glass slide); (b) Cr(III)-IIM, SA-AuNPs (layer on glass slide); Figure S5: Langmuir (a) and Freundlich (b) isotherms for adsorption of Cr(III) on the Cr(III)-IIM and NIIM; Figure S6: Adsorption kinetics of Cr(III) ions onto the Cr(III)-IIM at concentration 5 mg/L, pH 6, temperature 40 °C, and adsorbent dose (one membrane) 0.140 g: (a) pseudo-first order; (b) pseudo-second order; (c) Elovich, and (d) intra-particle diffusion model; Figure S7: Schematic representation of SA-AuNPs synthesis process; Table S1: Interference studies on the degree of sorption and the selectivity of Cr(III)-IIM hydrogel in the presence of different cations and anions in model solutions. (three parallel determinations); Table S2: Comparison of analytical figures of merit of analytical procedures using different sorbent materials for Cr(III)/Cr(VI) speciation.

Author Contributions: Conceptualization, I.D., P.V. and I.K.; methodology, I.D., P.V. and I.K.; investigation, I.D. and P.V.; writing—original draft preparation, I.D., P.V. and I.K.; writing—review and editing, I.D., P.V. and I.K.; data curation, I.D., P.V. and I.K.; project administration, I.D. All authors have read and agreed to the published version of the manuscript.

Funding: This research was funded by the Bulgarian Scientific Fund, Project, DN19/10, "Smart speciation".

Institutional Review Board Statement: Not applicable.

Informed Consent Statement: Not applicable.

Data Availability Statement: Not applicable.

Acknowledgments: The authors gratefully acknowledge the financial support from the Bulgarian National Science Fund (Grant DN19/10 "Smart speciation").

Conflicts of Interest: The authors declare no conflict of interest.

References

1. Costa, M.; Klein, C.B. Toxicity and carcinogenicity of chromium compounds in humans. *Crit. Rev. Toxicol.* **2006**, *36*, 155–163. [CrossRef] [PubMed]
2. Karimi-Maleh, H.; Ayati, A.; Ghanbari, S.; Orooji, Y.; Tanhaei, B.; Karimi, F.; Alizadeh, M.; Rouhi, J.; Fu, L.; Sillanpää, M. Recent advances in removal techniques of Cr(VI) toxic ion from aqueous solution: A comprehensive review. *J. Mol. Liq.* **2021**, *329*, 115062. [CrossRef]
3. Arain, M.B.; Ali, I.; Yilmaz, E.; Soylak, M. Nanomaterial's based chromium speciation in environmental samples: A review. *Trends Anal. Chem.* **2018**, *103*, 44–55. [CrossRef]

4. Filik, H.; Avan, A.A. Magnetic nanostructures for preconcentration, speciation and determination of chromium ions: A review. *Talanta* **2019**, *203*, 168–177. [CrossRef]
5. Herrero-Latorre, C.; Barciela-García, J.; García-Martin, S.; Peña-Crecente, R.M. Graphene and carbon nanotubes as solid phase extraction sorbets for the speciation of chromium: A review. *Anal. Chim. Acta* **2018**, *1002*, 1–17. [CrossRef] [PubMed]
6. Rakhunde, R.; Deshpande, L.; Juneja, H.D. Chemical speciation of chromium in water: A review. *Crit. Rev. Environ. Sci. Technol.* **2012**, *42*, 776–810. [CrossRef]
7. Trzonkowska, L.; Leśniewska, B.; Godlewska-Żyłkiewicz, B. Recent advances in on-line methods based on extraction for speciation analysis of chromium in environmental matrices. *Crit. Rev. Anal. Chem.* **2016**, *46*, 305–322. [CrossRef]
8. Vieira, M.A.; Grinberg, P.; Bobeda, C.R.; Reyes, M.N.; Campos, R.C. Non-chromatographic atomic spectrometric methods in speciation analysis: A review. *Spectrochim. Acta B At. Spectrosc.* **2009**, *64*, 459–476. [CrossRef]
9. Karadjova, I.; Yordanova, T.; Dakova, I.; Vasileva, P. Smart Materials in Speciation Analysis. In *Handbook of Smart Materials in Analytical Chemistry*, 1st ed.; de la Guardia, M., Esteve-Turrillas, F.A., Eds.; Wiley: New York, NY, USA, 2019; Volume 2, pp. 757–794.
10. Birlik, E.; Ersöz, A.; Açıkkalp, E.; Denizli, A.; Say, R. Cr(III)-imprinted polymeric beads: Sorption and preconcentration studies. *J. Hazard. Mater.* **2007**, *140*, 110–116. [CrossRef]
11. He, Q.; Chang, X.; Zheng, H.; Jiang, N.; Wang, X. Determination of chromium(III) and total chromium in natural waters using a surface ion imprinted silica gel as selective adsorbent. *Int. J. Environ. Anal. Chem.* **2008**, *88*, 373–384. [CrossRef]
12. Liu, Y.; Meng, X.; Han, J.; Liu, Z.; Meng, M.; Wang, Y.; Chen, R.; Tian, S. Speciation, adsorption and determination of chromium(III) and chromium(VI) on a mesoporous surface imprinted polymer adsorbent by combining inductively coupled plasma atomic emission spectrometry and UV spectrophotometry. *J. Sep. Sci.* **2013**, *36*, 3949–3957. [CrossRef]
13. Leśniewska, B.; Godlewska-Żyłkiewicz, B.; Wilczewska, A.Z. Separation and preconcentration of trace amounts of Cr(III) ions on ion-imprinted polymer for atomic absorption determinations in surface water and sewage samples. *Microchem. J.* **2012**, *105*, 88–93. [CrossRef]
14. Leśniewska, B.; Trzonkowska, L.; Zambrzycka, E.; Godlewska-Żyłkiewicz, B. Multi-commutation flow system with on-line solid phase extraction exploiting the ion-imprinted polymer and FAAS detection for chromium speciation analysis in sewage samples. *Anal. Methods* **2015**, *7*, 1517–1526. [CrossRef]
15. Zhang, N.; Suleiman, J.S.; He, M.; Hu, B. Chromium(III)-imprinted silica gel for speciation analysis of chromium in environmental water samples with ICP-MS detection. *Talanta* **2008**, *75*, 536–543. [CrossRef] [PubMed]
16. Yazdi, M.K.; Vatanpour, V.; Taghizadeh, A.; Taghizadeh, M.; Ganjali, M.R.; Munir, M.T.; Habibzadeh, S.; Saeb, M.R.; Ghaedi, M. Hydrogel membranes: A review. *Mater. Sci. Eng. C* **2020**, *114*, 111023. [CrossRef] [PubMed]
17. Lu, J.; Qin, Y.; Wu, Y.; Meng, M.; Yan, Y.; Li, C. Recent Advances in Ion-Imprinted Membranes: Separation and Detection via Ion-Selective Recognition. *Environ. Sci. Water Res. Technol.* **2019**, *5*, 1626–1653. [CrossRef]
18. Torres-Cartas, S.; Catalá-Icardo, M.; Meseguer-Lloret, S.; Simó-Alfonso, E.F.; Herrero-Martínez, J.M. Recent Advances in Molecularly Imprinted Membranes for Sample Treatment and Separation. *Separations* **2020**, *7*, 69. [CrossRef]
19. Chen, J.H.; Li, G.P.; Liu, Q.L.; Ni, J.C.; Wu, W.B.; Lin, J.M. Cr (III) ionic imprinted polyvinyl alcohol/sodium alginate (PVA/SA) porous composite membranes for selective adsorption of Cr (III) ions. *Chem. Eng. J.* **2010**, *165*, 465–473. [CrossRef]
20. Chen, J.H.; Xing, H.T.; Guo, H.X.; Li, G.P.; Weng, W.; Hu, S.R. Preparation, characterization and adsorption properties of a novel 3-aminopropyltriethoxysilane functionalized sodium alginate porous membrane adsorbent for Cr(III) ions. *J. Hazard. Mater.* **2013**, *248–249*, 285–294. [CrossRef]
21. Li, P.; Wang, X.; Wang, G.; Zhao, L.; Hong, Y.; Hu, X.; Zi, F.; Cheng, H. Synthesis and evaluation of ion-imprinted composite membranes of Cr(VI) based on β-diketone functional monomers. *RSC Adv.* **2021**, *11*, 38915–38924. [CrossRef]
22. Liu, Y.; Hu, D.; Hu, X.; Chen, S.; Zhao, L.; Chen, Y.; Yang, P.; Qin, X.; Cheng, H.; Zi, F. Preparation and Characterization of Chromium(VI) Ion-Imprinted Composite Membranes with a Specifically Designed Functional Monomer. *Anal. Lett.* **2020**, *53*, 1113–1139. [CrossRef]
23. Kumar, A.; Sood, A.; Han, S.S. Poly(vinyl alcohol)-alginate as potential matrix for various applications: A focused review. *Carbohydr. Polym.* **2022**, *277*, 118881. [CrossRef] [PubMed]
24. Daniel, F.B.; Robinson, M.; Olson, G.R.; Page, N.P. Toxicity Studies of Epichlorohydrin in Sprague-Dawley Rats. *Drug Chem. Toxicol.* **1996**, *19*, 41–58. [CrossRef]
25. Smith, D.R.; Wang, R.-S. Glutaraldehyde Exposure and its Occupational Impact in the Health Care Environment. *Environ. Health Prev. Med.* **2006**, *11*, 3–10. [CrossRef] [PubMed]
26. Zhou, Z.; Cui, K.; Mao, Y.; Chai, W.; Wang, N.; Ren, Z. Green preparation of D-tryptophan imprinted selfsupported membrane for ultrahigh enantioseparation of racemic tryptophan. *RSC Adv.* **2016**, *6*, 109992. [CrossRef]
27. Sani, A.; Cao, C.; Cui, D. Toxicity of gold nanoparticles (AuNPs): A review. *Biochem. Biophys. Rep.* **2021**, *26*, 100991. [CrossRef] [PubMed]
28. Thoniyot, P.; Tan, M.J.; Karim, A.A.; Young, D.J.; Loh, X.J. Nanoparticle–hydrogel composites: Concept, design, and applications of these promising, multi-functional materials. *Adv. Sci.* **2015**, *2*, 1400010. [CrossRef]
29. Ouardi, Y.E.; Giove, A.; Laatikainen, M.; Branger, C.; Laatikainen, K. Benefit of ion imprinting technique in solid-phase extraction of heavy metals, special focus on the last decade. *J. Environ. Chem. Eng.* **2021**, *9*, 106548. [CrossRef]

30. Borse, S.D.; Joshi, S.S. Optical and Structural Properties of PVA Capped Gold Nanoparticles and Their Antibacterial Efficacy. *Adv. Chem. Lett.* **2013**, *1*, 15–23. [CrossRef]
31. Zou, Z.; Zhang, B.; Nie, X.; Cheng, Y.; Hu, Z.; Liao, M.; Li, S. A sodium alginate-based sustained-release IPN hydrogel and its applications. *RSC Adv.* **2020**, *10*, 39722. [CrossRef] [PubMed]
32. Hassan, R.M.; Takagi, H.D. Degradation Kinetics of Some Coordination Biopolymers of Transition Metal Complexes of Alginates: Influence of Geometrical Structure and Strength of Chelation on the Thermal Stability. *Mater. Sci.* **2019**, *1*, 3. [CrossRef]
33. Zahakifar, F.; Keshtkar, A.R.; Talebi, M. Synthesis of sodium alginate (SA)/polyvinyl alcohol (PVA)/polyethylene oxide (PEO)/ZSM-5 zeolite hybrid nanostructure adsorbent by casting method for uranium (VI) adsorption from aqueous solutions. *Prog. Nucl. Energy* **2021**, *134*, 103642. [CrossRef]
34. Li, T.-T.; Yan, M.; Xu, W.; Shiu, B.-C.; Lou, C.-W.; Lin, J.-H. Mass-Production and Characterizations of Polyvinyl Alcohol/Sodium Alginate/Graphene Porous Nanofiber Membranes Using Needleless Dynamic Linear Electrospinning. *Polymers* **2018**, *10*, 1167. [CrossRef] [PubMed]
35. Sahu, D.; Sarkar, N.; Sahoo, G.; Mohapatra, P.; Swain, S.K. Nano silver imprinted polyvinyl alcohol nanocomposite thin films for Hg^{2+} sensor. *Sens. Actuators B Chem.* **2017**, *246*, 96–107. [CrossRef]
36. Park, J.H.; Karim, M.R.; Kim, I.K.; Cheong, I.W.; Kim, J.W.; Bae, D.G.; Cho, J.W.; Yeum, J.H. Electrospinning fabrication and characterization of poly (vinylalcohol)/montmorillonite/silver hybrid nanofibers for antibacterialapplications. *Colloid Polym. Sci.* **2010**, *288*, 115–121. [CrossRef]
37. Ahmed, H.B.; Abdel-Mohsen, A.M.; Emam, H.E. Green-assisted tool for nanogold synthesis based on alginate as a biological macromolecule. *RSC Adv.* **2016**, *6*, 73974–73985. [CrossRef]
38. Di Bello, M.P.; Lazzoi, M.R.; Mele, G.; Scorrano, S.; Mergola, L.; Del Sole, R. A new ion-imprinted chitosan-based membrane with an azo-derivative ligand for the efficient removal of Pd(II). *Materials* **2017**, *10*, 1133. [CrossRef] [PubMed]
39. Bertoni, F.A.; Bellú, S.E.; González, J.C.; Sala, L.F. Reduction of hypervalent chromium in acidic media by alginic acid. *Carbohydr. Polym.* **2014**, *114*, 1–11. [CrossRef] [PubMed]
40. Almeida, J.C.; Cardoso, C.E.D.; Tavares, D.S.; Freitas, R.; Trindade, T.; Vale, C.; Pereira, E. Chromium removal from contaminated waters using nanomaterials—A review. *Trends Anal. Chem.* **2019**, *118*, 277–291. [CrossRef]
41. Foo, K.Y.; Hameed, B.H. Insights into the modeling of adsorption isotherm systems. *Chem. Eng. J.* **2010**, *156*, 2–10. [CrossRef]
42. Al-Ghouti, M.A.; Da'ana, D.A. Guidelines for the use and interpretation of adsorption isotherm models: A review. *J. Hazard. Mater.* **2020**, *393*, 122383. [CrossRef] [PubMed]
43. Gao, X.; Guo, C.; Hao, J.; Zhao, Z.; Long, H.; Li, M. Adsorption of heavy metal ions by sodium alginate based adsorbent—A review and new perspectives. *Int. J. Biol. Macromol.* **2020**, *164*, 4423–4434. [CrossRef] [PubMed]
44. Tang, Q.; Li, N.; Lu, Q.; Wang, X.; Zhu, Y. Study on Preparation and Separation and Adsorption Performance of Knitted Tube Composite β-Cyclodextrin/Chitosan Porous Membrane. *Polymers* **2019**, *11*, 1737. [CrossRef]
45. Yang, H.; Liu, H.B.; Tang, Z.S.; Qiu, Z.D.; Zhu, H.X.; Song, Z.X.; Jia, A.L. Synthesis, performance, and application of molecularly imprinted membranes: A review. *J. Environ. Chem. Eng.* **2021**, *9*, 106352. [CrossRef]
46. Salehi, E.; Madaeni, S.S.; Vatanpour, V. Thermodynamic investigation and mathematical modeling of ion-imprinted membrane adsorption. *J. Membr. Sci.* **2012**, *389*, 334–342. [CrossRef]
47. Chen, N.; Zhang, Z.; Feng, C.; Zhu, D.; Yang, Y.; Sugiura, N. Preparation and characterization of porous granular ceramic containing dispersed aluminum and iron oxides as adsorbents for fluoride removal from aqueous solution. *J. Hazard. Mater.* **2011**, *186*, 863–868. [CrossRef] [PubMed]
48. Du, Q.J.; Sun, J.K.; Li, H. Highly enhanced adsorption of congo red onto graphene oxide/chitosan fibers by wet-chemical etching off silica nanoparticles. *Chem. Eng. J.* **2014**, *245*, 99–106. [CrossRef]
49. Tan, K.L.; Hameed, B.H. Insight into the adsorption kinetics models for the removal of contaminants from aqueous solutions. *J. Taiwan Inst. Chem. Eng.* **2017**, *74*, 25–48. [CrossRef]
50. Djerahov, L.; Vasileva, P.; Karadjova, I. Self-standing chitosan film loaded with silver nanoparticles as a tool for selective determination of Cr(VI) by ICP-MS. *Microchem. J.* **2016**, *129*, 23–28. [CrossRef]

Article

Design of Economical and Achievable Aluminum Carbon Composite Aerogel for Efficient Thermal Protection of Aerospace

Yumei Lv [1], Fei He [1,*], Wei Dai [2], Yulong Ma [1], Taolue Liu [1], Yifei Liu [1] and Jianhua Wang [1]

[1] CAS Key Laboratory of Mechanical Behavior and Design of Materials, Department of Thermal Science and Energy Engineering, University of Science and Technology of China, Hefei 230027, China
[2] Beijing Institute of Astronautical Systems Engineering, No.1 South Dahongmen Road, Fengtai District, Beijing 100071, China
* Correspondence: hefeihe@ustc.edu.cn; Tel.: +86-551-6360-0945

Citation: Lv, Y.; He, F.; Dai, W.; Ma, Y.; Liu, T.; Liu, Y.; Wang, J. Design of Economical and Achievable Aluminum Carbon Composite Aerogel for Efficient Thermal Protection of Aerospace. *Gels* 2022, 8, 509. https://doi.org/10.3390/gels8080509

Academic Editors: Esmaiel Jabbari, Daxin Liang, Ting Dong, Yudong Li and Caichao Wan

Received: 27 July 2022
Accepted: 13 August 2022
Published: 17 August 2022

Publisher's Note: MDPI stays neutral with regard to jurisdictional claims in published maps and institutional affiliations.

Copyright: © 2022 by the authors. Licensee MDPI, Basel, Switzerland. This article is an open access article distributed under the terms and conditions of the Creative Commons Attribution (CC BY) license (https://creativecommons.org/licenses/by/4.0/).

Abstract: Insulation materials play an extremely important role in the thermal protection of aerospace vehicles. Here, aluminum carbon aerogels (AlCAs) are designed for the thermal protection of aerospace. Taking AlCA with a carbonization temperature of 800 °C (AlCA–800) as an example, scanning electron microscopy (SEM) images show an integrated three-dimensional porous frame structure in AlCA–800. In addition, the thermogravimetric test (TGA) reveals that the weight loss of AlCA–800 is only ca. 10%, confirming its desirable thermal stability. Moreover, the thermal conductivity of AlCA–800 ranges from 0.018 W m^{-1} K^{-1} to 0.041 W m^{-1} K^{-1}, revealing an enormous potential for heat insulation applications. In addition, ANSYS numerical simulations are carried out on a composite structure to forecast the thermal protection ability of AlCA–800 acting as a thermal protection layer. The results uncover that the thermal protective performance of the AlCA–800 layer is outstanding, causing a 1185 K temperature drop of the structure surface that is exposed to a heat environment for ten minutes. Briefly, this work unveils a rational fabrication of the aluminum carbon composite aerogel and paves a new way for the efficient thermal protection materials of aerospace via the simple and economical design of the aluminum carbon aerogels under the guidance of ANSYS numerical simulation.

Keywords: aluminum carbon composite aerogel; simple fabrication method; thermal stability; thermal conductivity; heat insulation; numerical simulation

1. Introduction

With the increasing speed and flight time of aerospace vehicles, the high heat fluxes arising from aerodynamics and combustion have created a growing demand for insulation materials to protect the key components of aircrafts [1,2]. However, the heat resistance, thermal stability, heat insulation ability and weight of current insulation materials are still unable to satisfy the growing needs of aerospace thermal protection [3]. Therefore, it is necessary to explore novel materials with extremely light weight and outstanding thermal insulation performance at ultrahigh temperatures [4].

A tremendous amount of research is currently focused on carbon-based materials, which have the ability to withstand ultrahigh temperatures up to 3000 °C [5], making them the most promising candidates for lightweight aerospace materials [6]. Carbon aerogel (CA) possesses an abundant mesoporous and superimposed nanoparticle network, which confers it various unique properties, such as high specific surface area and extremely low density [7]. Besides, the carbon skeleton of the carbon aerogel has ultralow thermal conductivity [8], and the heat transfer through the gases is suppressed due to the nanoscale pores in the carbon aerogel [9]. Compared with other aerogels, the higher specific extinction coefficient of the carbon aerogel could significantly reduce the radiation heat transfer between it and the high temperature environment [10]. Overall, the carbon aerogels provide enormous potential for acting as a barrier in convection, conduction and radiation heat transfer. Furthermore, carbon aerogels are able to keep their mesoporous structure in an inert atmosphere above 2000 °C [11], so it appears to be one of the high temperature insulating materials with excellent thermal stability. Hence, developing the carbon aerogel is an effective way to address the problem of thermal protection under high temperatures [12].

Traditional carbon aerogels are fabricated by chemical organic precursors, such as hydroxybenzene, aldehyde, polyimide and polyimide, etc., which does not conform to the concept of green production [13]. Besides, the surface tension of the carbon aerogels is large during the preparation; thus, the pores easily collapse in the drying process [14]. There will also be the obvious expansion or contraction of the carbon aerogel in the carbonization process, which may lead to cracks in the aerogel [15]. Therefore, traditional carbon aerogel fabrication usually requires a complicated solution exchange process and high-cost supercritical drying process to reduce the surface tension and prevent the pore collapse [16], severely limiting the large-scale production and practical applications. To overcome this issue, researchers try to strengthen the gel skeleton with carbon fibers, carbon nanotubes and ceramic fibers as a remedy of those hindrances [17]. Unfortunately, the strong shrinkage mismatch between different structures will generate internal tensile stress [11,18], which still inevitably results in microcracks. Meanwhile, the addition of some fibers may lead to an increase in the thermal conductivity of the carbon aerogel [19], causing a contradiction between the heat insulation ability and mechanical properties. Hence, under the purpose of widely employing the carbon aerogel as thermal protection materials for aerospace applications, the raw materials with low cost, large-scale production and simple fabrication methods should be adopted in the design of the carbon aerogel [20,21]. Moreover, some materials ought to be added to address the problems of cracks and collapse in the production process of carbon aerogel, and at the same time the addition of these materials should not bring about other problems [22].

Starch is an abundant resource in the leaves and seeds, which could be widely produced from the nature [23]. In addition, the starch easily forms colloidal solutions at high temperatures under the high temperature [24]. Subsequently, hydrogels can be converted from the colloidal solution by reforming the hydrogen bonds during the decrease in temperature [25]. Hence, the starch exhibits prominent advantages as the raw material to fabricate the carbon aerogels, since it does not require a complex pre-treatment or additional post-processing step [26,27]. Moreover, the introduction of aluminum ions in the process of the fabrication of carbon aerogel has been proposed as the way to promote the polymerization of organic precursors [28]. Meanwhile, aluminum oxide has a certain effect on restraining the expansion or contraction of the carbon aerogel in the carbonization process [29]. Hence, the aluminum could promote the development of a three-dimensional porous frame structure in the carbon aerogel.

Herein, carbon aerogels are designed and fabricated using starch as a raw material for availability and economy. Meanwhile, aluminum ions are anchored on the carbon skeleton to improve the cracking during the process of drying or the carbonization process and ensure the integrity of the pores of the carbon aerogel. Taking an aluminum carbon composite aerogel (AlCA) with a carbonization temperature of 800 °C (AlCA–800) as an example, the scanning electron microscopy (SEM) image displays an unbroken three-dimensional porous frame structure in AlCA–800, which demonstrates that the aerogel is successfully prepared. In addition, elemental mapping images (EDS) exhibit a homogeneous distribution of C and Al_2O_3 in AlCA–800. Furthermore, the thermogravimetric test (TGA) indicates that the weight loss of AlCA–800 is only ca. 10%, further confirming its thermal stability. Moreover, the thermal conductivity of AlCA–800 ranges from 0.018 W m^{-1} K^{-1} to 0.041 W m^{-1} K^{-1}, revealing a superior heat insulation ability. In addition, ANSYS numerical simulations are carried out on a plate protected with a thermal protection layer made of AlCA–800. The results demonstrate that the thermal protection performance of AlCA–800 layer is desirable, causing a 1185 K temperature drop to the plate surface that was exposed to heat environment for ten minutes. Overall, this work unveils a rational fabrication of the aluminum carbon composite aerogel and paves a way for thermal protection materials with light weight and low thermal conductivity for aerospace applications. Meanwhile, the ANSYS numerical simulation is brought into the design of the materials for predicting the effect of thermal protection in practical applications, which makes the work more reliable and economical.

2. Results and Discussion

2.1. The Structure and Thermochemical Property of Carbon Aerogels

In this work, carbon aerogels (CA) and aluminum carbon composite aerogels (AlCA) were designed and fabricated as effective thermal protection material for aerospace applications, and by altering the carbonization temperature and time, the most suitable aerogel was finally selected. Scanning electron microscopy (SEM) images in Figure 1 revealed the three-dimensional porous frame structure of the CAs and AlCAs. One can observe that the pore structures of CAs and AlCAs can be readily adjusted by increasing the carbonization temperature and time. In addition, the integrity of micropores in CAs was not a patch on AlCAs due to the structural shrinkage and cracking during the carbonization process [30]. Among AlCAs, the AlCA–800 has the most excellent three-dimensional porous frame structure, while some holes were sightless and not fully formed in the AlCA–600 due to insufficient carbonization temperature and time [31]. Despite the slight cracks and fractures, AlCA–1000 still retained the three-dimensional porous frame, which indicated the thermal stability of the carbon skeleton. The BET surface, adsorption average pore diameter and quantity adsorbed were displayed in Table 1, further confirming the three-dimensional porous framework in the CAs and AlCAs [32]. Meanwhile, the quantity adsorbed of AlCA–800 was largest, corresponding to the result of SEM images.

To study the elemental composition of the samples, the X-ray diffraction (XRD) pattern was employed for the characterization of the CAs and AlCAs. As shown in Figure 2a, it can be observed that the XRD peak positions of the CAs and AlCAs were consistent but the peak intensities differed greatly, indicating that the substrate of aerogels was graphite [33] (JCPDS card No. 87-0722). However, the XRD peak of the aluminum was not detected due to the small quantity of aluminum. In addition, the Raman spectra of CA–600 and AlCA–600 in Figure 2b demonstrated that there were two characteristic peaks at wavelength 1347 cm^{-1} and 1585 cm^{-1}, which could be assigned to graphite D peak and G peak, respectively [34]. The peaks of CA–600 did not demonstrate a detectable difference from those of AlCA–600 due to the weak interaction between the lower temperature of Al and C, corresponding to the result of the XRD patterns (Figure 2a). In addition, Figure 2c revealed the discrepancy of the D peak in CA–800 and AlCA–800, which could be ascribed to the interaction of the Al and C and the decrease in the graphitization degree in AlCA–800 [34,35]. Moreover, as discovered in Figure 2d, the difference appeared in the CA–1000 and AlCA–1000 due to the same reason. In addition, the elemental mapping images (EDS) in Figure 2e further revealed the carbon skeleton in CA–800. A homogeneous distribution of C and Al_2O_3 in AlCA–800 was depicted in Figure 2f, confirming that the aluminum was successfully anchored on carbon. As mentioned above, all the results suggested that the CAs and AlCAs were successfully synthesized through the simple sol-gel method.

Figure 1. SEM images for (**a**) CA–600, (**b**) CA–800, (**c**) CA–1000, (**d**) AlCA–600, (**e**) AlCA–800 and (**f**) AlCA–1000.

Table 1. The BET parameters of CAs and AlCAs.

as	BET Surface /$m^2\ g^{-1}$	Adsorption Average Pore Diameter /nm	Quantity Adsorbed /($cm^3\ g^{-1}\ STP^{-1}$)
AlCA–600	291.9096	2.9517	150.4746
AlCA–800	356.1491	4.9803	286.6782
AlCA–1000	292.8679	3.0761	160.2679
CA–600	306.9739	2.6727	148.4311
CA–800	287.8915	3.3322	170.8282
CA–1000	300.4066	2.4997	133.4855

Thermogravimetric tests (TGA) were carried out to reveal the thermal stability of CAs and AlCAs. As shown in Figure 3, in the heating process from 30 °C to 300 °C, the removal of absorbed H_2O trapped in the samples led to a significant weight loss [36]. Then, there was a slight weight loss of the samples from 300 °C to 600 °C, which was mainly because of the oxidation of a small amount of organic residues [37]. After this stage, a significant weight loss of the carbon CAs and AlCAs occurred in the heating process from 600 °C to 1000 °C, which was attributed to the loss of carbonaceous residuals during depolymerization and decomposition [38]. Based on the TGA curves, the thermal stability of AlCAs was superior to the CAs, which indicated that the Al_2O_3 could promote the thermal stability of the aerogel [39]. In addition, the weight loss of AlCA–800 was only ca. 10%, and its thermal stability was better than that of other AlCAs. Moreover, the AlCA–1000 exhibited a favorable thermal stability, which meant that the AlCAs could be employed under higher temperatures. Nevertheless, due to the deficiency of carbonization temperature and time, there was not a complete three-dimensional porous frame structure in AlCA–600, and thus the thermal stability of AlCA–600 was poorer. Overall, the AlCAs had higher thermal stability; in particular, the AlCA–800 possessed excellent thermal stability, demonstrating a great potential to serve as a kind of efficient thermal protective material.

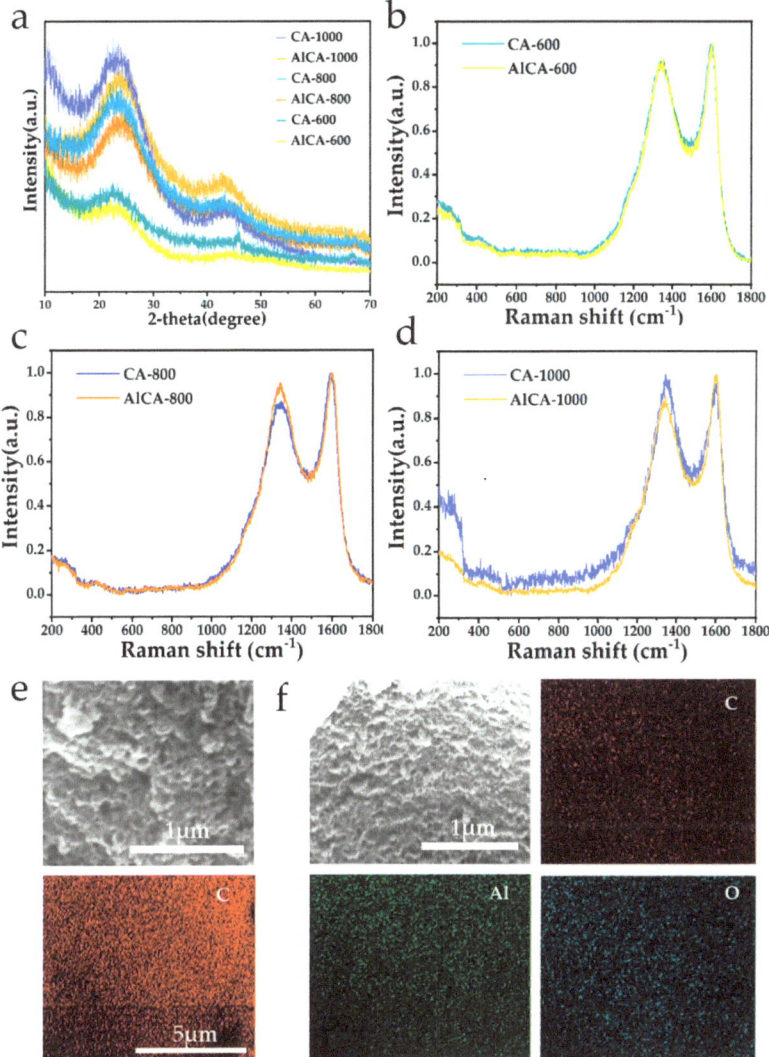

Figure 2. (a) XRD pattern for CAs and AlCAs, Raman spectra for (b) CA–600 and AlCA–600, (c) CA–800 and AlCA–800, (d) CA–1000 and AlCA–1000, EDS mapping for (e) CA–800 and (f) AlCA–800.

To investigate the thermal protection ability of the AlCAs, the thermal conductivities at different temperatures were tested. The density of AlCAs was displayed in Table 2. As shown in Figure 4, the thermal conductivity of AlCA–800 was the lowest, ranging from 0.018 W m^{-1} K^{-1} to 0.041 W m^{-1} K^{-1}, which could be attributed to the integrated three-dimensional porous frame structure. In addition, the AlCA–1000 demonstrated higher thermal conductivity at low temperatures, but the thermal conductivity presented a slow upward trend as the temperature increases, revealing a fine heat insulation ability under higher temperature. Due to the incomplete three-dimensional network structure, the AlCA–600 always demonstrated a relatively higher thermal conductivity. To sum up, the

AlCA–800 possessed desirable thermal stability and heat insulation capacity simultaneously, and hence AlCA–800 could be adopted in the thermal protection of aerospace.

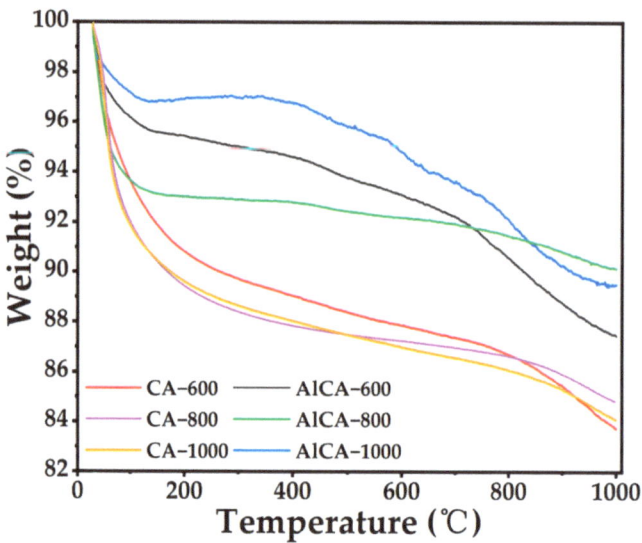

Figure 3. TGA curves of aerogels.

Table 2. Density of AlCAs.

Aerogels	Thickness /mm	Diameter /mm	Weight /g	Density /g cm^{-3}
AlCA–600	0.875	10	0.0157	0.229
AlCA–800	1.78	10.14	0.0401	0.279
AlCA–1000	0.519	10	0.0119	0.294

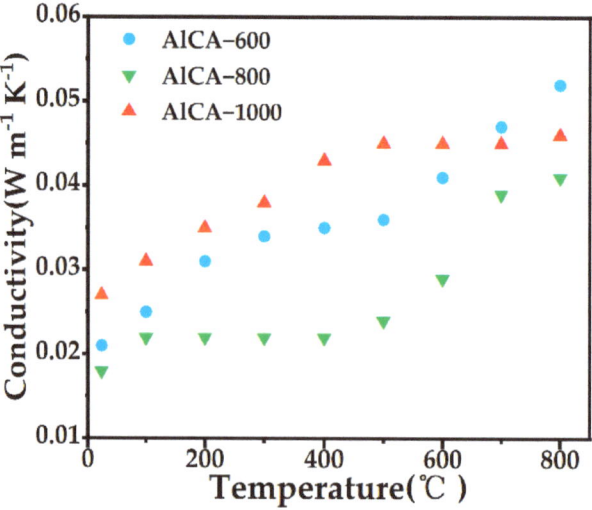

Figure 4. Thermal conductivity of AlCAs.

2.2. The Thermal Protection Performance of Carbon Aerogel

To explore the thermal protection performance of AlCA–800 as the thermal protection layer, the numerical simulations were performed on a plate with the thermal protection layer (combined structure) under the practical working condition of the scramjet engine (Figure 5a). Meanwhile, the plate without a thermal protection layer (single structure) was also numerical simulated for comparison. The combined structure of the plate to be protected and the thermal protection layer made of AlCA–800 were illustrated in Figure 5a. The plate was initially placed at the ambient environment. Then, its upper surface was exposed to a mainstream with 3000 K high temperature, and its lower surface was cooled by the air with 300 K. The thickness of the plate was 2 mm, while the thickness of carbon aerogel thermal protection layer was 0.5 mm. The TC4 titanium alloy with the thermal conductivity of 7.4 W m^{-1} K^{-1} was adopted as the material of the plate. For the thermal protection layer, the material parameters were customized using the experimental values of AICA–800. In order to reduce the computational region and thereby save computing resource, the periodic boundary conditions were employed on both sides. The numerical simulation was carried out through ANSYS Fluent, and the plate and the hot mainstream were coupled and solved with a transient state pressure-based solver. A second order upwind scheme was employed for the discretization. When the residual of continuity, mass, energy and momentum equations were all lower than 10^{-6}, the numerical calculations were considered convergent.

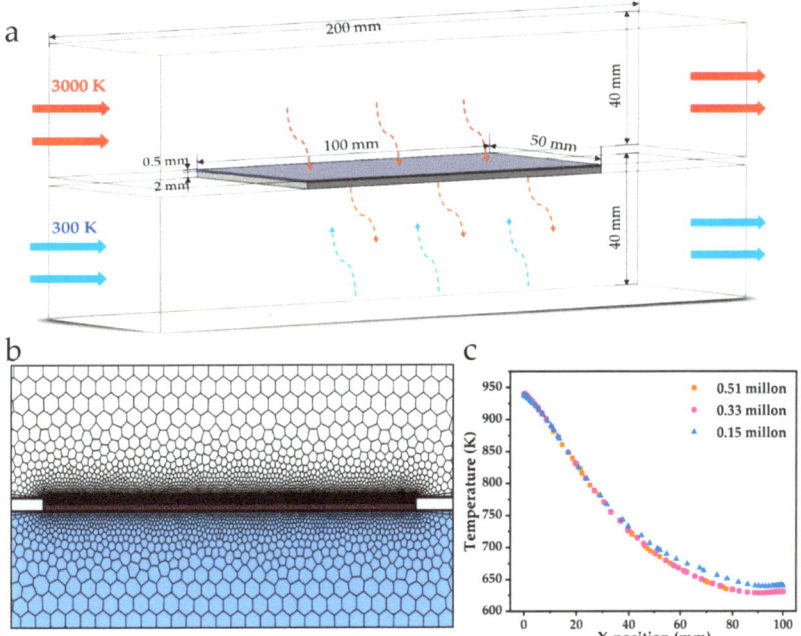

Figure 5. (**a**) Calculation domain and boundary conditions of structure; (**b**) unstructured meshes; (**c**) mesh independence verification.

Table 3 displayed the mathematical models describing the heat transfer between the high temperature mainstream and the structure. The fluid flow was calculated by the Menter's two-equation SST k-ω model [40], of which the veracity was confirmed by many practical applications. Meanwhile, the heat conduction appeared in the thermal protection layer and the lower plate, and thus the Fourier's law was used to describe the energy balance. Mainstream and cooling air were regarded as ideal gases, and the corresponding viscosity was determined by Sutherland's formula. Besides, the NIST could serve as the reasonable method to calculate other gas property parameters and fit them with a polynomial formula.

Table 3. Mathematical models.

Computational Domain	Conservation Equation
Mainstream	The continuity equation: $\frac{\partial \rho}{\partial t} + \frac{\partial}{\partial x_i}(\rho u_i) = S_m$ The momentum equation: $\frac{\partial}{\partial t}(\rho u_i) + \frac{\partial}{\partial x_i}(\rho u_i u_j) = -\frac{\partial p}{\partial x_i} + -\frac{\partial \tau}{\partial c_j} + \rho g_i + F_i$ The SST k-ω turbulence model was used to solve the Reynolds stress term The energy equation: $\frac{\partial}{\partial t}(\rho E) + \frac{\partial}{\partial x_i}(u_i(\rho E + p)) = \frac{\partial}{\partial x_i}\left(k_f \frac{\partial T}{\partial x_i}\right) + \sum_j h_j J_j + u_j(\tau_{ij})_f + S_h$
Solid wall	Fourier's law of heat conduction: $\frac{\partial}{\partial t}\rho h + \frac{\partial}{\partial x_i}(u_i \rho h) = \frac{\partial}{\partial x_i}\left(k_s \frac{\partial T}{\partial x_i}\right)$
Thermodynamic model	Ideal gas law: $P = \rho R_g T$ Sutherland formula: $\frac{\mu}{\mu_0} = \left(\frac{T}{T_0}\right)^{1.5} \frac{T_0 + T_s}{T + T_s}$

Figure 5b showed the unstructured grid generated in the local calculation region. The grid near the wall was refined due to strong energy exchanges in the fluid near the wall. Besides, the height of the first layer of the grids was set as 1×10^{-5} m to ensure the dimensionless wall distance y+ was less than 1, and the grid growth rate was set to 1.2 in this paper. The grid establishment method and distribution directly affected the accuracy of the results, and hence the grid independence of the numerical results was tested through three grid strategies, respectively. The temperature on the centerline of the plate after heating for 100 s was selected for monitoring and comparison. As shown in Figure 5c, the results obtained by the grid sizes of 0.33 million and 0.51 million were similar. Considering the accuracy of the calculation results and the calculation load, a grid size with 0.33 million was adopted in the following calculations.

The temperature distributions on the lengthwise section at the centerline of the two structures were shown in Figure 6a, the heat transfer from the high temperature mainstream to the plate was greatly hindered in the combined structure due to the extremely low thermal conductivity of the AlCA–800. Thus, there was a large temperature drop between the thermal protection layer and the plate. Figure 6b displayed the contours of temperature on the surface of the plates in two structures after 60 s of heating; it could be obviously observed that the surface temperature of the plate in the combined structure was lower, and the temperature distributed more uniformly. Besides, Figure 6c showed the variation of the temperature on the plate surface with heating time, one could see that the rate of the temperature rise slowed down for both structures, while the temperature differences between the two structures were becoming more and more significant. Notably, the temperature on the plate surface of the combined structure was only 1268 K after heating for 10 min under the protection of thermal protection layer, whereas that of the single structure bore the high temperature up to 2453 K, indicating the excellent heat insulation ability of AlCA–800 thermal protection layer. Therefore, the AlCA–800 could be employed as a competitive material for the thermal protection layer, and provided efficient thermal protection for the aerospace vehicles.

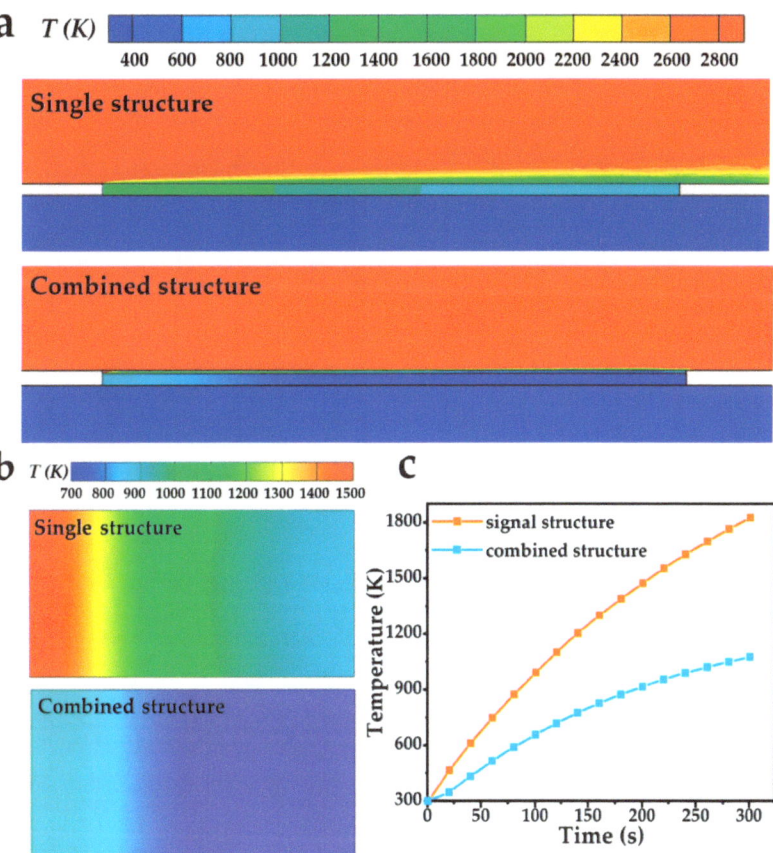

Figure 6. (a) Temperature distribution on the lengthwise section at the centerline of two structures; (b) temperature distribution on the surface of the plates; (c) temperature variation of the plate surfaces.

3. Conclusions

In summary, the aluminum carbon composite aerogels (AlCAs) were designed for realizing the effective thermal protection of aerospace vehicles. The AlCAs were fabricated with the starch, which had a large quantity and low price, being raw, and the preparation process was simple and achievable, guaranteeing the economy of the materials. In addition, the carbon aerogels (CAs) were also fabricated for comparison. The SEM images demonstrated a tightly three-dimensional porous frame structure in the AlCAs, which indicated the successful synthesis of the aerogel. Notably, the AlCA–800 possessed the most desirable three-dimensional porous frame structure due to appropriate carbonization temperature and time. Elemental mapping (EDS) images unveiled a homogeneous distribution of C and Al_2O_3 in AlCAs. Moreover, the TGA revealed the lower weight loss of AlCAs than CAs, which could be attribute to the addition of Al. In particular, the weight loss of the AlCA–800 was only ca. 10%, powerfully confirming its thermal stability. Importantly, the thermal conductivity of the AlCA–800 ranged from 0.018 W m^{-1} K^{-1} to 0.041 W m^{-1} K^{-1}, which was far below the existing thermal protection materials, meaning that the AlCA–800 had an outstanding heat insulation ability under high temperature. Furthermore, the numerical simulation was carried out on the plate with thermal protection layer made of AlCA–800, aiming at evaluating the thermal protection performance of the AlCA–800. The results uncovered that the thermal protective performance of the AlCA–800 layer was extraordinary, causing a 1185 K temperature drop to the plate surface, which was exposed to a heat environment for ten minutes. Consequently, this work not only paved a way for the simple and low-cost fabrication of thermal protection materials with light weight and low thermal conductivity, but also brought ANSYS numerical simulation for predicting its protection performance in practical application situation, which made the work more reliable and economical.

4. Materials and Methods

4.1. Preparation of Aluminum Carbon Composite Aerogel

Aluminium chloride ($AlCl_3$) and soluble starch were obtained from Sinopharm Chem. Reagent Co. Ltd. (Shanghai, China).

Taking carbon aerogel with a heating temperature of 800 °C as an example, the preparation process was as follows: Initially, 50 mL ultrapure water was heated to 80 °C and the 15 g/20 mL starch aqueous solution was added to the water under vigorous stirring for 5 min, and the hydrogels were obtained. Afterwards, the aerogels were obtained after freeze drying. Finally, the aerogels were carbonized at 800 °C for 4 h with a ramping rate of 2 °C min^{-1} under an Ar atmosphere.

Taking aluminum carbon composite aerogel with heating temperature 800 °C as an example, the preparation process was as follows: Initially, 10 mmol $AlCl_3$ was dissolved in the 50 mL ultrapure water, the suspension was heated to 80 °C under vigorous stirring and the hydrosols were obtained. Then, the 15 g/20 mL starch aqueous solution was added to the above suspension under vigorous stirring for 5 min, and the hydrogels were obtained. Afterwards, the aerogels were obtained after freeze drying. Finally, the aerogels were carbonized at 800 °C for 4 h with a ramping rate of 2 °C min^{-1} under the Ar atmosphere.

4.2. Characterization of Aluminum Carbon Composite Aerogel

SEM images were measured on a FEI Sirion-200 (FEI NanoPorts, Hillsboro, OR, USA). XRD patterns were performed on a Rigaku D/MAX-TTRIII diffractometer (Rigaku Corporatio, Tokyo, Japan) with Cu Kα radiation (λ = 1.54178 Å). Raman spectra was acquired by a JY LabRamHR Evolution (HORIBA Jobin Yvon, Palaiseau, France) with a 532 nm laser. SDT Q600 (TA Instruments, New Castle, DE, USA) thermal analyzer was employed to acquire TGA curves under nitrogen atmosphere. Thermal conductivity was determined by the NETZSCH LFA457 thermal analyzer (NETZSCH, Selb, Germany). BET surface area was acquired by automatic microporous gas adsorption analyzer system on ASAP 2020 PLUS (Micromeritics, Norcross, GA, USA).

Author Contributions: Conceptualization, F.H. and Y.L. (Yumei Lv); methodology, Y.L. (Yifei Liu); software, T.L.; validation, J.W., F.H.; investigation, Y.M.; resources, W.D.; data curation, Y.M.; writing—original draft preparation, Y.L. (Yumei Lv); writing—review and editing, F.H. All authors have read and agreed to the published version of the manuscript.

Funding: This research is funded by the National Natural Science Foundation of China, and the funding number is 51806206.

Institutional Review Board Statement: Not applicable.

Informed Consent Statement: Not applicable.

Data Availability Statement: The data that support the findings of this study are available from the corresponding author upon reasonable request.

Acknowledgments: This work is supported by the Supercomputing Center in the University of Science and Technology of China.

Conflicts of Interest: The authors declare no conflict of interest.

Nomenclature

k	Conductivity [$W \cdot m^{-1} \cdot K^{-1}$]
T	Temperature [K]
u	Velocity [$m \cdot s^{-1}$]
P	Pressure [Pa]
h	Coefficient of heat transfer [$W \cdot m^{-2} \cdot K^{-1}$]
R_g	Universal or ideal gas constant [$J \cdot kg^{-1} \cdot K^{-1}$]
Greek	
μ	Dynamic viscosity [$N \cdot s \cdot m^{-2}$]
ρ	Fluid density [$kg \cdot m^{-3}$]
τ	Shear stress
Subscripts	
s	Solid
f	Fluid

References

1. Padture, N.P. Advanced structural ceramics in aerospace propulsion. *Nat. Mater.* **2016**, *15*, 804–809. [CrossRef] [PubMed]
2. Uyanna, O.; Najafi, H. Thermal protection systems for space vehicles: A review on technology development, current challenges and future prospects. *Acta Astronaut.* **2020**, *176*, 341–356. [CrossRef]
3. Poloni, E.; Bouville, F.; Schmid, A.L.; Pelissari, P.I.; Pandolfelli, V.C.; Sousa, M.L.; Tervoort, E.; Christidis, G.; Shklover, V.; Leuthold, J. Carbon ablators with porosity tailored for aerospace thermal protection during atmospheric re-entry. *Carbon* **2022**, *195*, 80–91. [CrossRef]
4. Sanoj, P.; Kandasubramanian, B. Hybrid carbon-carbon ablative composites for thermal protection in aerospace. *J. Compos.* **2014**, *2014*, 825607. [CrossRef]
5. Li, J.; Guo, P.; Hu, C.; Pang, S.; Ma, J.; Zhao, R.; Tang, S.; Cheng, H.-M. Fabrication of Large Aerogel-Like Carbon/Carbon Composites with Excellent Load-Bearing Capacity and Thermal-Insulating Performance at 1800 °C. *ACS Nano* **2022**, *16*, 6565–6577. [CrossRef]
6. Zhang, Q.Z.; Zhang, D.; Miao, Z.C.; Zhang, X.L.; Chou, S.L. Research progress in MnO$_2$–carbon based supercapacitor electrode materials. *Small* **2018**, *14*, 1702883. [CrossRef]
7. Li, Y.; Liu, X.; Nie, X.; Yang, W.; Wang, Y.; Yu, R.; Shui, J. Multifunctional organic–inorganic hybrid aerogel for self-cleaning, heat-insulating, and highly efficient microwave absorbing material. *Adv. Funct. Mater.* **2019**, *29*, 1807624. [CrossRef]
8. Sun, W.; Du, A.; Gao, G.; Shen, J.; Wu, G. Graphene-templated carbon aerogels combining with ultra-high electrical conductivity and ultra-low thermal conductivity. *Microporous Mesoporous Mater.* **2017**, *253*, 71–79. [CrossRef]
9. Hu, L.; He, R.; Lei, H.; Fang, D. Carbon aerogel for insulation applications: A review. *Int. J. Thermophys.* **2019**, *40*, 39. [CrossRef]
10. Wiener, M.; Reichenauer, G.; Braxmeier, S.; Hemberger, F.; Ebert, H.-P. Carbon aerogel-based high-temperature thermal insulation. *Int. J. Thermophys.* **2009**, *30*, 1372–1385. [CrossRef]
11. Feng, J.; Zhang, C.; Feng, J.; Jiang, Y.; Zhao, N. Carbon aerogel composites prepared by ambient drying and using oxidized polyacrylonitrile fibers as reinforcements. *ACS Appl. Mater. Interfaces* **2011**, *3*, 4796–4803. [CrossRef]
12. Ding, J.; Wu, X.; Shen, X.; Cui, S.; Chen, X. A promising form-stable phase change material composed of C/SiO$_2$ aerogel and palmitic acid with large latent heat as short-term thermal insulation. *Energy* **2020**, *210*, 118478. [CrossRef]
13. Gu, W.; Sheng, J.; Huang, Q.; Wang, G.; Chen, J.; Ji, G. Environmentally friendly and multifunctional shaddock peel-based carbon aerogel for thermal-insulation and microwave absorption. *Nano-Micro Lett.* **2021**, *13*, 102. [CrossRef]
14. Jia, X.; Dai, B.; Zhu, Z.; Wang, J.; Qiao, W.; Long, D.; Ling, L. Strong and machinable carbon aerogel monoliths with low thermal conductivity prepared via ambient pressure drying. *Carbon* **2016**, *108*, 551–560. [CrossRef]
15. Al-Muhtaseb, S.A.; Ritter, J.A. Preparation and properties of resorcinol–formaldehyde organic and carbon gels. *Adv. Mater.* **2003**, *15*, 101–114. [CrossRef]
16. ElKhatat, A.M.; Al-Muhtaseb, S.A. Advances in tailoring resorcinol-formaldehyde organic and carbon gels. *Adv. Mater.* **2011**, *23*, 2887–2903. [CrossRef]
17. Feng, J.; Zhang, C.; Feng, J. Carbon fiber reinforced carbon aerogel composites for thermal insulation prepared by soft reinforcement. *Mater. Lett.* **2012**, *67*, 266–268. [CrossRef]
18. Drach, V.; Wiener, M.; Reichenauer, G.; Ebert, H.-P.; Fricke, J. Determination of the anisotropic thermal conductivity of a carbon aerogel–fiber composite by a non-contact thermographic technique. *Int. J. Thermophys.* **2007**, *28*, 1542–1562. [CrossRef]
19. Guo, P.; Li, J.; Pang, S.; Hu, C.; Tang, S.; Cheng, H.-M. Ultralight carbon fiber felt reinforced monolithic carbon aerogel composites with excellent thermal insulation performance. *Carbon* **2021**, *183*, 525–529. [CrossRef]
20. Yang, Z.; Li, J.; Xu, X.; Pang, S.; Hu, C.; Guo, P.; Tang, S.; Cheng, H.-M. Synthesis of monolithic carbon aerogels with high mechanical strength via ambient pressure drying without solvent exchange. *J. Mater. Sci. Technol.* **2020**, *50*, 66–74. [CrossRef]
21. Zhai, Z.; Ren, B.; Xu, Y.; Wang, S.; Zhang, L.; Liu, Z. Green and facile fabrication of Cu-doped carbon aerogels from sodium alginate for supercapacitors. *Org. Electron.* **2019**, *70*, 246–251. [CrossRef]

22. Chen, Y.; Zhang, L.; Yang, Y.; Pang, B.; Xu, W.; Duan, G.; Jiang, S.; Zhang, K. Recent progress on nanocellulose aerogels: Preparation, modification, composite fabrication, applications. *Adv. Mater.* **2021**, *33*, 2005569. [CrossRef]
23. Chang, X.; Chen, D.; Jiao, X. Starch-derived carbon aerogels with high-performance for sorption of cationic dyes. *Polymer* **2010**, *51*, 3801–3807. [CrossRef]
24. Zhai, Z.; Zheng, Y.; Du, T.; Tian, Z.; Ren, B.; Xu, Y.; Wang, S.; Zhang, L.; Liu, Z. Green and sustainable carbon aerogels from starch for supercapacitors and oil-water separation. *Ceram. Int.* **2021**, *47*, 22080–22087. [CrossRef]
25. Chen, Y.; Hao, Y.; Li, S.; Luo, Z.; Gao, Q. Preparation of hydroxybutyl starch with a high degree of substitution and its application in temperature-sensitive hydrogels. *Food Chem.* **2021**, *355*, 129472. [CrossRef]
26. Kubicka, M.; Bakierska, M.; Chudzik, K.; Świętosławski, M.; Molenda, M. Nitrogen-doped carbon aerogels derived from starch biomass with improved electrochemical properties for Li-ion batteries. *Int. J. Mol. Sci.* **2021**, *22*, 9918. [CrossRef]
27. Yuan, D.; Zhang, T.; Guo, Q.; Qiu, F.; Yang, D.; Ou, Z. Recyclable biomass carbon@ SiO_2 @ MnO_2 aerogel with hierarchical structures for fast and selective oil-water separation. *Chem. Eng. J.* **2018**, *351*, 622–630. [CrossRef]
28. Peng, F.; Jiang, Y.; Feng, J.; Cai, H.; Feng, J.; Li, L. Thermally insulating, fiber-reinforced alumina–silica aerogel composites with ultra-low shrinkage up to 1500 °C. *Chem. Eng. J.* **2021**, *411*, 128402. [CrossRef]
29. Gülgün, M.A.; Nguyen, M.H.; Kriven, W.M. Polymerized organic-inorganic synthesis of mixed oxides. *J. Am. Ceram. Soc.* **1999**, *82*, 556–560. [CrossRef]
30. Feng, J.; Feng, J.; Zhang, C. Shrinkage and pore structure in preparation of carbon aerogels. *J. Sol-Gel Sci. Technol.* **2011**, *59*, 371–380. [CrossRef]
31. Xu, F.; Xu, J.; Xu, H.; Lu, Y.; Yang, H.; Tang, Z.; Lu, Z.; Fu, R.; Wu, D. Fabrication of novel powdery carbon aerogels with high surface areas for superior energy storage. *Energy Storage Mater.* **2017**, *7*, 8–16. [CrossRef]
32. Buttersack, C. Modeling of type IV and V sigmoidal adsorption isotherms. *Phys. Chem. Chem. Phys.* **2019**, *21*, 5614–5626. [CrossRef] [PubMed]
33. Holder, C.F.; Schaak, R.E. Tutorial on Powder X-ray Diffraction for Characterizing Nanoscale Materials. *J. Acs Nano.* **2019**, *13*, 7359–7365. [CrossRef] [PubMed]
34. Wang, C.; Wang, L.; Liang, W.; Liu, F.; Wang, S.; Sun, H.; Zhu, Z.; Li, A. Enhanced light-to-thermal conversion performance of all-carbon aerogels based form-stable phase change material composites. *J. Colloid Interface Sci.* **2022**, *605*, 60–70. [CrossRef]
35. Eckmann, A.; Felten, A.; Mishchenko, A.; Britnell, L.; Krupke, R.; Novoselov, K.S.; Casiraghi, C. Probing the nature of defects in graphene by Raman spectroscopy. *Nano Lett.* **2012**, *12*, 3925–3930. [CrossRef]
36. Wang, S.; Liu, Q.; Luo, Z.; Wen, L.; Cen, K. Mechanism study on cellulose pyrolysis using thermogravimetric analysis coupled with infrared spectroscopy. *Front. Energy Power Eng. China* **2007**, *1*, 413–419. [CrossRef]
37. Cheng, B.; Zhao, L.; Yu, J.; Zhao, X. Facile fabrication of SiO_2/Al_2O_3 composite microspheres with a simple electrostatic attraction strategy. *Mater. Res. Bull.* **2008**, *43*, 714–722. [CrossRef]
38. Meng, Y.; Young, T.M.; Liu, P.; Contescu, C.I.; Huang, B.; Wang, S. Ultralight carbon aerogel from nanocellulose as a highly selective oil absorption material. *Cellulose* **2015**, *22*, 435–447. [CrossRef]
39. Zhang, R.; Jiang, N.; Duan, X.-J.; Jin, S.-L.; Jin, M.-L. Synthesis and characterization of Al_2O_3-C hybrid aerogels by a one-pot sol-gel method. *New Carbon Mater.* **2017**, *32*, 258–264. [CrossRef]
40. Lv, Y.; Liu, T.; Huang, X.; He, F.; Tang, L.; Zhou, J.; Wang, J. Numerical investigation and optimization of flat plate transpiration-film combined cooling structure. *Int. J. Therm. Sci.* **2022**, *179*, 107673. [CrossRef]

Article

Self-Assembling Peptide-Based Magnetogels for the Removal of Heavy Metals from Water

Farid Hajareh Haghighi [1,†], Roya Binaymotlagh [1,†], Laura Chronopoulou [1,2,*], Sara Cerra [1], Andrea Giacomo Marrani [1], Francesco Amato [1], Cleofe Palocci [1,2,*] and Ilaria Fratoddi [1]

1. Department of Chemistry, Sapienza University of Rome, Piazzale Aldo Moro 5, 00185 Rome, Italy; farid.hajarehhaghighi@uniroma1.it (F.H.H.); roya.binaymotlagh@uniroma1.it (R.B.); sara.cerra@uniroma1.it (S.C.); andrea.marrani@uniroma1.it (A.G.M.); francesco.amato@uniroma1.it (F.A.); ilaria.fratoddi@uniroma1.it (I.F.)
2. Research Center for Applied Sciences to the Safeguard of Environment and Cultural Heritage (CIABC), Sapienza University of Rome, Piazzale Aldo Moro 5, 00185 Rome, Italy
* Correspondence: laura.chronopoulou@uniroma1.it (L.C.); cleofe.palocci@uniroma1.it (C.P.); Tel.: +39-06-4991-3317 (C.P.)
† These authors contributed equally to this work.

Abstract: In this study, we present the synthesis of a novel peptide-based magnetogel obtained through the encapsulation of γ-Fe_2O_3-polyacrylic acid (PAA) nanoparticles (γ-Fe_2O_3NPs) into a hydrogel matrix, used for enhancing the ability of the hydrogel to remove Cr(III), Co(II), and Ni(II) pollutants from water. Fmoc-Phe (*Fluorenylmethoxycarbonyl*-Phenylalanine) and diphenylalanine (Phe_2) were used as starting reagents for the hydrogelator (Fmoc-Phe_3) synthesis via an enzymatic method. The PAA-coated magnetic nanoparticles were synthesized in a separate step, using the co-precipitation method, and encapsulated into the peptide-based hydrogel. The resulting organic/inorganic hybrid system (γ-Fe_2O_3NPs-peptide) was characterized with different techniques, including FT-IR, Raman, UV-Vis, DLS, ζ-potential, XPS, FESEM-EDS, swelling ability tests, and rheology. Regarding the application in heavy metals removal from aqueous solutions, the behavior of the obtained magnetogel was compared to its precursors and the effect of the magnetic field was assessed. Four different systems were studied for the separation of heavy metal ions from aqueous solutions, including (1) γ-Fe_2O_3NPs stabilized with PAA, (γ-Fe_2O_3NPs); (2) Fmoc-Phe_3 hydrogel (HG); (3) γ-Fe_2O_3NPs embedded in peptide magnetogel (γ-Fe_2O_3NPs@HG); and (4) γ-Fe_2O_3NPs@HG in the presence of an external magnetic field. To quantify the removal efficiency of these four model systems, the UV-Vis technique was employed as a fast, cheap, and versatile method. The results demonstrate that both Fmoc-Phe_3 hydrogel and γ-Fe_2O_3NPs peptide magnetogel can efficiently remove all the tested pollutants from water. Interestingly, due to the presence of magnetic γ-Fe_2O_3NPs inside the hydrogel, the removal efficiency can be enhanced by applying an external magnetic field. The proposed magnetogel represents a smart multifunctional nanosystem with improved absorption efficiency and synergic effect upon applying an external magnetic field. These results are promising for potential environmental applications of γ-Fe_2O_3NPs-peptide magnetogels to the removal of pollutants from aqueous media.

Keywords: magnetogels; magnetic nanoparticles; peptide-based hydrogels; hydrogel composites; water purification; Cr(III); Co(II); Ni(II)

1. Introduction

Worldwide, due to the increase in industrialization levels, the protection of the environment from industrial wastes has become more and more important over the past decades. The presence of toxic metal ions in industrial wastewaters can result in adverse effects on both human health and the environment even at low concentrations because these inorganic species are generally nondegradable in nature [1–3]. Some of the most common metal pollutants are cadmium (Cd), lead (Pb), mercury (Hg), nickel (Ni), arsenic (As), copper (Cu), chromium (Cr), cobalt (Co), and zinc (Zn). Table 1 summarizes the primary industrial sources, health side effects and the permitted quantity of these heavy metals, based on recent literature data [4–8].

Table 1. Heavy metals commonly found in industrial wastewaters, along with their health side effects and the permitted quantity in drinking water based on the World Health Organization (WHO) recommendations. Adapted with permission from Ref. [4]. Copyright 2021, Springer Nature.

Heavy Metal	Sources	Main Organ and System Affected	Permitted Amounts (µg)
Lead (Pb)	Lead-based batteries, solder, alloys, cable sheathing pigments, rust inhibitors, ammunition, glazes, plastic stabilizers	Bones, liver, kidneys, brain, lungs, spleen, immunological system, hematological system, cardiovascular system, reproductive system	10
Arsenic (As)	Electronics and glass production	Skin, lungs, brain, kidneys, metabolic system, cardiovascular system, immunological system, endocrine system	10
Copper (Cu)	Corroded plumbing systems, electronic and cables industry	Liver, brain, kidneys, cornea, gastrointestinal system, lungs, immunological system, hematological system	2000
Zinc (Zn)	Brass coating, rubber products, some cosmetics and aerosol deodorants	Stomach cramps, skin irritations, vomiting, nausea, anemia, convulsions	3000
Chromium (Cr)	Steel and pulp mills, tanneries	Skin, lungs, kidneys, liver, brain, pancreas, tastes, gastrointestinal system, reproductive system	50
Cadmium (Cd)	Batteries, paints, steel industry, plastic industries, metal refineries, corroded galvanized pipes	Bones, liver, kidneys, lungs, testes, brain, immunological system, cardiovascular system	3
Mercury (Hg)	Electrolytic production of caustic soda and chlorine, electrical appliances, runoff from landfills and agriculture, industrial and control instruments, laboratory apparatus, refineries	Brain, lungs, kidneys, liver, immunological system, cardiovascular system, endocrine and reproductive system	6
Nickel (Ni)	Nickel alloy production, stainless steel	Skin, gastrointestinal distress, lung, pulmonary fibrosis, kidney	70
[1] Cobalt (Co)	Cement industries, polishing disc used in diamond polishing, mobile batteries, televisions (TVs), liquid crystal display TVs, computer monitors	High concentrations cause vomiting, nausea, vision problems, thyroid gland damage	N/A

[1] The data for cobalt (Co) were taken from references [5,6].

In the past decades, there have been admirable efforts to develop fast and efficient techniques to remove metal ions from aqueous solutions, and different approaches have been studied and developed for water remediation, including ion exchange, electrochemical treatments, membrane separation, and flocculation [1,9]. However, these methods often suffer from their own limitations in terms of complexity, cost, and efficiency. In recent years, sol- and gel-based adsorption strategies have attracted increasing attention for scaling-up the removal process because of their simplicity, high efficiency, and low-cost operation. To this aim, polymers, clay minerals, and carbon-based materials have been investigated as adsorbents, and promising results have been obtained in this regard [10–14]. On this basis, an ideal adsorbent should have high adsorption capacity, fast adsorption kinetics, good chemical stability, and easy preparation method [2].

Hydrogels have been recently introduced as suitable alternatives as absorbent materials due to their highly porous structure and adsorbing functional groups, as well as their large surface area and high swelling ability [15–21]. They are colloidal materials that possess a three-dimensional (3D) network based on amphipathic polymer building blocks. The polymers link to each other, forming an insoluble 3D matrix that can absorb and entrap a significant amount of water [22,23]. For water treatment applications, it is highly recommended to use biocompatible absorbents, and among them, peptide-based hydrogels for example can be suitable candidates [1,15]. In general, hydrogels can absorb large amounts of metal ions inside their matrices, due to the presence of suitable functional groups. As is well-known, on this basis, hydrogels physically or chemically interact with pollutants via one or more mechanisms. One of the major mechanisms of heavy metals removal by hydrogels is based on electrostatic interactions that usually occur when, as a function of pH, hydroxyl and carboxyl groups of hydrogels are deprotonated, imparting negative charges to the hydrogel. Other possible interactions involve H-bond formation. These types of hydrogels can be easily used and scaled up for industrial applications of wastewater treatment because of their cost-effectiveness and standard synthetic methods. In particular, they can be prepared by using an enzymatic approach and under mild conditions, which are highly suitable for large-scale production [19,24].

To enhance hydrogels adsorption efficiency and reduce their operational costs, they can be hybridized (or combined) with magnetic nanoparticles (NPs) to form "magnetogels", which are promising organic-inorganic nanohybrids for environmental and biological applications, as they combine benefits of both hydrogels and magnetic nanoparticles into a single inorganic/organic hybrid [25–32]. In fact, magnetic sorbents entrapped within hydrogel-based materials can promote heavy metals removal thanks to the large number of functional groups on hydrogel surfaces, thus improving adsorption selectivity as well as sorbent capacity. These magnetogels are considered as soft smart multifunctional nanosystems, which provide the possibility for enhancing the adsorption efficiency of hydrogels upon applying an external magnetic field. In addition, the presence of magnetic nanoparticles can modify hydrogels' structure through covalent or non-covalent interactions, which offers the possibility to finely tune their physico-chemical properties. The porous magnetogels present in their structure active functional groups (e.g., carboxyl, hydroxyl and amino groups) that are able to remove contaminants by means of electrostatic interactions, ionic exchange, or complexation with contaminants such as heavy metal ions. More importantly, the interaction of magnetogels with an external magnetic field can promote the separation, collection, and reuse of hydrogel adsorbents and also have an enhancing effect on the adsorption of magnetogels [28]. Among the various types of magnetic NPs, maghemite (γ-Fe_2O_3) and magnetite (Fe_3O_4) nanoparticles have been employed for biological and environmental applications due to their high magnetization, biocompatibility, and well-assessed synthesis methods [25]. These two types of magnetic nanoparticles both exhibit strong on/off superparamagnetic properties in the presence/absence of a magnetic field, respectively. Although maghemite NPs (γ-Fe_2O_3NPs) show a slightly smaller magnetic moment compared to magnetite NPs (Fe_3O_4NPs), they are more stable in air and have the benefits of possessing a much lower optical absorption in the visible region [33], suitable for biotechnologies. For water remediation applications, γ-Fe_2O_3NPs can act as photocatalysts to break down and remove various organic contaminants. To date, various nanocomposite-based γ-Fe_2O_3NPs have been used as adsorbents to remove different metallic- and organic-based contaminants from aqueous solutions (drinking water, groundwater, wastewater, and acid mine drainage) with significant adsorption efficiency. The pollutants include heavy metal ions and several dyes (e.g., methyl orange (MO), methylene blue (MB), rose bengal, Congo red (CR), brilliant cresyl blue, thionine, and Janus green B) [34,35]. Although several worthwhile water purification studies have been performed using magnetogels [26,36–39], only a few water remediation applications of peptide-based magnetogels are known. Thus, it is worth studying this topic to gain more insight toward the development of advanced and smart biocompatible adsorbents.

Concerning magnetogels synthesis, several methods have been introduced, such as blending [40], grafting [41], in situ precipitation [42], and swelling [43]. Among these methods, the swelling strategy is considered as an in situ methodology, and the magnetogels are prepared in one step. On the other side, blending and grafting methods are known as "ex situ" strategies, and the magnetogels are synthesized in more than one step [27].

Regarding the blending method, it is based on sequential syntheses of the components, starting with the magnetic NPs, which are then blended with hydrogel precursors to make the resultant magnetogels. Despite the simplicity of the blending method, the magnetic nanoparticles may have an interfering effect on hydrogel formation and also have a negative influence on the final structure of the gel. Moreover, lack of proper stabilization of nanoparticles can result in their heterogeneous distribution or diffusion out of the gel upon swelling [27]. Based on the interaction of MNPs with the hydrogel network, magnetogels are classified into class I and class II, as discussed by Weeber et al. [27]. In class I magnetogels, nanoparticles have a weak interaction with the hydrogel network (also called blends) through physical interactions, e.g., nanoparticles embedded in the aqueous compartments or adsorbed onto the fibers. On the other hand, class II magnetogels exhibit strong interactions among MNPs and hydrogel fibers, through covalent bonding or strong physical forces [27]. From an industrial point of view, the synthesis of novel magnetic hybrid sorbents with improved properties in terms of sorbent efficiency or ability to simultaneously introduce different extraction approaches is highly desirable.

Hence, due to the importance of magnetogels, we studied the synthesis, characterization, and potential water remediation application of novel peptide-based magnetogel nanocomposites by ex situ encapsulation of polyacrylic acid (PAA)-modified iron oxide magnetic nanoparticles (γ-Fe$_2$O$_3$NPs) into a peptide hydrogel matrix (γ-Fe$_2$O$_3$NPs@HG). The γ-Fe$_2$O$_3$NPs were synthesized by the well-known co-precipitation method, and for the hydrogel matrix, Fmoc-Phe and diphenylalanine (Phe$_2$) were used as starting materials for the hydrogelator synthesis through an enzymatic reaction [44]. The magnetogel nanocomposite was characterized using different spectroscopic, morphological, and structural techniques and applied to the separation of Cr(III), Co(II), and Ni(II) from aqueous solutions. The effects of contact time and an external magnetic field on the adsorption efficiency of the different contaminants were investigated.

2. Results and Discussion

2.1. Preparation of γ-Fe$_2$O$_3$NPs and γ-Fe$_2$O$_3$NPs@HG Magnetogel

Co-precipitation is known as a simple and cheap method to synthesize magnetic iron oxide NPs (γ-Fe$_2$O$_3$NPs) from aqueous solutions of Fe(II) and Fe(III) by the addition of a base as a precipitating agent at mild temperature, and a large amount of NPs can be prepared by this method. The co-precipitation process, schematized in Figure 1a, does not require organic solvents or toxic precursor iron complexes and proceeds at temperatures below 100 °C. More importantly, it can be developed and scaled up from lab to industry due to its simplicity, reproducibility, and eco-friendly reaction conditions. However, this method sometimes suffers from a lack of control over particle size distribution, probably because of the complicated set of pathways that lead to the formation of NPs [45]. The general mechanism for the formation of MNPs first involves hydroxylation of the ferrous and ferric ions to form Fe(OH)$_2$ and Fe(OH)$_3$, respectively. These two low-soluble hydroxides (K_{ps} (25 °C) = 7.9 × 10^{-15} and 6.3 × 10^{-38}, for ferrous and ferric hydroxide, respectively [46]) can be obtained at alkaline pHs (pH > 8), and when NaOH is used as the precipitating agent, a black colloidal solution of iron containing NPs is formed instantaneously. By applying a 2:1 molar ratio of Fe(III):Fe(II) and an oxygen-free environment, magnetite NPs (Fe$_3$O$_4$NPs) are the main product of this reaction through the following possible reactions (Equations (1)–(4)) [45]:

$$Fe(III) + 3OH^- \rightarrow Fe(OH)_3 \tag{1}$$

$$Fe(OH)_3 \rightarrow FeOOH + H_2O \tag{2}$$

$$Fe(II) + 2OH^- \rightarrow Fe(OH)_2 \tag{3}$$

$$2FeOOH + Fe(OH)_2 \rightarrow Fe_3O_4NPs\downarrow + 2H_2 \tag{4}$$

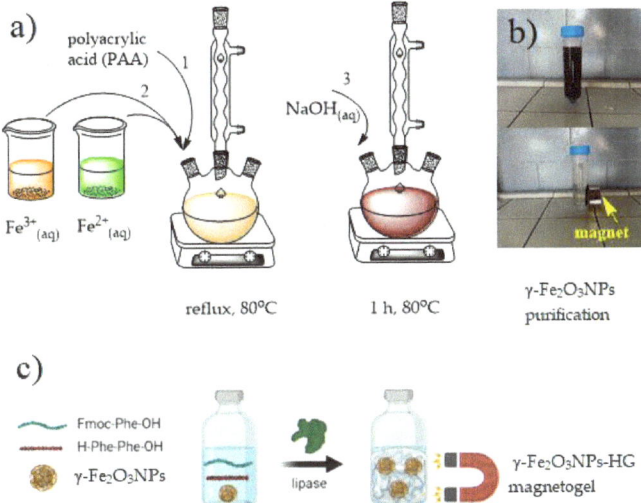

Figure 1. (**a**) Schematic in situ synthesis of PAA-stabilized γ-Fe$_2$O$_3$NPs; (**b**) separation of γ-Fe$_2$O$_3$NPs with an external magnetic field (the enlarged image is provided in SI as Figure S1); (**c**) illustration of the ex situ incorporation of γ-Fe$_2$O$_3$NPs into the peptide-based hydrogels γ-Fe$_2$O$_3$NPs@HG (blending method).

Magnetite shows an inverse (or normal) spinel crystal structure, and its unit cell contains 32 O^{2-} anions, 8 Fe(II), and 16 Fe(III) cations. Due to the presence of reduced iron (Fe(II)) in this crystal structure, Fe$_3$O$_4$NPs are easily subject to oxidation and are transformed to a more stable maghemite phase (γ-Fe$_2$O$_3$NPs) by the following equation (Equation (5)) [47]:

$$3Fe_3O_4NPs + 0.5O_2 + 2H^+ \rightarrow 4\gamma\text{-}Fe_2O_3NPs + Fe^{2+} + H_2O \tag{5}$$

Another important aspect of iron-based NPs is their colloidal stability after the synthesis. Due to their magnetic properties, iron-based NPs are more vulnerable to agglomeration because of the magnetic attraction among particles. In general, colloidal stability is the result of a balance between repulsive interactions (steric and electrostatic) and attractive forces (Van der Waals, dipolar, and magnetic), which can be influenced by the medium parameters, including composition, pH, and ionic strength [48]. To enhance their colloidal stability, MNPs should be stabilized by steric, electrostatic, or a combination of these repulsive forces. In the electrostatic stabilization, the repulsive forces between the NPs originate from likewise charges [49], and for the steric stabilization, the presence of large molecules provides a repulsive hindrance for the surface of NPs [22]. Steric stabilization is usually favored because it is less sensitive to medium parameters and therefore more suitable when MNPs are in contact with complex media [50–52]. To this aim, several small and large stabilizing agents have been applied for the surface functionalization of MNPs, such as polymers (polytrolox ester, PAA, and polyacrylic-co-maleic acid), natural antioxidants (green tea polyphenols, curcumin, quercetin, and anthocyanins) and organic or inorganic acids (gallic, ascorbic, citric, and humic acid) [53]. Among these stabilizers, the functionalization of magnetic NPs with PAA provides both steric and electrostatic effects on the NPs' surface [54]. The electrostatic effect of PAA originates from its carboxylate groups, and the steric effect from its polymeric nature. For water purification applications, PAA is known for its ability to absorb a large amount of water and is used as a superabsorbent. Another advantage of PAA is its biocompatibility which is highly desirable [40]. For the PAA functionalization of NPs, two methods are generally used: in situ and post (ex situ) surface coating. For in situ functionalization, PAA is used simultaneously with the iron precursors during the synthesis of magnetic NPs, and both synthesis and functionalization occur simultaneously in one step. For the post (ex situ) method, PAA is added to pre-synthesized NPs in a separate step (next step) from the synthesis. Generally, the in situ method is more preferable due to the inhibition of particle growth in a high concentration of PAA. Also, the high hydrophilicity and colloidal stability induced by PAA stabilization of magnetic NPs can decrease the long gelation time of supramolecular magnetogels. For instance, it was reported that the stabilization of iron oxide NPs with polyacrylic acid allowed homogeneous encapsulation of NPs up to 30 m/m% in both Npx-L-Asp-Z-ΔPhe-OH and Npx-L-Tyr-Z-ΔPhe-OH hydrogels containing non-canonical amino acids [27]. These are some important benefits of PAA for its applications as composite adsorbents. For instance, PAA-functionalized magnetic magnetite particles have been used as an adsorbent for basic dyes [53,55,56]. For the above reasons, we synthesized the magnetic NPs stabilized by PAA using the in situ strategy to obtain small and colloidally stable γ-Fe$_2$O$_3$NPs, followed by preparation of γ-Fe$_2$O$_3$NPs@HG magnetogel.

The isolation of γ-Fe$_2$O$_3$NPs was first confirmed by the visual magnetic behavior of the purified precipitates. As can be seen in Figure 1b and Figure S1, the particles are strongly attracted to the external magnet and easily redispersed after removing the magnet. To confirm the stabilizing effect of PAA, uncoated magnetic NPs were also synthesized, using the same reaction conditions but without the presence of PAA. For the synthesis of γ-Fe$_2$O$_3$NPs@HG magnetogels, we have used the blending strategy mentioned in the introduction section as a well-known method to synthesize magnetogels [25,27]. This method is based on sequential syntheses of the components, starting with the magnetic NPs, which are then blended with hydrogel precursors to make the resulting magnetogels. The formation of the hydrogel followed a well-assessed procedure, using a microbial lipase to catalyze the synthesis in water of self-assembling peptides generated by the peptide bond formation between 9-fluorenylmethoxycarbonyl-phenylalanine (Fmoc–Phe) and the dipeptide diphenylalanine (Phe$_2$) (Figure 1c) [57].

2.2. Raman, FT-IR/ATR, and XPS Characterization of γ-Fe$_2$O$_3$NPs

Fourier transform-infrared (FT-IR) and Raman spectroscopies are two commonly used techniques to characterize iron oxide nanoparticles, as they can provide information on the oxide phase through detection of phonon modes [58]. The FT-IR spectra of bare and PAA-stabilized γ-Fe$_2$O$_3$NPs are presented in Figure 2a in the 1000–400 cm^{-1} region, containing information about the phase of both NPs. Based on literature, the magnetite phase has a sharp and symmetrical vibration at around 571 cm^{-1}, assigned to the Fe–O bonds present in tetrahedral and octahedral sites of the spinel structure. It is a general characteristic band of iron oxide NPs, and its sharpness clarifies the pure defect-free phase. In maghemite, this band splits into two characteristic vibrations, due to the creation of vacancy defects and vanishing of Fe(II) ion from the octahedral sites upon the formation of γ-Fe$_2$O$_3$, causing a decrease in Fe–O bond length, and hence, corresponding splitting of the band occurs [59]. It should be mentioned that both phases (magnetite or maghemite) show a weak band in the 440–460 cm^{-1} region [58]. Regarding the crystal structure, also group theory theoretically predicts that if the γ-Fe$_2$O$_3$NPs have a spinel crystal, then there are four T_1 modes expected at 212, 362, 440, and 553 cm^{-1} [60], of which we were able to only detect the highest two frequencies with our experimental setup. Both spectra of γ-Fe$_2$O$_3$NPs clearly show the two T_1 modes around 460 and 570 cm^{-1} for uncoated γ-Fe$_2$O$_3$NPs, and 462 and 565 cm^{-1} for the PAA stabilized γ-Fe$_2$O$_3$NPs, confirming a spinel crystal structure for both NPs of our experiment. Hence, in both spectra, the formation of γ-Fe$_2$O$_3$NPs was confirmed from the broadening and splitting of the band into 628 and 570 cm^{-1} for uncoated γ-Fe$_2$O$_3$NPs, and 626 and 565 cm^{-1} for PAA-stabilized γ-Fe$_2$O$_3$NPs. Regarding the PAA coating and its interaction with the surface of NPs, the FT-ATR data are shown in Figure 2b for PAA-stabilized γ-Fe$_2$O$_3$NPs (red line) and pristine PAA (black line) in the 4000–600 cm^{-1} region. The FT-ATR spectrum of PAA stabilized γ-Fe$_2$O$_3$NPs is not a simple superposition of the PAA spectrum, and the relative intensities of the main vibrational bands show some changes, which suggests that PAA alters its symmetry when it attaches to the NPs. The main features of these spectra are summarized in Table 2, which shows the characteristic frequencies of free PAA, including the carbonyl stretching (−C=O) at 1669 cm^{-1}, −CH$_2$ scissoring at 1446 cm^{-1}, and the −C−O stretching at 1236 cm^{-1}, as well as the symmetric stretching frequencies of the carboxylate ions (−COO$^-$) at 1402 cm^{-1}. Compared to the PAA spectrum, the PAA-stabilized γ-Fe$_2$O$_3$NPs spectrum shows all these characteristic bands (except −C−O stretching) with a maximum of 6 cm^{-1} shift. There is also an additional band for the PAA-stabilized γ-Fe$_2$O$_3$NPs at around 1556 cm^{-1}, assigned to −COO$^-$ asymmetric stretching. These results confirm the surface functionalization of NPs with PAA, and based on the small shifts observed in frequencies of PAA-stabilized γ-Fe$_2$O$_3$NPs (compared to the PAA spectrum), a physical interaction can be proposed between the negatively charged PAA (–COO$^-$) and positive surface Fe(III) ions of NPs [61].

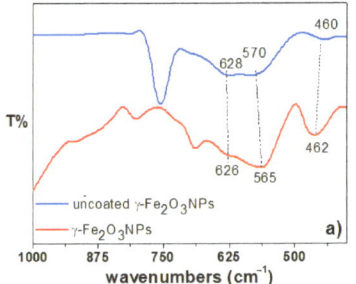 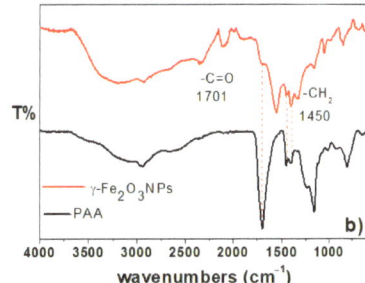

Figure 2. (**a**) FT-IR spectra of uncoated γ-Fe$_2$O$_3$NPs NPs (blue line) and PAA-stabilized γ-Fe$_2$O$_3$NPs (red line); (**b**) FT-ATR of PAA-stabilized γ-Fe$_2$O$_3$NPs (red line) and PAA (black line).

Table 2. Comparison of FT-ATR peak assignments (in cm^{-1}) for polyacrylic acid (PAA) in solid form and after attachment onto the γ-Fe$_2$O$_3$NPs surface.

PAA	PAA-Stabilized γ-Fe$_2$O$_3$NPs	Peak Assignment
1699	1705	−C=O (free COOH)
-	1556	−COO$^-$ (asymmetric)
1446	1444	−CH$_2$ scissor
1402	1408	−COO$^-$ (symmetric)
1236	-	−C−O

As mentioned above, Raman spectroscopy can discriminate iron oxide phases because they exhibit distinct Raman signatures originating from their different oxidation states [62]. Hence, Raman spectroscopy was used as a complementary technique to better understand the structure of the synthesized γ-Fe$_2$O$_3$NPs.

The Raman spectra of the synthesized uncoated γ-Fe$_2$O$_3$NPs and PAA-coated γ-Fe$_2$O$_3$NPs are shown in Figure 3a,b, respectively. The results clearly match Raman spectra for maghemite previously reported in the literature [58,63], with the three broad observed Raman active phonon modes at around 350 cm^{-1} (T$_1$), 500 cm^{-1} (E), and 700 cm^{-1} (A$_1$). Regarding the PAA-coated sample, we clearly see the Raman modes at higher wavenumbers, with two main bands centered at around 1400 and 2927 cm^{-1}, assigned to −COO$^-$ (symmetric) and −CH/−CH$_2$ stretching bands, respectively [64]. These results are in accordance with the FT-IR spectra and confirm the presence of maghemite phase for the PAA-coated γ-Fe$_2$O$_3$NPs [58,62,65–68].

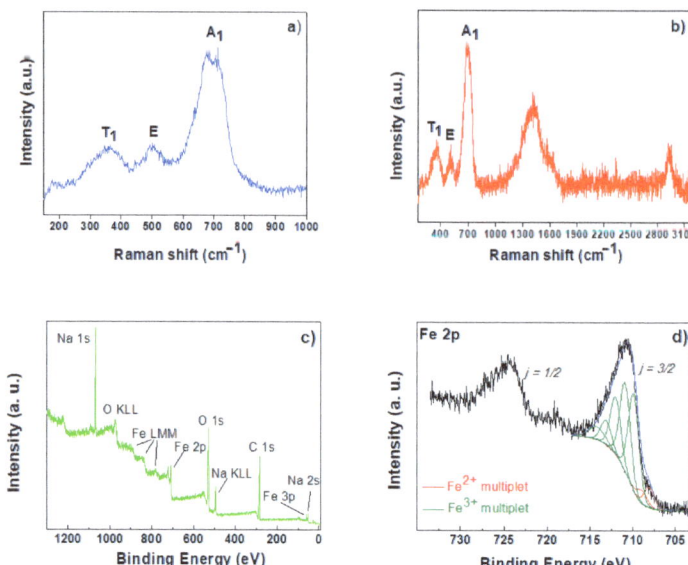

Figure 3. (a) Raman spectra of uncoated γ-Fe$_2$O$_3$NPs; and (b) PAA-coated γ-Fe$_2$O$_3$NPs; (c) XPS survey spectrum of PAA-coated γ-Fe$_2$O$_3$NPs; (d) XPS Fe2p spectrum of PAA-coated γ-Fe$_2$O$_3$NPs.

The maghemite phase is also confirmed by XPS results (Figure 3c,d) [62,69–73], and the PAA-coated γ-Fe$_2$O$_3$NPs sample was analyzed with XPS in order to ascertain the chemical composition of the inorganic core structure. Figure 3c shows the XPS spectrum, where the signals due to ionization of Fe, C, O, and Na are visible, the latter resulting from the use of NaOH in the synthesis of PAA-coated γ-Fe$_2$O$_3$NPs. Figure 3d shows the Fe 2p spectrum of PAA-coated γ-Fe$_2$O$_3$NPs, which is composed of two rather broad spin-orbit split components (j = 3/2 and 1/2), whose maxima are separated by a ΔE_{so} ~ 13.6 eV. This spectrum was curve-fitted in order to determine the oxidation state of the iron species. The curve-fitting procedure was conducted following the work by Grosvenor et al. [69,70], who applied a Shirley background removal to the 2p$_{3/2}$ envelope and successfully used the Gupta and Sen (GS) multiplets calculated for free metal ions [71,72] to account for electrostatic interactions in high-spin Fe(II) and Fe(III) compounds [69,70]. Also in the present case, a five-fold GS multiplet for Fe(III) compounds was used in the curve-fitting (green curves in Figure 3d, first component at 710.05 eV binding energy), achieving a good match with the experimental data. This agreement strongly supports the attribution of this signal to γ-Fe$_2$O$_3$, as reported by Grosvenor et al. A further two-fold multiplet was added at low binding energy (red curves in Figure 3d, first component at 708.17 eV binding energy), in order to better reproduce the experimental signal. This multiplet appears also in the spectrum by Grosvenor et al. as a "pre-peak" and might stem from residual Fe(II) high-spin components possibly present in γ-Fe$_2$O$_3$ [69].

2.3. DSL and UV-Vis Characterization of γ-Fe$_2$O$_3$NPs

The hydrodynamic size and surface charges of the two γ-Fe$_2$O$_3$NPs systems (bare and coated with PAA) were studied by DLS and ζ-potential techniques (Figure 4a,b). Compared to uncoated γ-Fe$_2$O$_3$NPs, PAA-coated γ-Fe$_2$O$_3$NPs show smaller size and higher negative surface charge, which can be a good indication of PAA coating and, more importantly, PAA's stabilizing effect on the NPs. Using PAA in the in situ synthesis of NPs allows for controlling the size of γ-Fe$_2$O$_3$NPs because the attached PAA moiety provides steric (by its polymeric nature) and electrostatic (by its –COO$^-$ group) stabilizations, thus permitting γ-Fe$_2$O$_3$NPs dispersion in aqueous media [54]. In fact, the presence of PAA on the surface of γ-Fe$_2$O$_3$NPs prolongs the colloidal stability by slowing down the agglomeration process. Another major advantage of PAA is that the carboxylic acid-enriched surfaces of PAA-coated γ-Fe$_2$O$_3$NPs may provide a platform for attaching these NPs to other systems to prepare multifunctional nanohybrids. UV-Vis spectra of PAA-coated γ-Fe$_2$O$_3$NPs (1.6–0.005 mg/mL) were monitored over a 7-day period to characterize their colloidal stability against agglomeration and sedimentation [74–76]. As can be seen in Figure 4c, results show no significant changes in the intensities, indicating the stability of the preparations against aggregation. The images of three PAA-coated γ-Fe$_2$O$_3$NPs suspensions (1.6, 0.16, and 0.08 mg/mL) are shown in Figure S2a, in which no precipitation was observed, consistent with the UV-Vis spectra (Figure 4c). Also, the colloidal stability of higher concentrations of PAA-coated γ-Fe$_2$O$_3$NPs (10 mg/mL) was visually monitored over one week, and a remarkable stability of the NPs against sedimentation was observed. Conversely, uncoated γ-Fe$_2$O$_3$NPs showed a fast aggregation and sedimentation that were visually detected, even at low concentrations (Figure S2c). These results are consistent with previous studies on the use of organic molecules as stabilizing agents for MNPs. However, the uncoated NPs showed a limited amount of precipitation in the vials (Figure S2c), demonstrating the positive long-term stabilizing effect of PAA-coated γ-Fe$_2$O$_3$NPs (Figure 4c). These results are comparable and consistent with the literature [27,40,53–56] about the stabilizing effect of PAA on γ-Fe$_2$O$_3$NPs, explained in Section 2.1.

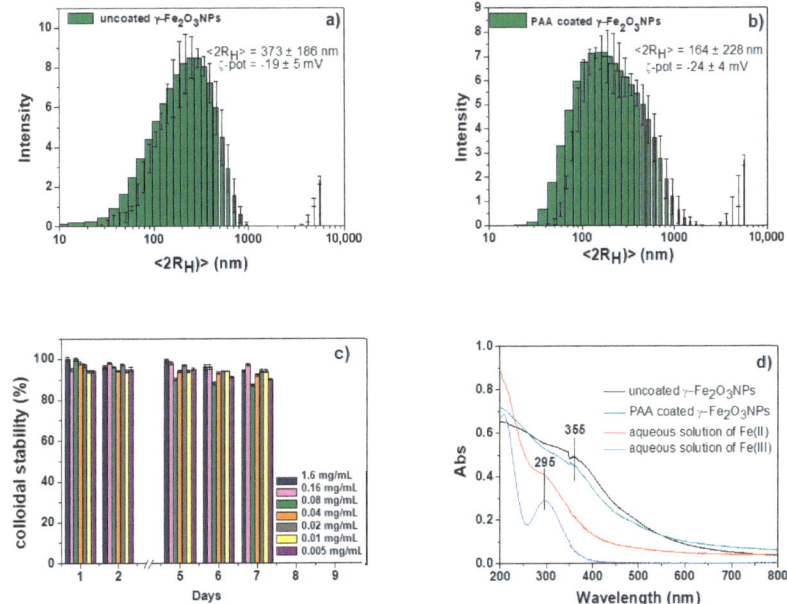

Figure 4. (**a**,**b**) DLS results of the two bare and PAA-coated γ-Fe$_2$O$_3$NPs; (**c**) stability tests of PAA-coated γ-Fe$_2$O$_3$NPs assessed by following the maximum absorption peak at 355 nm; (**d**) UV-Vis spectra of freshly prepared precursors ions (Fe^{2+}$_{(aq)}$ and Fe^{3+}$_{(aq)}$) and the two NPs.

The UV-Vis spectra of both NPs systems show the presence of a new absorption band at around 355 nm (Figure 4d), which is different compared to the spectra of precursors ions (Fe^{2+}$_{(aq)}$ and Fe^{3+}$_{(aq)}$) and assigned to the band gap of maghemite derived from O(2p) → Fe(3d) transitions [77,78]. In the freshly prepared samples of both NPs, shoulder peaks centered at around 450 nm are observed due to the presence of a minor amount of Fe$_3$O$_4$NPs, which then oxidized to the maghemite phase, and are not seen in the results of previous section (Section 2.2).

The peaks of $Fe^{3+}_{(aq)}$ and $Fe^{2+}_{(aq)}$ are assigned to charge-transfer electronic transition in octahedral aquo complexes [79–81]. Regarding these UV-Vis spectra of the $Fe^{3+}_{(aq)}/Fe^{2+}_{(aq)}$, it is worthy to explain that their electronic transition is largely governed by their d-electron configuration. The charge-transfer transitions are both spin- and Laporte-allowed and occur in the UV region, which is closely related to the strength of the applied ligand field (10Dq) [82].

2.4. FESEM-EDS Characterization of γ-Fe$_2$O$_3$NPs and γ-Fe$_2$O$_3$NPs@HG Magnetogel

The solid-state morphology and size of the γ-Fe$_2$O$_3$NPs were evaluated using Field Emission Scanning Electron Microscopy (FESEM) (Figure 5a,b). The FESEM image of uncoated γ-Fe$_2$O$_3$NPs demonstrates a grain-like morphology with a size range of 15–65 nm (Figure 5a,c), and a little aggregation is detected in the image. In agreement with DLS results, the FESEM images of PAA-coated γ-Fe$_2$O$_3$NPs show smaller particles (Figure 5b,d, 10–40 nm). This result is consistent with the general mechanism of γ-Fe$_2$O$_3$NPs formation including a nucleation step in the beginning of co-precipitation, followed by nuclei growth and coalescence. This mechanism is supported by our results, in which both DLS and FESEM analyses indicate that the γ-Fe$_2$O$_3$NPs obtained without PAA have a larger average size compared to the particles in the presence of PAA. EDS elemental analyses showed the presence of C element, which is due to the presence of PAA, previously confirmed by the FT-IR/ATR and Raman results (Figure S3). For both γ-Fe$_2$O$_3$NPs (with and without PAA), we see smaller distribution particles, compared with the DLS results, relating to the differences between these two techniques. In fact, FESEM probes the electron-rich part of the particle in the solid state; then only the inner core can be seen, and the result obtained would be smaller. On the other hand, the DLS measures a hydrodynamic diameter based on the diffusion of the particles in the solutions, and in most cases, we see larger nanoparticles due to the hydrodynamic layer. FESEM images of the hydrogel and magnetogel are shown in Figure 5e,f, both exhibiting the typical fibrillar structure, suggesting that the presence of PAA-coated γ-Fe$_2$O$_3$NPs does not change the macromolecular structure of the gel. Also, the small particles seen in the magnetogel image can be assigned to γ-Fe$_2$O$_3$NPs.

2.5. Rheological Studies and Swelling Ability

The viscoelastic behavior and the adsorption properties of hydrogel materials are tightly correlated properties. The goal of the rheological analyses was to understand how the presence of PAA-coated γ-Fe$_2$O$_3$NPs and their concentration could modulate the viscoelastic behavior of the hydrogels. As can be seen in Figure 6, for all systems, the experimental curves show a typical trend of a viscoelastic gel-like material, characterized by G' values much larger than G'' values [83]. All the magnetogels showed lower mechanical strength compared to the hydrogel alone, and a concentration-dependent trend is observed for the magnetogels' mechanical strength, in which the highest concentration of γ-Fe$_2$O$_3$NPs (30 mg/mL) had the lowest storage modulus. The decrease in the strength of the magnetogels may be due to the presence of PAA-coated γ-Fe$_2$O$_3$NPs with different sizes, morphologies, and surface charges. The negatively charged PAA-coated γ-Fe$_2$O$_3$NPs could expand the hydrogel network and increase the number of pores and free spaces, lowering the mechanical strength. The PAA-coated γ-Fe$_2$O$_3$NPs can interact with some hydroxyl and amine groups of the hydrogel using Fe(III) and PAA moieties expanding the internal network structure [84–88]. The swelling abilities of hydrogel and magnetogel samples were measured and are summarized in Table 3. All magnetogels showed higher swelling behavior in comparison to the native peptide-based hydrogel. This result may be attributed to the interaction of the hydrogel networks with PAA-coated γ-Fe$_2$O$_3$NPs, neutralizing the repulsions in the networks and resulting in the penetration of more water in order to compensate for the buildup of osmotic ion pressure [44,89]. Another water-adsorbing moiety is the attached PAA [40]. These results are consistent with the rheological studies, and there is an inverse relationship between them, meaning that the higher mechanical strength causes the lower swelling ability of the gels [38]. For all successive removal applications, we used the 10 mg/mL PAA-coated γ-Fe$_2$O$_3$NPs magnetogels because of their higher mechanical strength compared to the other two magnetogels.

Table 3. Swelling abilities of the hydrogel systems.

Samples	Swelling Degree (q)
HG	62.18 ± 0.35
γ-Fe$_2$O$_3$NPs@HG (10 mg/mL)	73.82 ± 0.99
γ-Fe$_2$O$_3$NPs@HG (20 mg/mL)	81.28 ± 0.22
γ-Fe$_2$O$_3$NPs@HG (30 mg/mL)	88.24 ± 0.31

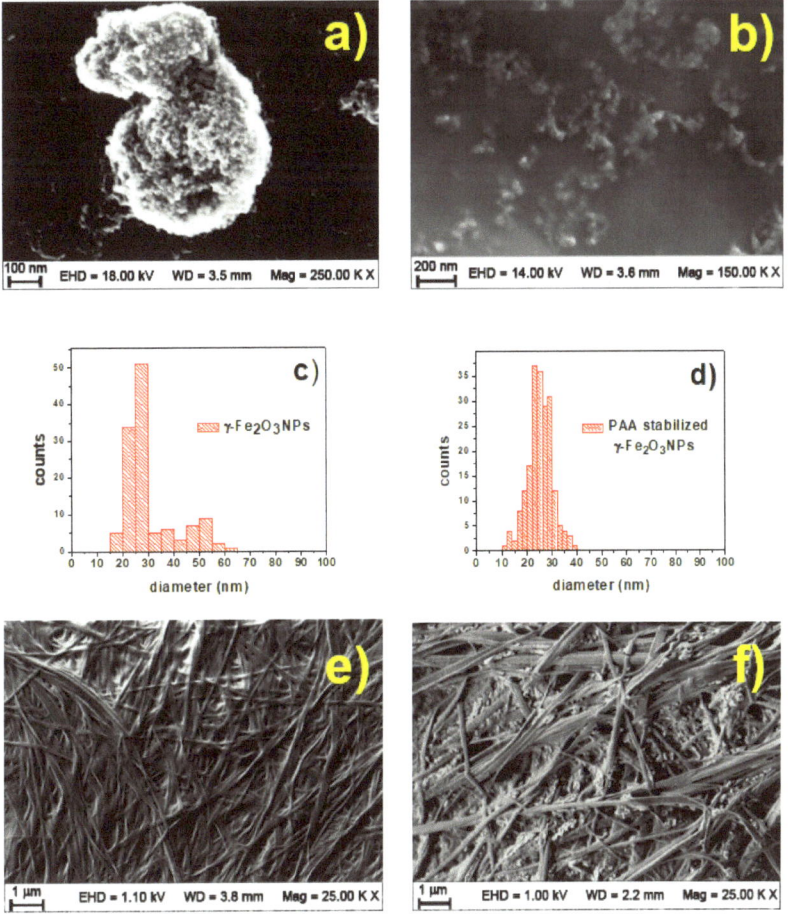

Figure 5. (**a**) FESEM results of uncoated γ-Fe$_2$O$_3$NPs and (**b**) PAA-coated γ-Fe$_2$O$_3$NPs; (**c**) size histogram of γ-Fe$_2$O$_3$NPs and (**d**) PAA-coated γ-Fe$_2$O$_3$NPs; (**e**) FESEM of peptide hydrogel alone and (**f**) γ-Fe$_2$O$_3$NPs@HG magnetogel. EDS data of PAA-coated γ-Fe$_2$O$_3$NPs are reported in the supporting section (Figure S3).

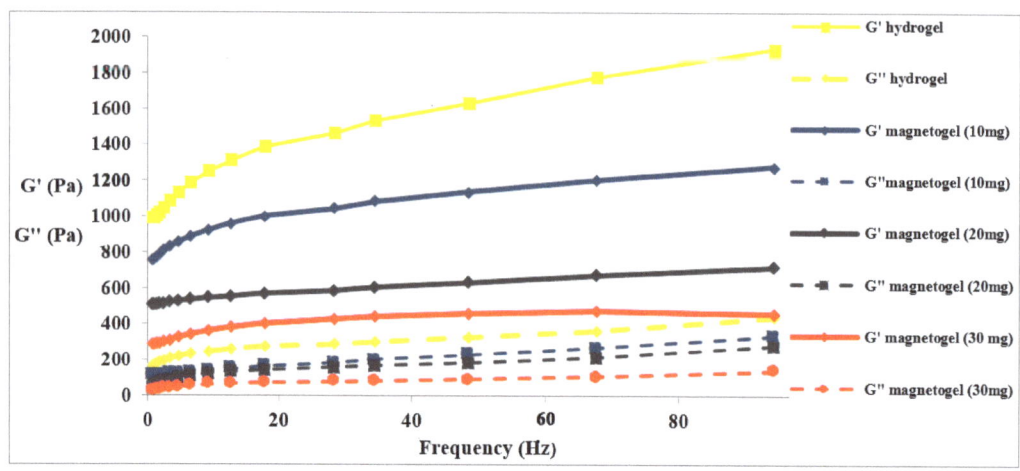

Figure 6. Frequency sweep of the Fmoc-Phe$_3$ hydrogel (HG) and γ-Fe$_2$O$_3$NPs embedded in peptide magnetogel (γ-Fe$_2$O$_3$NPs@HG) at different concentrations.

2.6. Magnetogels Application in the Removal of Metallic Cations

A simple and straightforward procedure of absorption was applied to Co(II), Ni(II), and Cr(III) aqueous solutions in the presence of fixed quantities of HG and γ-Fe$_2$O$_3$NPs@HG, following the UV-Vis peak arising from the metal ions over time, also comparing the effect of an external magnetic field (1.42–1.47 T). In the following paragraphs, the results of the studies are reported.

2.6.1. Co(II) Removal Studies

Co(II)$_{(aq)}$ has a broad metal-based absorption peak in the visible region centered at 512 nm [90]. Over the whole removal process, the intensity of this peak decreases without significant change in the wavelength for all the three adsorbents (Figures 7a and S4) tested, i.e., HG, γ-Fe$_2$O$_3$NPs@HG, and γ-Fe$_2$O$_3$NPs@HG upon magnetic field application. In particular, in Figure 7a, the UV-Vis spectra of γ-Fe$_2$O$_3$NPs@HG upon magnetic field application is reported. The adsorption capacity of these systems was monitored as a function of contact time (Co(II)$_{initial}$ = 61 mg/mL), and q_t (in mg g^{-1}) plots are shown in Figure 7b, with contact times ranging from 0 to 480 min. The adsorption capacities follow the trend γ-Fe$_2$O$_3$NPs@HG upon magnetic field application > γ-Fe$_2$O$_3$NPs@HG > HG, confirming the enhancing effect of γ-Fe$_2$O$_3$NPs on the adsorption. The interaction of magnetogel with an external magnet further increases the capacity, compared to that of the magnetogel. For each adsorbent, a fast Co(II) removal was observed in the first 15 min, which is due to the large number of active sites available on the hydrogel [91]. This rapid adsorption might be due to chemical rather than physical adsorption because of the possible complexation between the suitable functional groups of hydrogels and Co(II) ions. This phenomenon was already reported in several studies of Co(II) adsorption [16,92,93]. Then, their adsorption gradually slowed until reaching equilibrium after 480 min for all systems. At equilibrium, the capacity of γ-Fe$_2$O$_3$NPs@HG upon magnetic field application is about 1.25 times higher than that of the other two adsorbents. Total adsorption efficiencies were estimated, and the results show that γ-Fe$_2$O$_3$NPs@HG and γ-Fe$_2$O$_3$NPs@HG upon magnetic field application increase the removal efficiency by 0.3% and 5.3%, respectively (Table 4), in comparison with hydrogel alone. Under these experimental conditions, the external magnetic field has a relevant effect on the adsorption capacity and efficiency of the hydrogel.

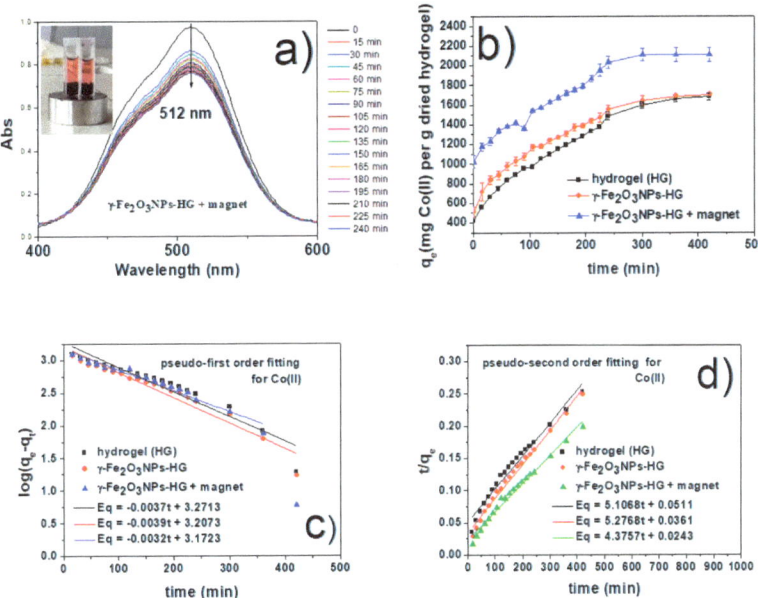

Figure 7. (a) UV-Vis study of Co(II) adsorption by γ-Fe$_2$O$_3$NPs@HG upon magnetic field application (the UV-vis spectra for the peptide HG and γ-Fe$_2$O$_3$NPs@HG are reported in the Supporting Information section, together with the calibration curve for Co(II) aqueous solutions); (**b**) adsorption capacity of HG, γ-Fe$_2$O$_3$NPs@HG, and γ-Fe$_2$O$_3$NPs@HG upon magnetic field application versus time; (**c**) fit of kinetic data to pseudo-first order model and (**d**) pseudo-second order model for Co(II).

Table 4. Co(II) adsorption efficiencies and capacities, obtained from the UV−Vis measurements at the equilibrium, at room temperature.

Adsorbent	Removal% (RE%)	Experimental q_e (mg g^{-1})
HG	20.4 ± 0.3	1680 ± 34
γ-Fe$_2$O$_3$NPs@HG	20.7 ± 0.4	1703 ± 42
γ-Fe$_2$O$_3$NPs@HG + magnet	25.7 ± 0.6	2111 ± 72

As time passes, the removal speeds decrease, owing to more active sites occupied by Co(II) ions. To better understand the adsorption kinetics, two well-known kinetic models, pseudo-first order and pseudo-second order [91], were applied and evaluated using the given linearized equations (Equations (9) and (10)), as provided in Section 4.8. In general, the pseudo-first order model describes a reversible adsorption between solid and liquid phases [94], claiming physisorption rather than chemisorption. Conversely, the second-order model mainly suggests a chemical adsorption of adsorbates onto the adsorbents [95], in which chemical bonding occurs among the metal ions and polar functional groups of the adsorbents. Two kinetic plots are shown in Figure 7c,d, and the subsequently calculated kinetics parameters are summarized in Table 5. Considering the correlation coefficient, R^2 better-fitted straight lines were obtained from the pseudo-second order relation for all the adsorbents, compared to those obtained for the pseudo-first order plots. Moreover, the calculated equilibrium adsorption capacities, q_e, obtained from the pseudo-second order equation were much closer to the experimental trend seen in Figure 7b. The results suggest that the adsorption of Co(II) onto our synthesized HG and γ-Fe$_2$O$_3$NPs@HG systems occurs through chemisorption, in agreement with previous works [96]. Also, we can clearly see the significant effect of magnetic interaction on the speed of adsorption, as shown in Table 5.

Table 5. Co(II) adsorption rate constant obtained from pseudo-first order and pseudo-second order models, at room temperature.

Adsorbent	Pseudo-First Order			Pseudo-Second Order		
	k_1 (min^{-1})	q_e (mg g^{-1})	R^2	k_2 (g mg^{-1} min^{-1})	q_e (mg g^{-1})	R^2
HG	0.0036	1867	0.9057	0.0005	1958	0.9722
γ-Fe$_2$O$_3$NPs@HG	0.0036	1611	0.9363	0.0005	1895	0.9872
γ-Fe$_2$O$_3$NPs@HG + magnet	0.0032	1486	0.9667	0.0004	2285	0.9834

The chemisorption of Co(II) can be interpreted based on the coordination chemistry of Co(II) aquo complexes, which are kinetically labile species. For substitution reactions, the transition metal kinetics are governed by their electronic configurations. For octahedral complexes, the d orbitals split into high energy e_g ($d_z{}^2$ and $d_{x^2-y^2}$) and low energy T_{2g} (d_{xz}, d_{yz}, d_{xy}) levels, of which e_g orbitals show anti-bonding characters. Due to the high-spin d^7 configuration of Co(II) aquo complexes having the $T_{2g}{}^5 e_g{}^2$, they also exhibit the Jahn-Teller effect (z-out), which is a geometric distortion of a non-linear molecular system that reduces its symmetry and energy, seen in the complexes having the occupied e_g levels. Therefore, the Co(II) ions in the dissolution process have high lability and are vulnerable to the chemical substitution reaction to achieve their stable electronic configurations. In fact, the presence of Fmoc-Phe and diphenylalanine (Phe$_2$) in the hydrogel network can provide the chelating moiety (through their nitrogen and oxygen atoms) for the Co(II) ions, resulting in stable Co(II) complexes [97].

Regarding the enhancing effect of γ-Fe$_2$O$_3$NPs nanoparticles, it is consistent with previous studies on magnetogels [98,99] and is related to the fact that NPs embedded in hydrogel can increase the cross-linking degree and porosity of the gel, providing a channel for the entry, exit, and adsorption of some substances [28]. More importantly, the interaction of γ-Fe$_2$O$_3$NPs@HG with the external magnet further enhances the adsorption. As is well-known, magnetogels can exhibit an on/off effect on the hydrogel pores [27]. In fact, swelling or shrinking states of γ-Fe$_2$O$_3$NPs@HG can be influenced by the magnetic dipole-dipole orientation of γ-Fe$_2$O$_3$NPs toward the external magnetic field [100] and can increase the permeability of Co(II) into the hydrogel network for the chemisorption [89,101].

2.6.2. Ni(II) Removal Studies

Ni(II)$_{(aq)}$ complexes display a typical octahedral structure with six water ligands in the first coordination shell [102] and an absorption maximum at 394 nm due to spin-allowed transitions. Upon interaction, a decrease in the absorption band was observed without significant changes in the wavelength (see supporting information, Figure S5). The γ-Fe$_2$O$_3$NPs@HG dramatically enhances the adsorption of Ni(II), with and without an external magnet. All the three systems show very fast adsorptions in the early 30 min, and applying an external magnet results in reaching the equilibrium earlier, after around 150 min for γ-Fe$_2$O$_3$NPs@HG + magnet. Considering the effect of an external magnet on γ-Fe$_2$O$_3$NPs@HG adsorption ability, a plateau is reached after 150 min. Without applying the external magnet, the equilibrium is reached later (after 360 min) but with 10% higher adsorption capacity (Figure 8a). The adsorption efficiencies are summarized in Table 6, showing that the γ-Fe$_2$O$_3$NPs@HG and γ-Fe$_2$O$_3$NPs@HG + magnet systems can increase the removal efficiency by 7.3% and 5.1%, respectively, compared to the native hydrogel.

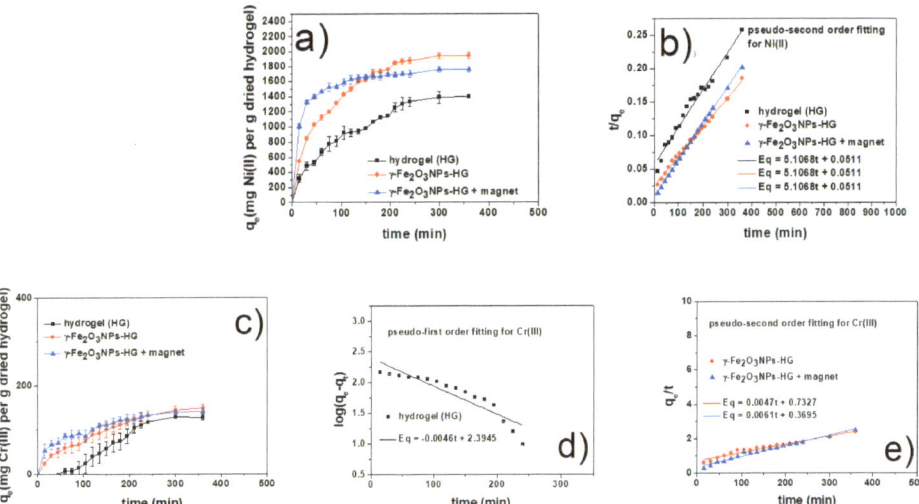

Figure 8. (**a**) Ni(II) adsorption capacity of HG, γ-Fe$_2$O$_3$NPs@HG, and γ-Fe$_2$O$_3$NPs@HG + magnet versus time; (**b**) fit of kinetic data to the pseudo-second-order model for Ni(II); (**c**) Cr(III) adsorption capacity of HG, γ-Fe$_2$O$_3$NPs@HG, and γ-Fe$_2$O$_3$NPs@HG + magnet versus time; (**d**) fit of kinetic data to the pseudo-first and (**e**) second-order models for Cr(III). (The complete UV-Vis study of Ni(II) and Cr(III) adsorption are reported in the supporting information section together with the Ni(II) and Cr(III) calibration curves.)

Table 6. Ni(II) adsorption efficiencies and capacities, obtained from the UV−Vis measurements at the equilibrium, at room temperature.

Adsorbent	Removal% (RE%)	Experimental q_e (mg g^{-1})
HG	18.6 ± 0.1	1399 ± 12
γ-Fe$_2$O$_3$NPs@HG	25.9 ± 0.4	1945 ± 41
γ-Fe$_2$O$_3$NPs@HG + magnet	23.7 ± 0.4	1758 ± 39

For all the three adsorbents, the speed of adsorption is higher in the beginning and slows down with time. Also, the adsorption speed of magnetogels (with and without an external magnet) is higher than that of the native hydrogel. Applying the kinetics models for the Ni(II) removal, it can be seen (Figure 8b) that the pseudo-second order model was more consistent for all these three systems (all second-order coefficients R^2 are higher than 0.97), supporting a chemisorption mechanism for Ni(II) (Table 7).

Table 7. Ni(II) adsorption rate constant obtained from pseudo-first order and pseudo-second order models, at room temperature.

Adsorbent	Pseudo-First Order			Pseudo-Second Order		
	k_1 (min^{-1})	q_e (mg g^{-1})	R^2	k_2 (g mg^{-1} min^{-1})	q_e (mg g^{-1})	R^2
HG	0.0115	1948	0.8548	0.000005	1785	0.9735
γ-Fe$_2$O$_3$NPs@HG	0.0161	2998	0.8247	0.000007	2325	0.9954
γ-Fe$_2$O$_3$NPs@HG + magnet	0.0092	613	0.9457	0.000036	1851	0.9993

2.6.3. Cr(III) Removal Studies

Cr(III) removal showed a different pattern, compared to Co(II) and Ni(II) adsorption. It is known that the aqueous Cr(III)$_{(aq)}$ can show three absorption peaks due to both the d→d electronic transitions [103]. In the present study, the absorption peak of 420 nm was monitored and assigned to spin-allowed transition. The UV-Vis spectra (reported in the supporting information, Figure S6) show the decrease in the 420 nm absorption band due to the removal of Cr(III). As can be seen in Figure 8c, the adsorption rate of both magnetogels γ-Fe$_2$O$_3$NPs@HG with and without the applied magnet is higher than that of the hydrogel HG in the first half of the experiments. However, at equilibrium, all adsorbents reach almost similar values of removal capacity (Figure 8c) (120–130 mg/g). Similar to the Co(II) and Ni(II) results, here also we clearly see the higher speed of magnetogels in removing the contaminant, further enhanced by the presence of a magnetic field. Table 8 reports the removal efficiencies for all the systems.

Table 8. Cr(III) adsorption efficiencies and capacities, obtained from the UV−Vis measurements at the equilibrium, at room temperature.

Adsorbent	Removal% (RE%)	Experimental q_e (mg g^{-1})
HG	13.2 ± 0.1	127 ± 6
γ-Fe$_2$O$_3$NPs@HG	15.5 ± 0.1	149 ± 8
γ-Fe$_2$O$_3$NPs@HG + magnet	14.7 ± 0.1	142 ± 5

Kinetic studies provided us further detailed information on Cr(III) removal. The fitting results of the kinetic models are given in Table 9. For the hydrogel, the correlation coefficient (R^2) provided by the pseudo-first order model is much higher than that of the pseudo-second order, suggesting a physical mechanism for the adsorption of Cr(III) (Figure 8d). This can be explained by the electronic configuration of Cr(III) aqua complexes (d^3), which are kinetically inert and have a low rate for the chemical substitution reaction in normal conditions. The physical mechanism might be due to the electrostatic interaction of positively charged Cr(III) ions with the negatively charged hydrogel network [24,44]. Conversely, the data obtained with both magnetogels (with and without magnet) fit with a pseudo-second order model, suggesting that the presence of γ-Fe$_2$O$_3$NPs@HG and magnetic field can both change the hydrogel network and porosity, providing chelating conditions inside the hydrogel suitable for the formation of the chemical bonds among hydrogel hyteroatoms and Cr(III) ions [104].

Table 9. Cr(III) adsorption rate constant obtained from pseudo-first-order and pseudo-second-order models, at room temperature.

Adsorbent	Pseudo-First Order			Pseudo-Second Order		
	k_1 (min^{-1})	q_e (mg g^{-1})	R^2	k_2 (g mg^{-1} min^{-1})	q_e (mg g^{-1})	R^2
HG	0.0105	248	0.8355	-	27	0.0040
γ-Fe$_2$O$_3$NPs@HG	0.0101	184	0.9052	0.00003	212	0.9538
γ-Fe$_2$O$_3$NPs@HG + magnet	0.0112	134	0.9197	0.0001	163	0.9794

3. Conclusions

The goal of this work was to direct attention to emerging and novel research involving magnetogel nanohybrid materials that might be relevant in future applications for the treatment of wastewater, as well as in other fields.

Generally, composite hydrogels are promising adsorbents with tunable features, and we demonstrated that the addition of effective functional groups in nanohybrid materials through chemical conjugation is a promising strategy to further improve the adsorption abilities of hydrogels. In fact, the results achieved pointed out that the presence of γ-Fe$_2$O$_3$NPs provides magnetic properties to the resulting nanohybrids, which can be applied for magnetic-based removal applications of contaminants, such as heavy metal ions, from aqueous phases. The results of the removal studies demonstrate that the presence of γ-Fe$_2$O$_3$NPs in combination with the application of an external magnetic field increases the adsorption efficiency of the hydrogel matrix for all the metal ions tested in this study; in particular, the γ-Fe$_2$O$_3$NPs@HG + magnet was effective to absorb up to 2111 ± 72 mg/g for Co(II), 1758 ± 39 mg/g for Ni(II), and 142 ± 5 mg/g for Cr(III).

The kinetic models showed the chemisorption of these cations onto the γ-Fe$_2$O$_3$NPs@HG (with and without the magnetic field). Regarding the native HG, Co(II) and Ni(II) showed chemisorption, but for the Cr(III), the results were fitted with a physical adsorption mechanism. This work showed that the peptide-based magnetogels can be introduced as promising adsorbing materials for wastewater treatment to remove heavy metals from aqueous solutions. In the future, this study could be expanded to test the recovering ability of the three adsorbing systems for the recycling of metal ions, and extensive efforts should be directed to scale up the applications and test the developed materials in practical scenarios.

4. Materials and Methods

4.1. Materials

L-Phenylalanyl-L-phenylalanine (H-Phe-Phe-OH, 98%, 312.36 g/mol) and N-(9-Fluorenylmethoxycarbonyl)-L-phenylalanine (Fmoc-L-phenylalanine: Fmoc-Phe-OH, 99%, 387.44 g/mol) were purchased from Bachem GmbH (Weil am Rhein, Germany) and used as received. FeCl$_2$·4H$_2$O (198.75 g/mol) and FeCl$_3$ (162.20 g/mol) were purchased from Fluka. NaOH (39.99 g/mol) and CrCl$_3$·6H$_2$O (158.36 g/mol) were purchased from Carlo Erba Reagents (Cornaredo (MI), Italy). NiCl$_2$·6H$_2$O (129.59 g/mol) and CoCl$_2$·6H$_2$O (129.83 g/mol) were obtained from Alfa Aesar. Lipase from *Pseudomonas fluorescens* (PFL ≥ 20,000 U/mg) was purchased from Sigma-Aldrich (Milan, Italy) and used as received. Ultra-pure water (H$_2$O$_{up}$) was obtained using a Zeneer Power I Scholar-UV (Full Tech Instruments, Rome, Italy) apparatus. In this study, the external magnetic field was provided by a commercial neodymium-based magnet possessing 39.270 cm^3 volume, magnetization quality of N5, and 1.42–1.47 T of magnetic strength.

4.2. Synthesis of γ-Fe$_2$O$_3$NPs@HG Magnetogels

γ-Fe$_2$O$_3$NPs were synthesized via the co-precipitation method in which 40 mg of FeCl$_3$ and 25 mg of FeCl$_2$·4H$_2$O were dissolved in 25 mL of an aqueous solution of polyacrylic acid (PAA) and degassed with Ar$_{(g)}$ for 15 min, followed by increasing the temperature to 80 °C [87]. Then, pH of this solution was increased to 11 by a fast addition of NaOH (10 M). The mixture was stirred at 80 °C for 1 h (with constant monitoring of the pH) and then cooled down to room temperature. The dark-brown colloidal solution of MNPs was collected by a strong magnet (1.42–1.47 T) and washed three times with a total volume of 150 mL of ultra-pure water to remove the excess amount of NaOH and other non-magnetic species. The resultant NPs were freeze-dried and stored at room temperature. Regarding PAA, it was synthesized by radical polymerization of acrylic acid (2 mL) in the presence of the initiator potassium persulfate (50 mg) in a total volume of 25 mL of water at 80 °C for 5 h. FmocPhe$_3$ hydrogel synthesis was performed according to our previous work [44]. In brief, Phe$_2$ and FmocPhe peptides were added in equimolar amounts to a colloidal solution containing 2 mL aqueous solution of the as-synthesized MNPs-PAA (10–30 mg/mL) and 420 μL of 0.5 M NaOH, stirring magnetically for 10 min. Then pH was adjusted to 7 by adding 1.5 mL of 0.1 M HCl, followed by the addition of 100 μL lipase aqueous solution (50 mg/mL), leaving the resultant mixture at 30 °C for 30 min.

4.3. UV-Vis, FT-IR/ATR and Raman Spectroscopies and Dynamic Light Scattering (DLS)

All UV-Vis spectra of the MNPs and Fe^{2+}/Fe^{3+} ions were recorded in 1.00 cm optical path quartz cells using a Cary 100 Varian spectrophotometer. For the removal studies, we used the plastic PMMA cells. FTIR-ATR data were collected with a Bruker Vertex 70 instrument (Bruker Optics, Ettlingen, Germany) using KRS-5 cells in the 4000–400 cm^{-1} range or in ATR mode on a diamond crystal in the 4000–600 cm^{-1} spectral region. Dynamic light scattering (DLS) measurements were performed using a Malvern Zetasizer with a minimum of 10 replicates. All measurements were carried out at least three times with reporting the average value ± standard deviation. Raman spectra were run at room temperature in backscattering geometry with an inVia Renishaw micro-Raman spectrometer equipped with an air-cooled CCD detector and super-Notch filters. An Ar$^+$ ion laser (λ_{laser} = 514 nm) was used, coupled to a Leica DLML microscope with a 20× objective. The resolution was 2 cm^{-1}, and spectra were calibrated using the 520.5 cm^{-1} line of a silicon wafer [105].

4.4. X-ray Photoelectron Spectroscopy (XPS)

The XPS analysis was conducted using a modified Omicron NanoTechnology MXPS system.(Scienta Omicron GmbH, Taunusstein, Germany) Samples were excited by achromatic AlKα photons (hν = 1486.6 eV), operating the anode at 14–15 kV and 10–20 mA. The take-off angle and pass energy were fixed at 21° and 20 eV, respectively. Samples were prepared by casting onto a hydrogenated Si(100) wafer a 20 µL drop of MNPs-PAA, and the obtained Si-supported sample was left to dry overnight and then mounted on a stainless steel sample holder for measurement [106].

4.5. Electron Microscopy Studies

FESEM images were performed using an Auriga Zeiss field emission scanning electron microscope supported by an energy dispersive X-ray spectroscopy detector (FESEM-EDS) instrument (Zeiss, Oberkochen, Germany). Samples were deposited onto conducting silicon stubs without the need of a conductive coating and analyzed at an accelerating voltage that avoided radiation damage.

4.6. Rheology Measurements

The rheological properties of three magnetogels containing different concentration of NPs and of the native hydrogel were studied by using an Anton Paar MCR 302 rotational rheometer in frequency sweep experiments, (Anton Paar, Turin, Italy) as reported previously [44].

4.7. Swelling Test

Swelling tests of the hydrogel samples were conducted as reported previously [44]. The swelling degree (q) was calculated by using the following equation:

$$q = (W_s - W_d)/W_d \tag{6}$$

where W_s is the weight of the hydrogel after removing the swelling solution and W_d is the weight of the freeze-dried sample.

4.8. Adsorption Experiments

Magnetogel samples were prepared in cuvettes, and 2 mL of the target solutions (Co(II), Ni(II), and Cr(III)) were cast on top of them, using different concentrations of the cations (see Figure S7). We only used the 10 mg type γ-Fe$_2$O$_3$NPs@HG magnetogel because of its higher mechanical strength, evidenced by the rheological characterization.

The removal studies were performed using UV-Vis spectroscopy, and the results are presented in Figures 7, 8 and S2–S4. For all the tested pollutants, we used three main hydrogel-based absorbents including (1) HG, (2) γ-Fe$_2$O$_3$NPs@HG magnetogel, and (3) γ-Fe$_2$O$_3$NPs@HG + magnet). We also studied the removal efficiency of γ-Fe$_2$O$_3$NPs alone (10 mg/mL), and no significant change was observed in the UV-Vis of solutions. The UV-

Vis absorbances of the solutions were monitored over time with 15 min intervals, and the removal efficiency (RE) was estimated by absorption spectra using Equation (7) as follows:

$$RE\,(\%) = (C_0 - C_f)/C_0 \times 100 \quad (7)$$

where C_0 is the initial concentration of pollutant and C_f is the concentration of pollutant in the eluted solution. The calibration curves were obtained and used for the calculations. Also, the adsorption capacities were estimated using the stock solutions of the pollutants (Co(II), Ni(II), and Cr(III)) prepared at pH 7, and their concentrations remaining in solutions after specific time intervals were determined by UV-Vis spectrophotometry. The adsorption capacity (q_e, mg g^{-1}) of the adsorbents was calculated using Equation (8) [107–109]:

$$q_e = (C_0 - C_e)/m \times V \quad (8)$$

where m (g) is the dried hydrogel mass, C_0 and C_e (mg L^{-1}) are the initial and equilibrium pollutant concentrations, and V (L) is the solution volume, respectively.

Kinetic behavior was studied using non-linear pseudo-first order and pseudo-second order kinetic models (Equations (9) and (10)) [108,110].

$$\log(q_e - q_t) = \log q_e - k_1 t/2.303 \quad (9)$$

$$t/q_t = 1/k^2 q_e^2 + t/q_e \quad (10)$$

Supplementary Materials: The following supporting information can be downloaded at: https://www.mdpi.com/article/10.3390/gels9080621/s1, Figure S1: Separation of γ-Fe$_2$O$_3$NPs with an external magnetic field (1.42–1.47 T); Figure S2: PAA-coated γ-Fe$_2$O$_3$NPs solutions after one week for (a) 1.6, 0.16 and 0.08 mg/mL and (b) 10 mg/mL; (c) sedimentation of uncoated γ-Fe$_2$O$_3$NPs after 24 h.; Figure S3: EDS spectrum of PAA-coated γ-Fe$_2$O$_3$NPs; Figure S4: UV-Vis study of Co(II) adsorption for (a) the peptide HG; (b) γ-Fe$_2$O$_3$NPs@HG, and (c) the calibration curve for Co(II) aqueous solutions.; Figure S5: UV-Vis study of Ni(II) adsorption for (a) the peptide HG; (b) γ-Fe$_2$O$_3$NPs@HG; (c) γ-Fe$_2$O$_3$NPs@HG upon magnetic field application; (d) the calibration curve for Ni(II) aqueous solutions, and (e) fit of kinetic data to pseudo-first order model for Ni(II).; Figure S6: UV-Vis study of Cr(III) adsorption for (a) the peptide HG; (b) γ-Fe$_2$O$_3$NPs@HG; (c) γ-Fe$_2$O$_3$NPs@HG upon magnetic field application, and (d) the calibration curve for Cr(III) aqueous solutions.; Figure S7: Methodology used for studying the removal efficiency of Co(II), as an example here.

Author Contributions: Conceptualization: F.H.H. and R.B.; Methodology: L.C. and S.C.; Measurements and elaboration data: F.H.H., R.B., L.C., S.C., A.G.M., F.A., C.P. and I.F.; Writing—original draft preparation: F.H.H. and R.B.; Writing—review and editing, L.C., S.C., C.P. and I.F. All authors have read and agreed to the published version of the manuscript.

Funding: The authors acknowledge the financial support of Sapienza Funding Grants Ateneo 2022 (RM1221867C322C1 and RM12218167B480E0).

Institutional Review Board Statement: Not applicable.

Informed Consent Statement: Not applicable.

Data Availability Statement: Data available on request.

Conflicts of Interest: The authors declare no conflict of interest.

References

1. Mondal, B.; Bairagi, D.; Nandi, N.; Hansda, B.; Das, K.S.; Edwards-Gayle, C.J.C.; Castelletto, V.; Hamley, I.W.; Banerjee, A. Peptide-Based Gel in Environmental Remediation: Removal of Toxic Organic Dyes and Hazardous Pb^{2+} and Cd^{2+} Ions from Wastewater and Oil Spill Recovery. *Langmuir* **2020**, *36*, 12942–12953. [CrossRef]
2. Zhao, X.; Wang, D.; Xiang, C.; Zhang, F.; Liu, L.; Zhou, X.; Zhang, H. Facile Synthesis of Boron Organic Polymers for Efficient Removal and Separation of Methylene Blue, Rhodamine B, and Rhodamine 6G. *ACS Sustain. Chem. Eng.* **2018**, *6*, 16777–16787. [CrossRef]

3. Sharma, G.; Kumar, A.; Ghfar, A.A.; García-Peñas, A.; Naushad, M.; Stadler, F.J. Fabrication and Characterization of Xanthan Gum-Cl-Poly(Acrylamide-Co-Alginic Acid) Hydrogel for Adsorption of Cadmium Ions from Aqueous Medium. *Gels* 2022, *8*, 23. [CrossRef] [PubMed]
4. Qasem, N.A.A.; Mohammed, R.H.; Lawal, D.U. Removal of Heavy Metal Ions from Wastewater: A Comprehensive and Critical Review. *Npj Clean Water* 2021, *4*, 36. [CrossRef]
5. Mahey, S.; Kumar, R.; Sharma, M.; Kumar, V.; Bhardwaj, R. A Critical Review on Toxicity of Cobalt and Its Bioremediation Strategies. *SN Appl. Sci.* 2020, *2*, 1279. [CrossRef]
6. Purushotham, D.; Rashid, M.; Lone, M.A.; Rao, A.N.; Ahmed, S.; Nagaiah, E.; Dar, F.A. Environmental Impact Assessment of Air and Heavy Metal Concentration in Groundwater of Maheshwaram Watershed, Ranga Reddy District, Andhra Pradesh. *J. Geol. Soc. India* 2013, *81*, 385–396. [CrossRef]
7. Fratoddi, I.; Cerra, S.; Salamone, T.A.; Fioravanti, R.; Sciubba, F.; Zampetti, E.; Macagnano, A.; Generosi, A.; Paci, B.; Scaramuzzo, F.A.; et al. Functionalized Gold Nanoparticles as an Active Layer for Mercury Vapor Detection at Room Temperature. *ACS Appl. Nano Mater.* 2021, *4*, 2930–2940. [CrossRef]
8. Cerra, S.; Salamone, T.A.; Bearzotti, A.; Hajareh Haghighi, F.; Mercurio, M.; Marsotto, M.; Battocchio, C.; Fioravanti, R.; Diociaiuti, M.; Fratoddi, I. Thiol-Functionalized Palladium Nanoparticles Networks: Synthesis, Characterization, and Room Temperature (Toxic) Vapor Detection. *Part. Part. Syst. Charact.* 2023, *40*, 2200189. [CrossRef]
9. Dakova, I.; Vasileva, P.; Karadjova, I. Cr(III) Ion-Imprinted Hydrogel Membrane for Chromium Speciation Analysis in Water Samples. *Gels* 2022, *8*, 757. [CrossRef]
10. Chowdhury, A.; Khan, A.A.; Kumari, S.; Hussain, S. Superadsorbent Ni–Co–S/SDS Nanocomposites for Ultrahigh Removal of Cationic, Anionic Organic Dyes and Toxic Metal Ions: Kinetics, Isotherm and Adsorption Mechanism. *ACS Sustain. Chem. Eng.* 2019, *7*, 4165–4176. [CrossRef]
11. Maiti, D.; Mukhopadhyay, S.; Devi, P.S. Evaluation of Mechanism on Selective, Rapid, and Superior Adsorption of Congo Red by Reusable Mesoporous α-Fe$_2$O$_3$ Nanorods. *ACS Sustain. Chem. Eng.* 2017, *5*, 11255–11267. [CrossRef]
12. Ramalingam, B.; Parandhaman, T.; Choudhary, P.; Das, S.K. Biomaterial Functionalized Graphene-Magnetite Nanocomposite: A Novel Approach for Simultaneous Removal of Anionic Dyes and Heavy-Metal Ions. *ACS Sustain. Chem. Eng.* 2018, *6*, 6328–6341. [CrossRef]
13. Minju, N.; Jobin, G.; Savithri, S.; Ananthakumar, S. Double-Silicate Derived Hybrid Foams for High-Capacity Adsorption of Textile Dye Effluent: Statistical Optimization and Adsorption Studies. *Langmuir* 2019, *35*, 9382–9395. [CrossRef]
14. Ray, S.; Das, A.K.; Banerjee, A. PH-Responsive, Bolaamphiphile-Based Smart Metallo-Hydrogels as Potential Dye-Adsorbing Agents, Water Purifier, and Vitamin B12 Carrier. *Chem. Mater.* 2007, *19*, 1633–1639. [CrossRef]
15. Fortunato, A.; Mba, M. A Peptide-Based Hydrogel for Adsorption of Dyes and Pharmaceuticals in Water Remediation. *Gels* 2022, *8*, 672. [CrossRef]
16. Seida, Y.; Tokuyama, H. Hydrogel Adsorbents for the Removal of Hazardous Pollutants—Requirements and Available Functions as Adsorbent. *Gels* 2022, *8*, 220. [CrossRef]
17. Godiya, C.B.; Ruotolo, L.A.M.; Cai, W. Functional Biobased Hydrogels for the Removal of Aqueous Hazardous Pollutants: Current Status, Challenges, and Future Perspectives. *J. Mater. Chem. A* 2020, *8*, 21585–21612. [CrossRef]
18. Okesola, B.O.; Smith, D.K. Applying Low-Molecular Weight Supramolecular Gelators in an Environmental Setting–Self-Assembled Gels as Smart Materials for Pollutant Removal. *Chem. Soc. Rev.* 2016, *45*, 4226–4251. [CrossRef]
19. Chronopoulou, L.; Margheritelli, S.; Toumia, Y.; Paradossi, G.; Bordi, F.; Sennato, S.; Palocci, C. Biosynthesis and Characterization of Cross-Linked Fmoc Peptide-Based Hydrogels for Drug Delivery Applications. *Gels* 2015, *1*, 179–193. [CrossRef]
20. Boni, R.; Regan, L. Modulating the Viscoelastic Properties of Covalently Crosslinked Protein Hydrogels. *Gels* 2023, *9*, 481. [CrossRef]
21. Chen, H.; Feng, R.; Xia, T.; Wen, Z.; Li, Q.; Qiu, X.; Huang, B.; Li, Y. Progress in Surface Modification of Titanium Implants by Hydrogel Coatings. *Gels* 2023, *9*, 423. [CrossRef] [PubMed]
22. Hajareh Haghighi, F.; Mercurio, M.; Cerra, S.; Salamone, T.A.; Bianymotlagh, R.; Palocci, C.; Romano Spica, V.; Fratoddi, I. Surface Modification of TiO$_2$ Nanoparticles with Organic Molecules and Their Biological Applications. *J. Mater. Chem. B* 2023, *11*, 2334–2366. [CrossRef] [PubMed]
23. Ningrum, E.O.; Gotoh, T.; Ciptonugroho, W.; Karisma, A.D.; Agustiani, E.; Safitri, Z.M.; Dzaky, M.A. Novel Thermosensitive-Co-Zwitterionic Sulfobetaine Gels for Metal Ion Removal: Synthesis and Characterization. *Gels* 2021, *7*, 273. [CrossRef] [PubMed]
24. Binaymotlagh, R.; Chronopoulou, L.; Hajareh Haghighi, F.; Fratoddi, I.; Palocci, C. Peptide-Based Hydrogels: New Materials for Biosensing and Biomedical Applications. *Materials* 2022, *15*, 5871. [CrossRef]
25. Veloso, S.R.S.; Ferreira, P.M.T.; Martins, J.A.; Coutinho, P.J.G.; Castanheira, E.M.S. Magnetogels: Prospects and Main Challenges in Biomedical Applications. *Pharmaceutics* 2018, *10*, 145. [CrossRef]
26. Salahuddin, B.; Aziz, S.; Gao, S.; Hossain, M.S.A.; Billah, M.; Zhu, Z.; Amiralian, N. Magnetic Hydrogel Composite for Wastewater Treatment. *Polymers* 2022, *14*, 5074. [CrossRef]
27. Veloso, S.R.S.; Andrade, R.G.D.; Castanheira, E.M.S. Review on the Advancements of Magnetic Gels: Towards Multifunctional Magnetic Liposome-Hydrogel Composites for Biomedical Applications. *Adv. Colloid Interface Sci.* 2021, *288*, 102351. [CrossRef]
28. Gang, F.; Jiang, L.; Xiao, Y.; Zhang, J.; Sun, X. Multi-Functional Magnetic Hydrogel: Design Strategies and Applications. *Nano Sel.* 2021, *2*, 2291–2307. [CrossRef]

29. Milakin, K.A.; Taboubi, O.; Acharya, U.; Lhotka, M.; Pokorný, V.; Konefał, M.; Kočková, O.; Hromádková, J.; Hodan, J.; Bober, P. Polypyrrole-Barium Ferrite Magnetic Cryogels for Water Purification. *Gels* **2023**, *9*, 92. [CrossRef]
30. Gonçalves, A.; Almeida, F.V.; Borges, J.P.; Soares, P.I.P. Incorporation of Dual-Stimuli Responsive Microgels in Nanofibrous Membranes for Cancer Treatment by Magnetic Hyperthermia. *Gels* **2021**, *7*, 28. [CrossRef]
31. Häring, M.; Schiller, J.; Mayr, J.; Grijalvo, S.; Eritja, R.; Díaz, D.D. Magnetic Gel Composites for Hyperthermia Cancer Therapy. *Gels* **2015**, *1*, 135–161. [CrossRef]
32. Zamora-Mora, V.; Soares, P.I.P.; Echeverria, C.; Hernández, R.; Mijangos, C. Composite Chitosan/Agarose Ferrogels for Potential Applications in Magnetic Hyperthermia. *Gels* **2015**, *1*, 69–80. [CrossRef]
33. De Melo, F.M.; Grasseschi, D.; Brandão, B.B.N.S.; Fu, Y.; Toma, H.E. Superparamagnetic Maghemite-Based CdTe Quantum Dots as Efficient Hybrid Nanoprobes for Water-Bath Magnetic Particle Inspection. *ACS Appl. Nano Mater.* **2018**, *1*, 2858–2868. [CrossRef]
34. Ali, A.F.; Atwa, S.M.; El-Giar, E.M. 6—Development of Magnetic Nanoparticles for Fluoride and Organic Matter Removal from Drinking Water. In *Water Purification*; Grumezescu, A.M.B.T.-W.P., Ed.; Academic Press: Cambridge, MA, USA, 2017; pp. 209–262; ISBN 978-0-12-804300-4.
35. Kunduru, K.R.; Nazarkovsky, M.; Farah, S.; Pawar, R.P.; Basu, A.; Domb, A.J. 2—Nanotechnology for Water Purification: Applications of Nanotechnology Methods in Wastewater Treatment. In *Water Purification*; Grumezescu, A.M., Ed.; Academic Press: Cambridge, MA, USA, 2017; pp. 33–74; ISBN 978-0-12-804300-4.
36. Asadi, S.; Eris, S.; Azizian, S. Alginate-Based Hydrogel Beads as a Biocompatible and Efficient Adsorbent for Dye Removal from Aqueous Solutions. *ACS Omega* **2018**, *3*, 15140–15148. [CrossRef]
37. Amiralian, N.; Mustapic, M.; Hossain, M.S.A.; Wang, C.; Konarova, M.; Tang, J.; Na, J.; Khan, A.; Rowan, A. Magnetic Nanocellulose: A Potential Material for Removal of Dye from Water. *J. Hazard. Mater.* **2020**, *394*, 122571. [CrossRef]
38. Trache, D.; Tarchoun, A.F.; Derradji, M.; Hamidon, T.S.; Masruchin, N.; Brosse, N.; Hussin, M.H. Nanocellulose: From Fundamentals to Advanced Applications. *Front. Chem.* **2020**, *8*, 392. [CrossRef]
39. Sanchez, L.M.; Actis, D.G.; Gonzalez, J.S.; Zélis, P.M.; Alvarez, V.A. Effect of PAA-Coated Magnetic Nanoparticles on the Performance of PVA-Based Hydrogels Developed to Be Used as Environmental Remediation Devices. *J. Nanoparticle Res.* **2019**, *21*, 64. [CrossRef]
40. Chełminiak, D.; Ziegler-Borowska, M.; Kaczmarek, H. Synthesis of Magnetite Nanoparticles Coated with Poly(Acrylic Acid) by Photopolymerization. *Mater. Lett.* **2016**, *164*, 464–467. [CrossRef]
41. Liang, Y.-Y.; Zhang, L.-M.; Jiang, W.; Li, W. Embedding Magnetic Nanoparticles into Polysaccharide-Based Hydrogels for Magnetically Assisted Bioseparation. *ChemPhysChem* **2007**, *8*, 2367–2372. [CrossRef]
42. Sang, J.; Wu, R.; Guo, P.; Du, J.; Xu, S.; Wang, J. Affinity-Tuned Peroxidase-like Activity of Hydrogel-Supported Fe_3O_4 Nanozyme through Alteration of Crosslinking Concentration. *J. Appl. Polym. Sci.* **2016**, *133*, 43065. [CrossRef]
43. Witt, M.U.; Hinrichs, S.; Möller, N.; Backes, S.; Fischer, B.; von Klitzing, R. Distribution of $CoFe_2O_4$ Nanoparticles Inside PNIPAM-Based Microgels of Different Cross-Linker Distributions. *J. Phys. Chem. B* **2019**, *123*, 2405–2413. [CrossRef] [PubMed]
44. Binaymotlagh, R.; Del Giudice, A.; Mignardi, S.; Amato, F.; Marrani, A.G.; Sivori, F.; Cavallo, I.; Di Domenico, E.G.; Palocci, C.; Chronopoulou, L. Green In Situ Synthesis of Silver Nanoparticles-Peptide Hydrogel Composites: Investigation of Their Antibacterial Activities. *Gels* **2022**, *8*, 700. [CrossRef] [PubMed]
45. Ahn, T.; Kim, J.H.; Yang, H.-M.; Lee, J.W.; Kim, J.-D. Formation Pathways of Magnetite Nanoparticles by Coprecipitation Method. *J. Phys. Chem. C* **2012**, *116*, 6069–6076. [CrossRef]
46. Patnaik, P. *Handbook of Inorganic Chemicals*; McGraw-Hill: New York, NY, USA, 2003; Volume 529.
47. Rebodos, R.L.; Vikesland, P.J. Effects of Oxidation on the Magnetization of Nanoparticulate Magnetite. *Langmuir* **2010**, *26*, 16745–16753. [CrossRef]
48. Cerra, S.; Carlini, L.; Salamone, T.A.; Hajareh Haghighi, F.; Mercurio, M.; Pennacchi, B.; Sappino, C.; Battocchio, C.; Nottola, S.; Matassa, R.; et al. Noble Metal Nanoparticles Networks Stabilized by Rod-Like Organometallic Bifunctional Thiols. *ChemistrySelect* **2023**, *8*, e202300874. [CrossRef]
49. Gutiérrez, L.; de la Cueva, L.; Moros, M.; Mazarío, E.; de Bernardo, S.; de la Fuente, J.M.; Morales, M.P.; Salas, G. Aggregation Effects on the Magnetic Properties of Iron Oxide Colloids. *Nanotechnology* **2019**, *30*, 112001. [CrossRef]
50. Harris, L.A.; Goff, J.D.; Carmichael, A.Y.; Riffle, J.S.; Harburn, J.J.; St Pierre, T.G.; Saunders, M. Magnetite Nanoparticle Dispersions Stabilized with Triblock Copolymers. *Chem. Mater.* **2003**, *15*, 1367–1377. [CrossRef]
51. Pardoe, H.; Chua-Anusorn, W.; Pierre, T.G.S.; Dobson, J. Structural and Magnetic Properties of Nanoscale Iron Oxide Particles Synthesized in the Presence of Dextran or Polyvinyl Alcohol. *J. Magn. Magn. Mater.* **2001**, *225*, 41–46. [CrossRef]
52. Di Corato, R.; Espinosa, A.; Lartigue, L.; Tharaud, M.; Chat, S.; Pellegrino, T.; Ménager, C.; Gazeau, F.; Wilhelm, C. Magnetic Hyperthermia Efficiency in the Cellular Environment for Different Nanoparticle Designs. *Biomaterials* **2014**, *35*, 6400–6411. [CrossRef]
53. Nahar, Y.; Rahman, M.A.; Hossain, M.K.; Sharafat, M.K.; Karim, M.R.; Elaissari, A.; Ochiai, B.; Ahmad, H.; Rahman, M.M. A Facile One-Pot Synthesis of Poly(Acrylic Acid)-Functionalized Magnetic Iron Oxide Nanoparticles for Suppressing Reactive Oxygen Species Generation and Adsorption of Biocatalyst. *Mater. Res. Express* **2020**, *7*, 16102. [CrossRef]
54. Rutnakornpituk, M.; Puangsin, N.; Theamdee, P.; Rutnakornpituk, B.; Wichai, U. Poly (Acrylic Acid)-Grafted Magnetic Nanoparticle for Conjugation with Folic Acid. *Polymer* **2011**, *52*, 987–995. [CrossRef]

55. Jain, N.; Wang, Y.; Jones, S.K.; Hawkett, B.S.; Warr, G.G. Optimized Steric Stabilization of Aqueous Ferrofluids and Magnetic Nanoparticles. *Langmuir* 2010, *26*, 4465–4472. [CrossRef]
56. Lin, C.-L.; Lee, C.-F.; Chiu, W.-Y. Preparation and Properties of Poly(Acrylic Acid) Oligomer Stabilized Superparamagnetic Ferrofluid. *J. Colloid Interface Sci.* 2005, *291*, 411–420. [CrossRef]
57. Chronopoulou, L.; Lorenzoni, S.; Masci, G.; Dentini, M.; Togna, A.R.; Togna, G.; Bordi, F.; Palocci, C. Lipase-Supported Synthesis of Peptidic Hydrogels. *Soft Matter* 2010, *6*, 2525–2532. [CrossRef]
58. Aaron, M.J.; Heather, C. Allen Vibrational Spectroscopic Characterization of Hematite, Maghemite, and Magnetite Thin Films Produced by Vapor Deposition. *ACS Appl. Mater. Interfaces* 2010, *2*, 2804–2812.
59. Yadav, B.S.; Singh, R.; Vishwakarma, A.K.; Kumar, N. Facile Synthesis of Substantially Magnetic Hollow Nanospheres of Maghemite (γ-Fe_2O_3) Originated from Magnetite (Fe_3O_4) via Solvothermal Method. *J. Supercond. Nov. Magn.* 2020, *33*, 2199–2208. [CrossRef]
60. Chamritski, I.; Burns, G. Infrared-and Raman-Active Phonons of Magnetite, Maghemite, and Hematite: A Computer Simulation and Spectroscopic Study. *J. Phys. Chem. B* 2005, *109*, 4965–4968. [CrossRef]
61. Kirwan, L.J.; Fawell, P.D.; van Bronswijk, W. In Situ FTIR-ATR Examination of Poly(Acrylic Acid) Adsorbed onto Hematite at Low PH. *Langmuir* 2003, *19*, 5802–5807. [CrossRef]
62. Testa-Anta, M.; Ramos-Docampo, M.A.; Comesaña-Hermo, M.; Rivas-Murias, B.; Salgueiriño, V. Raman Spectroscopy to Unravel the Magnetic Properties of Iron Oxide Nanocrystals for Bio-Related Applications. *Nanoscale Adv.* 2019, *1*, 2086–2103. [CrossRef]
63. De Faria, D.L.A.; Venâncio Silva, S.; De Oliveira, M.T. Raman Microspectroscopy of Some Iron Oxides and Oxyhydroxides. *J. Raman Spectrosc.* 1997, *28*, 873–878. [CrossRef]
64. Murli, C.; Song, Y. Pressure-Induced Polymerization of Acrylic Acid: A Raman Spectroscopic Study. *J. Phys. Chem. B* 2010, *114*, 9744–9750. [CrossRef] [PubMed]
65. de Faria, D.L.A.; Lopes, F.N. Heated Goethite and Natural Hematite: Can Raman Spectroscopy Be Used to Differentiate Them? *Vib. Spectrosc.* 2007, *45*, 117–121. [CrossRef]
66. Guo, C.; Hu, Y.; Qian, H.; Ning, J.; Xu, S. Magnetite (Fe_3O_4) Tetrakaidecahedral Microcrystals: Synthesis, Characterization, and Micro-Raman Study. *Mater. Charact.* 2011, *62*, 148–151. [CrossRef]
67. Shebanova, O.N.; Lazor, P. Raman Study of Magnetite (Fe3O4): Laser-induced Thermal Effects and Oxidation. *J. Raman Spectrosc.* 2003, *34*, 845–852. [CrossRef]
68. Slavov, L.; Abrashev, M.V.; Merodiiska, T.; Gelev, C.; Vandenberghe, R.E.; Markova-Deneva, I.; Nedkov, I. Raman Spectroscopy Investigation of Magnetite Nanoparticles in Ferrofluids. *J. Magn. Magn. Mater.* 2010, *322*, 1904–1911. [CrossRef]
69. Grosvenor, A.P.; Kobe, B.A.; Biesinger, M.C.; McIntyre, N.S. Investigation of Multiplet Splitting of Fe 2p XPS Spectra and Bonding in Iron Compounds. *Surf. Interface Anal.* 2004, *36*, 1564–1574. [CrossRef]
70. Biesinger, M.C.; Payne, B.P.; Grosvenor, A.P.; Lau, L.W.M.; Gerson, A.R.; Smart, R.S.C. Resolving Surface Chemical States in XPS Analysis of First Row Transition Metals, Oxides and Hydroxides: Cr, Mn, Fe, Co and Ni. *Appl. Surf. Sci.* 2011, *257*, 2717–2730. [CrossRef]
71. Gupta, R.P.; Sen, S.K. Calculation of Multiplet Structure of Core p-Vacancy Levels. *Phys. Rev. B* 1974, *10*, 71–77. [CrossRef]
72. Gupta, R.P.; Sen, S.K. Calculation of Multiplet Structure of Core p-Vacancy Levels. II. *Phys. Rev. B* 1975, *12*, 15–19. [CrossRef]
73. Dong, J.; Ozaki, Y.; Nakashima, K. Infrared, Raman, and Near-Infrared Spectroscopic Evidence for the Coexistence of Various Hydrogen-Bond Forms in Poly(Acrylic Acid). *Macromolecules* 1997, *30*, 1111–1117. [CrossRef]
74. Sharif, S.M.; Golestani Fard, F.; Khatibi, E.; Sarpoolaky, H. Dispersion and Stability of Carbon Black Nanoparticles, Studied by Ultraviolet–Visible Spectroscopy. *J. Taiwan Inst. Chem. Eng.* 2009, *40*, 524–527. [CrossRef]
75. Safaei, Y.; Aminzare, M.; Golestani-Fard, F.; Khorasanizadeh, F.; Salahi, E. Suspension Stability of Titania Nanoparticles Studied by UV-VIS Spectroscopy Method. *Iran. J. Mater. Sci. Eng.* 2012, *9*, 62–68.
76. Salzmann, C.G.; Chu, B.T.T.; Tobias, G.; Llewellyn, S.A.; Green, M.L.H. Quantitative Assessment of Carbon Nanotube Dispersions by Raman Spectroscopy. *Carbon* 2007, *45*, 907–912. [CrossRef]
77. Horia, F.; Easawi, K.; Khalil, R.; Abdallah, S.; El-Mansy, M.; Negm, S. Optical and Thermophysical Characterization of Fe_3O_4 Nanoparticle. *IOP Conf. Ser. Mater. Sci. Eng.* 2020, *956*, 12016. [CrossRef]
78. Jung, H.; Schimpf, A.M. Photochemical Reduction of Nanocrystalline Maghemite to Magnetite. *Nanoscale* 2021, *13*, 17465–17472. [CrossRef]
79. Sutherland, T.I.; Sparks, C.J.; Joseph, J.M.; Wang, Z.; Whitaker, G.; Sham, T.K.; Wren, J.C. Effect of Ferrous Ion Concentration on the Kinetics of Radiation-Induced Iron-Oxide Nanoparticle Formation and Growth. *Phys. Chem. Chem. Phys.* 2017, *19*, 695–708. [CrossRef]
80. Heinrich, C.A.; Seward, T.M. A Spectrophotometric Study of Aqueous Iron(II) Chloride Complexing from 25 to 200 °C. *Geochim. Cosmochim. Acta* 1990, *54*, 2207–2221. [CrossRef]
81. Zhao, R.; Pan, P. A Spectrophotometric Study of Fe(II)-Chloride Complexes in Aqueous Solutions from 10 to 100 °C. *Can. J. Chem.* 2001, *79*, 131–144. [CrossRef]
82. Po, H.N.; Sutin, N. Stability Constant of the Monochloro Complex of Iron(II). *Inorg. Chem.* 1968, *7*, 621–624. [CrossRef]
83. Kulshrestha, A.; Sharma, S.; Singh, K.; Kumar, A. Magnetoresponsive Biocomposite Hydrogels Comprising Gelatin and Valine Based Magnetic Ionic Liquid Surfactant as Controlled Release Nanocarrier for Drug Delivery. *Mater. Adv.* 2022, *3*, 484–492. [CrossRef]

84. Gils, P.S.; Ray, D.; Sahoo, P.K. Designing of Silver Nanoparticles in Gum Arabic Based Semi-IPN Hydrogel. *Int. J. Biol. Macromol.* **2010**, *46*, 237–244. [CrossRef] [PubMed]
85. Nagaraja, K.; Krishna Rao, K.S.V.; Zo, S.; Soo Han, S.; Rao, K.M. Synthesis of Novel Tamarind Gum-Co-Poly(Acrylamidoglycolic Acid)-Based PH Responsive Semi-IPN Hydrogels and Their Ag Nanocomposites for Controlled Release of Chemotherapeutics and Inactivation of Multi-Drug-Resistant Bacteria. *Gels* **2021**, *7*, 237. [CrossRef] [PubMed]
86. Jayaramudu, T.; Raghavendra, G.M.; Varaprasad, K.; Sadiku, R.; Ramam, K.; Raju, K.M. Iota-Carrageenan-Based Biodegradable Ag^0 Nanocomposite Hydrogels for the Inactivation of Bacteria. *Carbohydr. Polym.* **2013**, *95*, 188–194. [CrossRef] [PubMed]
87. Veloso, S.R.S.; Martins, J.A.; Hilliou, L.O.; Amorim, C.; Amaral, V.S.; Almeida, B.G.; Jervis, P.J.; Moreira, R.; Pereira, D.M.; Coutinho, P.J.G.; et al. Dehydropeptide-Based Plasmonic Magnetogels: A Supramolecular Composite Nanosystem for Multimodal Cancer Therapy. *J. Mater. Chem. B* **2020**, *8*, 45–64. [CrossRef] [PubMed]
88. Carvalho, A.; Gallo, J.; Pereira, D.M.; Valentão, P.; Andrade, P.B.; Hilliou, L.; Ferreira, P.M.T.; Bañobre-López, M.; Martins, J.A. Magnetic Dehydrodipeptide-Based Self-Assembled Hydrogels for Theragnostic Applications. *Nanomaterials* **2019**, *9*, 541. [CrossRef]
89. Nagireddy, N.R.; Yallapu, M.M.; Kokkarachedu, V.; Sakey, R.; Kanikireddy, V.; Pattayil Alias, J.; Konduru, M.R. Preparation and Characterization of Magnetic Nanoparticles Embedded in Hydrogels for Protein Purification and Metal Extraction. *J. Polym. Res.* **2011**, *18*, 2285–2294. [CrossRef]
90. Jørgensen, C.K.; De Verdier, C.-H.; Glomset, J.; Sörensen, N.A. Studies of Absorption Spectra. IV. Some New Transition Group Bands of Low Intensity. *Acta Chem. Scand.* **1954**, *8*, 1502–1512. [CrossRef]
91. Nonkumwong, J.; Ananta, S.; Srisombat, L. Effective Removal of Lead(Ii) from Wastewater by Amine-Functionalized Magnesium Ferrite Nanoparticles. *RSC Adv.* **2016**, *6*, 47382–47393. [CrossRef]
92. Li, Z.; Li, Y.; Chen, C.; Cheng, Y. Magnetic-Responsive Hydrogels: From Strategic Design to Biomedical Applications. *J. Control. Release* **2021**, *335*, 541–556. [CrossRef]
93. Marey, A.; Ahmed, D.F. Batch Adsorption Studies of Natural Composite Hydrogel for Removal of Co(II) Ions. *J. Appl. Membr. Sci. Technol.* **2022**, *26*, 13–18. [CrossRef]
94. Low, K.S.; Lee, C.K.; Liew, S.C. Sorption of Cadmium and Lead from Aqueous Solutions by Spent Grain. *Process Biochem.* **2000**, *36*, 59–64. [CrossRef]
95. Ho, Y.S.; McKay, G. A Comparison of Chemisorption Kinetic Models Applied to Pollutant Removal on Various Sorbents. *Process Saf. Environ. Prot.* **1998**, *76*, 332–340. [CrossRef]
96. Zhang, W.; Hu, L.; Hu, S.; Liu, Y. Optimized Synthesis of Novel Hydrogel for the Adsorption of Copper and Cobalt Ions in Wastewater. *RSC Adv.* **2019**, *9*, 16058–16068. [CrossRef]
97. Lawrance, G.A. Leaving Groups on Inert Metal Complexes with Inherent or Induced Lability. In *Advances in Inorganic Chemistry*; Academic Press: Cambridge, MA, USA, 1989; Volume 34, pp. 145–194; ISBN 0898-8838.
98. Facchi, D.P.; Cazetta, A.L.; Canesin, E.A.; Almeida, V.C.; Bonafé, E.G.; Kipper, M.J.; Martins, A.F. New Magnetic Chitosan/Alginate/Fe_3O_4@SiO_2 Hydrogel Composites Applied for Removal of Pb(II) Ions from Aqueous Systems. *Chem. Eng. J.* **2018**, *337*, 595–608. [CrossRef]
99. Yao, G.; Bi, W.; Liu, H. PH-Responsive Magnetic Graphene Oxide/Poly(NVI-Co-AA) Hydrogel as an Easily Recyclable Adsorbent for Cationic and Anionic Dyes. *Colloids Surfaces A Physicochem. Eng. Asp.* **2020**, *588*, 124393. [CrossRef]
100. Van Berkum, S.; Biewenga, P.D.; Verkleij, S.P.; van Zon, J.H.B.A.; Boere, K.W.M.; Pal, A.; Philipse, A.P.; Erné, B.H. Swelling Enhanced Remanent Magnetization of Hydrogels Cross-Linked with Magnetic Nanoparticles. *Langmuir* **2015**, *31*, 442–450. [CrossRef]
101. Saadli, M.; Braunmiller, D.L.; Mourran, A.; Crassous, J.J. Thermally and Magnetically Programmable Hydrogel Microactuators. *Small* **2023**, *19*, 2207035. [CrossRef]
102. Liu, W.; Migdisov, A.; Williams-Jones, A. The Stability of Aqueous Nickel(II) Chloride Complexes in Hydrothermal Solutions: Results of UV–Visible Spectroscopic Experiments. *Geochim. Cosmochim. Acta* **2012**, *94*, 276–290. [CrossRef]
103. Radoń, M.; Drabik, G. Spin States and Other Ligand–Field States of Aqua Complexes Revisited with Multireference Ab Initio Calculations Including Solvation Effects. *J. Chem. Theory Comput.* **2018**, *14*, 4010–4027. [CrossRef]
104. Staszak, K.; Kruszelnicka, I.; Ginter-Kramarczyk, D.; Góra, W.; Baraniak, M.; Lota, G.; Regel-Rosocka, M. Advances in the Removal of Cr(III) from Spent Industrial Effluents-A Review. *Materials* **2023**, *16*, 378. [CrossRef]
105. Amato, F.; Motta, A.; Giaccari, L.; Di Pasquale, R.; Scaramuzzo, F.A.; Zanoni, R.; Marrani, A.G. One-Pot Carboxyl Enrichment Fosters Water-Dispersibility of Reduced Graphene Oxide: A Combined Experimental and Theoretical Assessment. *Nanoscale Adv.* **2023**, *5*, 893–906. [CrossRef] [PubMed]
106. Marrani, A.G.; Motta, A.; Palmieri, V.; Perini, G.; Papi, M.; Dalchiele, E.A.; Schrebler, R.; Zanoni, R. A Comparative Experimental and Theoretical Study of the Mechanism of Graphene Oxide Mild Reduction by Ascorbic Acid and N-Acetyl Cysteine for Biomedical Applications. *Mater. Adv.* **2020**, *1*, 2745–2754. [CrossRef]
107. Dalalibera, A.; Vilela, P.B.; Vieira, T.; Becegato, V.A.; Paulino, A.T. Removal and Selective Separation of Synthetic Dyes from Water Using a Polyacrylic Acid-Based Hydrogel: Characterization, Isotherm, Kinetic, and Thermodynamic Data. *J. Environ. Chem. Eng.* **2020**, *8*, 104465. [CrossRef]
108. Li, H.; Cao, X.; Zhang, C.; Yu, Q.; Zhao, Z.; Niu, X.; Sun, X.; Liu, Y.; Ma, L.; Li, Z. Enhanced Adsorptive Removal of Anionic and Cationic Dyes from Single or Mixed Dye Solutions Using MOF PCN-222. *RSC Adv.* **2017**, *7*, 16273–16281. [CrossRef]

109. Zhang, X.; Li, F.; Zhao, X.; Cao, J.; Liu, S.; Zhang, Y.; Yuan, Z.; Huang, X.; De Hoop, C.F.; Peng, X.; et al. Bamboo Nanocellulose/Montmorillonite Nanosheets/Polyethyleneimine Gel Adsorbent for Methylene Blue and Cu(II) Removal from Aqueous Solutions. *Gels* 2023, *9*, 40. [CrossRef]
110. Aljar, M.A.A.; Rashdan, S.; Almutawah, A.; El-Fattah, A.A. Synthesis and Characterization of Biodegradable Poly(Vinyl Alcohol)-Chitosan/Cellulose Hydrogel Beads for Efficient Removal of Pb(II), Cd(II), Zn(II), and Co(II) from Water. *Gels* **2023**, *9*, 328. [CrossRef]

Disclaimer/Publisher's Note: The statements, opinions and data contained in all publications are solely those of the individual author(s) and contributor(s) and not of MDPI and/or the editor(s). MDPI and/or the editor(s) disclaim responsibility for any injury to people or property resulting from any ideas, methods, instructions or products referred to in the content.

Article

Superhydrophobic/Superoleophilic PDMS/SiO$_2$ Aerogel Fabric Gathering Device for Self-Driven Collection of Floating Viscous Oil

Feng Liu [1,†], Xin Di [1,†], Xiaohan Sun [1], Xin Wang [1], Tinghan Yang [1], Meng Wang [2], Jian Li [1], Chengyu Wang [1,*] and Yudong Li [1,*]

[1] Key Laboratory of Bio-Based Material Science & Technology of Ministry of Education, Northeast Forestry University, Harbin 150040, China
[2] State Key Laboratory of Urban Water Resource and Environment, School of Environment, Harbin Institute of Technology, Harbin 150001, China
* Correspondence: wangcy@nefu.edu.cn (C.W.); lydlnn0000@163.com (Y.L.)
† These authors contributed equally to this work.

Abstract: The persistent challenge of removing viscous oil on water surfaces continues to pose a major concern and requires immediate attention. Here, a novel solution has been introduced in the form of a superhydrophobic/superoleophilic PDMS/SiO$_2$ aerogel fabric gathering device (SFGD). The SFGD is based on the adhesive and kinematic viscosity properties of oil, enabling self-driven collection of floating oil on the water surface. The SFGD is able to spontaneously capture the floating oil, selectively filter it, and sustainably collect it into its porous fabric interior through the synergistic effects of surface tension, gravity, and liquid pressure. This eliminates the need for auxiliary operations such as pumping, pouring, or squeezing. The SFGD demonstrates exceptional average recovery efficiencies of 94% for oils with viscosities ranging from 10 to 1000 mPa·s at room temperature, including dimethylsilicone oil, soybean oil, and machine oil. With its facile design, ease of fabrication, high recovery efficiency, excellent reclaiming capabilities, and scalability for multiple oil mixtures, the SFGD represents a significant advancement in the separation of immiscible oil/water mixtures of various viscosities and brings the separation process one step closer to practical application.

Keywords: oil collection; oil–water separation; superhydrophobic; superoleophilic

Citation: Liu, F.; Di, X.; Sun, X.; Wang, X.; Yang, T.; Wang, M.; Li, J.; Wang, C.; Li, Y. Superhydrophobic/Superoleophilic PDMS/SiO$_2$ Aerogel Fabric Gathering Device for Self-Driven Collection of Floating Viscous Oil. *Gels* **2023**, *9*, 405. https://doi.org/10.3390/gels9050405

Academic Editors: Pavel Gurikov and Miguel Sanchez-Soto

Received: 18 January 2023
Revised: 19 March 2023
Accepted: 27 March 2023
Published: 12 May 2023

Copyright: © 2023 by the authors. Licensee MDPI, Basel, Switzerland. This article is an open access article distributed under the terms and conditions of the Creative Commons Attribution (CC BY) license (https://creativecommons.org/licenses/by/4.0/).

1. Introduction

With the development of the social economy and industrialization, oil pollution such as waste edible oil, mechanical abandoned oil, and industrial spilled oil increases rapidly, which results in a serious threat to the ecosystem and human health [1,2]. Conventional methods and technologies such as flotation, separators, centrifugation, oil containment booms, and skimmers have been developed for oil removal but are not effective for totally eliminating oil from water, especially oil with viscosity, thus making the separation incomplete with oil residual in water [3,4]. Moreover, these methods usually involve low selectivity, tedious operations, energy-consuming processes, low separation efficiency, and so on, which severely blocks the practical usage of these approaches. Therefore, novel materials with special selectivity, good mechanical stability, excellent separation efficiency, and reliable recyclability are urgently needed for the separation of oil/water mixtures [5,6].

Recently, various separation materials with special wettability such as membrane films, porous materials, and gelation have been widely developed [7–12]. Superhydrophilic materials with hierarchical structures and water-binding affinity could adsorb water and make water become trapped in the rough structures once contacting water during the separation process [13,14]. The adsorbed water forms a hydration layer, which definitely reduces the contact area between the oil and sample surface and thus decreases the oil adhesive property. During the separation process, these materials could easily attract and filter water content from oil/water mixtures under gravity or external force, often displaying outstanding oil/water separation efficiency [15,16]. However, besides the drawbacks of the rigorous fabrication process, complicated operation steps, and the process being energy-consuming, having low recyclability, and sometimes requiring pump assistance, the large-scale usage of these materials for oil/water separation still remains limited because of their separation style: the water content of the viscous-oil/water mixtures passes through the materials instead of gathering floating viscous oil content from the mixtures, requiring oil/water mixtures to be gathered first [17]. Meanwhile, most reported materials are mainly focused on organic solvents or light oils (e.g., gasoline and diesel) and are still easily polluted and fouled by sticky oils [18–20].

Porous absorbent materials with water repellency such as sponges, foam materials, rubber, carbon-based materials, and chemosynthesis adsorbents have drawn much attention to dealing with floating oil assigned to their prominent adsorption characteristics and extrusion property [21–23]. The materials could spontaneously and selectively adsorb oil from the water surface due to their lipophilicity and water resistance, usually possessing excellent adsorption capacity and exceptional separation efficiency. Polydimethylsiloxane (PDMS) sorbent has been widely used for oil–water separation due to its high selectivity and recovery rates for oil types, as well as its ease of use and cost-effectiveness compared to other sample preparation methods [24–26]. However, its use also has several disadvantages, including higher costs than other sample preparation materials, environmental impact due to the difficulty of biodegradation of the synthetic material, the need for reusability, and limited capacity for large sample volumes. Graphene/PDMS sponge has gained attention as a promising material for oil–water separation due to its high surface area and selective affinity for different types of oil. In addition, its reusability and durability make it a cost-effective solution for oil–water separation [27–29]. However, the synthesis and preparation of graphene/PDMS sponge can be challenging and requires advanced methods, resulting in high production costs. Furthermore, the stability of the material may not be consistent over time, leading to degradation and decreased efficiency in oil–water separation. However, it is of concern that these adsorbent materials, once used for viscous oil adsorption and removal, would show a dramatic decline in adsorption properties, leading to an inevitable decrease in recovery efficiencies after a limited number of uses [21]. The primary reason for the decline is due to the stickiness and kinematic viscosity of the oils, which would lead to accumulated contaminants, severe pore fouling, and an irreversible decrease in their adsorption properties. In the meantime, artificial squeezing or pump-driven procedures may require tedious separation operations and high energy consumption. Therefore, novel devices and methods with easy operation, self-driven property, good oil recovery, and stable recyclability should be designed for floating viscous oil collection and removal.

Herein, a superhydrophobic/superoleophilic $PDMS/SiO_2$ aerogel fabric gathering device (SFGD) is designed for collection and removal of floating viscous oil spills. Firstly, a layer of superhydrophobic/superoleophilic $PDMS/SiO_2$ composite aerogel coating is prepared on the surface of the burlap fabric using PDMS and nano-silica aerogel particles [30,31]. Subsequently, the burlap fabric is combined with porous plastic balls (i.e., by filling the porous plastic balls as an internal support frame inside the burlap sack) to form the SFGD. The device could spontaneously gather floating oil on a water surface and sustainably collect it into a porous sack relying on the effects of surface tension, gravity, and liquid pressure, requiring no auxiliary operations (e.g., pumping, pouring, and squeezing). During the separation process, the partially submerged fabric surface remains superhydrophobic even underwater, which mainly results from the adhered waterproof viscous oil, hierarchical surface structures, and the intrinsic water repellency of the textile surface. Therefore, an interesting phenomenon appears that the water tightly wraps the submerged part of the sack surface while being totally forbidden to pass through the textured surface; simultaneously, the oil could easily flow into the sack while being inevitably locked into the sack by the wrapped water. Finally, the floating viscous oil is collected into a container and the SGFD could be reutilized for the oil/water separation, thus demanding no clean-up treatment. Moreover, the fabrics display good water repellence, excellent wear resistance detected by an oscillating abrasion tester, and prominent reusability via a recycling experiment. The facile prepared SFGD with scalable fabrication, high recovery efficiency, and prominent reclamation for the separation of immiscible oil/water mixtures of various oil viscosities utilizing a self-driven approach is displayed to illustrate the necessity of the unique device designed herein.

2. Results and Discussion

2.1. Mechanism and Surface Wettability

Polydimethylsiloxane has been widely applied for the construction of superhydrophobic/superoleophilic surfaces due to its low surface energy and adhesive properties [32]. Nano-silica aerogel particles were prepared by the sol-gel method and used as the surface coating composition for providing additional nanoscale surface roughness and enhancing surface [30]. As shown in Figure 1a, PDMS prepolymer was dissolved into n-hexane solution under stirring and then the added nano-silica aerogel particles were dispersed under ultrasonic treatment because of the aggregation of the particles. After a simple dip-coating procedure and a subsequent drying process, the burlap fabric was coated with a uniform superhydrophobic PDMS/SiO$_2$ composite aerogel coating as exhibited in the SEM images (Figure 1b). During the preparation of the PDMS/SiO$_2$ composite aerogel coating on the fabric surface, the adhesive properties of the PDMS played a crucial role for both the fabric substrate and silica. The strong adhesion of the PDMS ensured firm and continuous bonding between the coating and the fabric surface, which was essential for the long-term stability and durability of the superhydrophobic coating. The adhesive properties of the PDMS also prevented the coating from peeling or flaking off the fabric substrate, ensuring that the coating remained intact even under high stress or wear conditions. In addition, the addition of nano-silica aerogel particles not only built a rough structure for the overall superhydrophobic coating, but also increased its mechanical stability and durability. The adhesive properties of the PDMS also helped to bond the silica particles to the fabric, forming a strong and uniform coating that could resist wear and tear. Therefore, the combination of the PDMS and silica enhanced the overall performance of the coating and made it more superhydrophobic. With the synergistic effect of hybridization and composite formation between PDMS and nano-silica aerogel particles, a layer of superhydrophobic/superoleophilic PDMS/SiO$_2$ composite aerogel coating is formed on the surface of burlap fabric.

Generally, surfaces with low surface energy showed a stronger affinity to oil than water [33,34]. The wettability of the coated burlap fabric was measured in air conditions with water contact angles (WCAs) of 156° (Figure 1c top) and an oil contact angle of 0° (Figure 1c bottom), respectively, showing excellent water repellency and superoleophilicity. Once it contacted the as-prepared sample, the oil droplet would spread on the modified surface (see the Supplementary Information, Figure S2). As demonstrated in Figure 1d, the sack coated by PDMS/SiO$_2$ composite aerogel coating displayed excellent water impact resistance (Figure 1d left) and outstanding water-holding properties (Figure 1d right).

As reported, the modified Young's equation could be not only applicable to analyze the wettability of an oil droplet on a solid surface underwater but also valid to a water droplet on a surface in oil [35,36]. The modified formula of water contact angle on an ideal smooth surface underoil (θ_{WO}) is displayed in Equation (1). According to the equation,

$$\cos \theta'_{WO} = r \cos \theta_{WO} \tag{1}$$

where θ'_{WO} represents the water contact angle of a water droplet on the rough surface and r is the roughness of the surface. the value of surface roughness (r) is greater than 1, illustrating that for material with underoil WCA (θ_{WO}) more than 90°, the real value of θ'_{WO} increase with the strengthening of surface roughness. Thus, the nano-silica aerogel-particles play an essential role in the construction of underoil superhydrophobicity. The wettability of the as-prepared rough surfaces with superhydrophobicity was explored via a contact angle meter (Figure 2 and Figure S3).

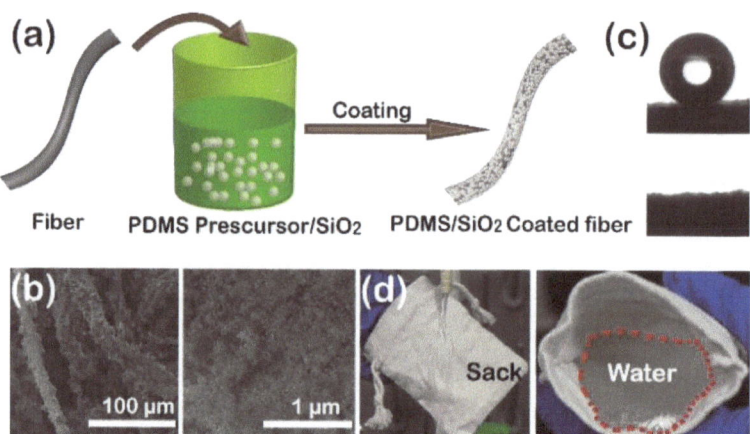

Figure 1. (a) Schematic illustration of the construction process for the PDMS/SiO$_2$-coated burlap fabric. (b) SEM images of the PDMS/SiO$_2$ composite aerogel coating on burlap fibers (left) with partial magnification (right). (c) Photographs of water contact angle (top) and oil contact angle (bottom) on the coated fabric surface in air conditions, respectively. (d) Photographs exhibiting the surface hydrophobicity of the water impacting test (left) and the water holding test (right).

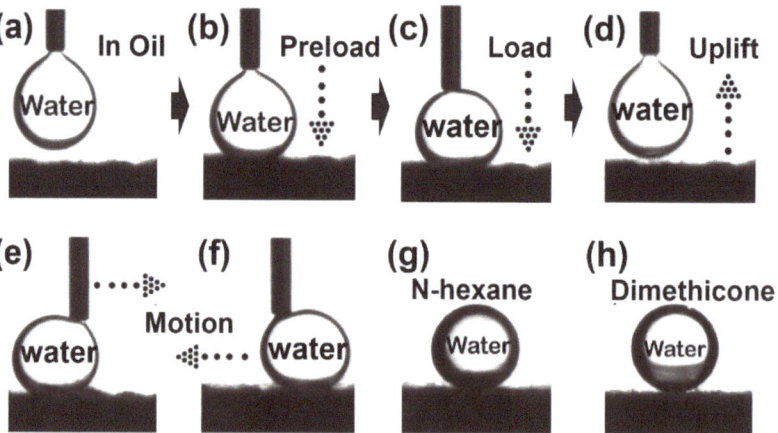

Figure 2. Photographs of (a–f) the underoil water-adhesion detection process on PDMS/SiO$_2$-coated fabric surface and (g,h) water contact angles on the prepared surfaces under n-hexane and under dimethyl silicone oil, respectively.

In oil, the superoleophilic PDMS/SiO$_2$ composite aerogel coating could quickly absorb oil and the oil could be tightly adhered to the textured surface and trapped in the rough structures, which was attributed to the excellent surface affinity to oil. The oil-coated fabric surface demonstrated remarkable superhydrophobicity with a WCA of 154.1° under n-hexane (Figure 2g). As shown in Figure 2a–d, when a water droplet came into contact with the oil/solid composite interface (Figure 2b) under n-hexane and was pressed downward (Figure 2c), the droplet was compressed but did not wet the interface, retaining its intact shape. When the water droplet was lifted, it remained non-adherent to the interface

(Figure 2d). Additionally, when the water droplet was brought into contact with the interface and pressed down a certain distance, upon moving the droplet to the right (Figure 2e) or left (Figure 2f), there was no lag or adhesion between the droplet and the interface. Furthermore, as shown in Figure 2h, the WCA on the textured surface under dimethylsilicone oil was 165.8°, providing a solid theoretical foundation for the collection and removal of viscous oils (the WCAs on the composite surface in the other viscous oils can be seen in Figure S3). In summary, the burlap fabric coated by PDMS/SiO$_2$ composite aerogel coating possesses predominantly excellent oil affinity and underoil water repellency.

2.2. Surface Chemical Component Analysis

As displayed in Figure 3, the chemical composition of the prepared burlap fabric was examined by X-ray photoelectron spectroscopy (XPS) (Figure 3a) and FT-IR spectroscopy (Figure 3b).

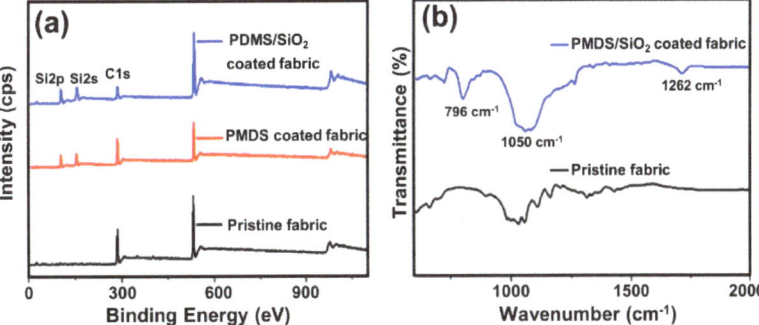

Figure 3. (a) XPS survey spectra of pristine burlap fabric, burlap fabric coated with PDMS only, and PDMS/SiO$_2$ composite aerogel coating. (b) FTIR spectra of pristine burlap fabric and burlap fabric coated with PDMS/SiO$_2$ composite aerogel coating.

Table S1 summarizes the quantitative data obtained for pristine burlap fabric, burlap fabric coated with PDMS only, and PDMS/SiO$_2$ composite aerogel coating. The XPS graphic of the burlap fabric exhibited C1s, O1s, Si2p, and Si2s peaks as shown in Figure 3a. Comparing the results, there was a significant increase in Si peaks after the fabrics were coated with PDMS and PDMS/SiO$_2$. Meanwhile, there was an apparent relatively higher ratio of O to C (43.3/24.1) or Si to C (32.6/24.1) content on the PDMS/SiO$_2$-coated fabric compared to the PDMS coated fabric (O to C = 27.6/48.8; Si to C = 23.6/48.8), which could be attributed to the nano-silica aerogel particles [37]. The contents increase could also be qualified through EDX as shown in Figure S4. Figure 3b showed that the absorption peaks at 1050 cm^{-1} and 1262 cm^{-1} were assigned to Si-O-Si stretching vibrations attributed to silicon dioxide and silicon rubber. The occurrence of prominent bands at 796 cm^{-1} represented the Si-C vibrations conforming that the PDMS adhered to the fabric surface. In summary, Figure 3 demonstrated that silicon dioxide and silicon rubber were deposited on the superhydrophobic burlap fabric surface.

2.3. Mechanical Robustness

The stability of micro- and nano-scale rough structures is an essential part of preparing durable special wettability surfaces which deeply affects the practicability of the materials [38]. The stickiness of the PDMS could coat and fix the nanoparticles on the fabric surfaces and the elasticity of the dried PDMS will disperse the forces once subjected to mechanical forces, which could protect the coated nano silica from facing the external force directly. To confirm the mechanical stability of the PDMS/SiO$_2$ composite aerogel coating on the fabric surface, the as-prepared burlap fabric was investigated by an oscillating abrasion tester as shown in Figure 4a. The fabric side of the sample was placed into the oscillating abrasion tester equipment with grit covering the whole sample. As exhibited in Figure 4b, the WCA of the sample remained above 152° even after 1500 abrasion cycles showing excellent durability. The stable hydrophobicity of the burlap fabric was primarily attributed to the remaining PDMS/SiO$_2$ rough structures (Figure 4b inserted images). Moreover, a blue-dyed water droplet could easily roll off the fabric surface after being treated for 1500 cycles (see the supplementary Information, Figure S6).

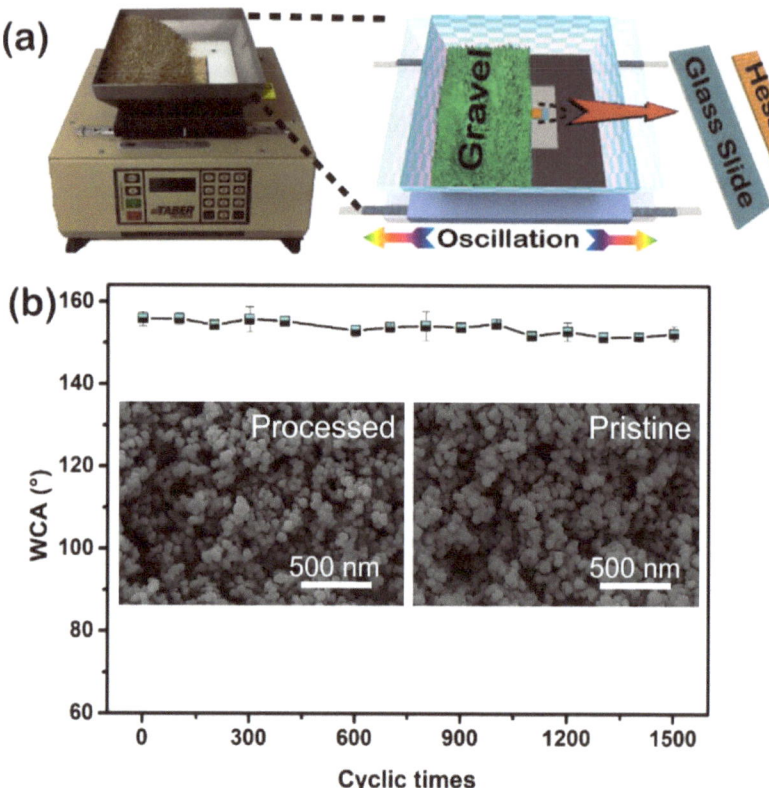

Figure 4. (**a**) Schematic Illustration of surface mechanical detection by an oscillating abrasion tester. (**b**) WCAs change of the burlap fabric by the PDMS/SiO$_2$ layer in the process of the abrasion test, and the SEM image after 1500 cycles and the corresponding insert SEM images of the pristine sample and the treated sample.

2.4. Separation of Viscous Oil/Water Mixture

For understanding the separation mechanism of the immiscible oil/water mixture via the SFGD, a schematic illustration for viscous oil collection and removal was provided as displayed in Figure 5. The SFGD was assembled by a superhydrophobic/superoleophilic burlap sack and a porous plastic ball (Figure 5(a,b$_1$)). The porous hollow ball was selected as an internal prop that could keep the sack owning a steady inner space. When dropping into a beaker containing viscous oil/water mixtures (Figure 5(b$_2$)), the SFGD would partially submerge in water due to gravity but not sink because of the buoyancy meaning that the average density of the device is lower than water. Subsequently, due to the superhydrophobic/superoleophilic properties of the SFGD surface fabric, the viscous oil on the water surface rapidly wetted the SFGD surface and slowly and self-drivenly passed through the outer fabric of the SFGD under the joint gravity of the SFGD and the oil as well as the liquid pressure, while the water was selectively blocked on the outer side due to the superhydrophobicity of the burlap fabric and the superhydrophobicity under the oil, finally completing the collection of the viscous oil on the water surface (Figure 5(b_3)). Thus, an interesting phenomenon occurs that the water tightly wraps the submerged part of the sack surface while being totally prevented from penetrating the textured surface; however, the oil could be easily attracted and flow into the sack while being inevitably locked into the sack by the wrapped water. As shown in Figure 5(b_4), when equilibrium (zero resultant force) was achieved, the oil-filled SFGD still partially submerged in water and remained unsinkable ascribed to its lower density than the water, and it maintained water repellency because of the water resistance of the collected oil and surface superhydrophobicity. Subsequently, the device was taken out and the oil was poured into a container. The leakage rate of the collected oil was rather low due to the relatively high kinematic viscosity of the inherent oil property, which could guarantee the oil recovery. Finally, after the pouring process, the SFGD was directly reused for viscous oil collection and removal without any wash treatment (Figure 5(b_1,b_5)).

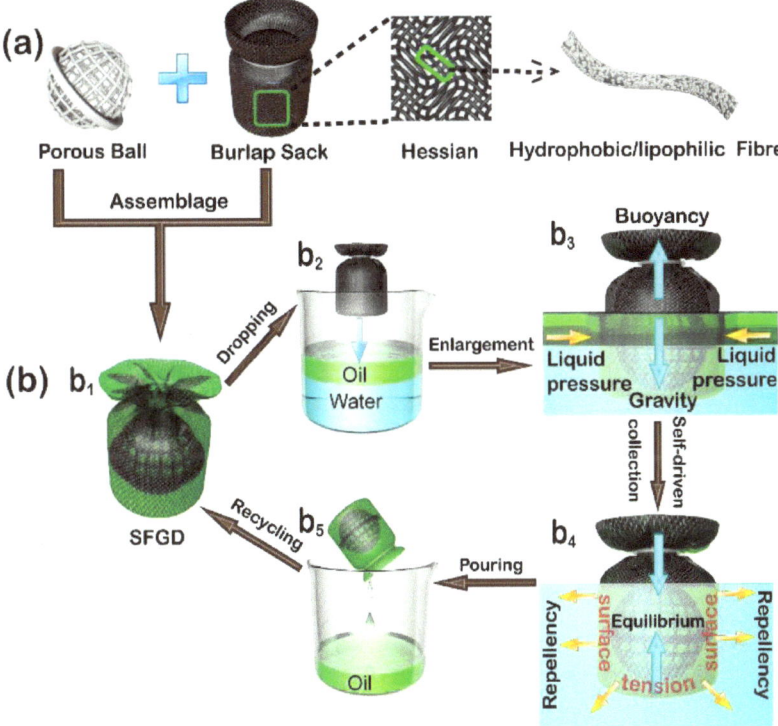

Figure 5. Scheme for (**a**) assembling the superhydrophobic/superoleophilic aerogel fabric gathering device (SFGD) and (**b**) the process (**b_1–b_5**) of separating the immiscible viscous oil/water mixture.

Figure 6a shows that floating silicon oil (100 ± 8 mPa·s) was collected and removed from the blue-dyed water surface via the SFGD following the steps in Figure 5b. It required 2 h for managing the oil collection. Obviously, on first use, the SFGD was gradually adsorbed with silicon oil during the gathering process which led to the mass increase of the oil-poured empty SFGD. Therefore, after the oil removal process, the weight of the uncleaned empty SFGD increased from 14.67 g to 38.09 g, and the weight difference of the device before and after usage was 23.42 g as shown in Figure 6b. The mass and recover efficiency of the collected oil was 68.9 g and 71.5%, respectively. However, when repeating the above separation via an uncleaned device, the weight difference of the SFGD could be controlled to ±2 g with the average recovery of 94.81% after 50 cycles (Figure 6b). Furthermore, during the long-time water resistance detection (see the Supplementary Information, Figure S6), the SFGD filled with the collected oil remained floating on the water for 15 days with an oil recovery efficiency of no less than 95.8%, showing excellent practicability. The SFGD could be applied to collect various viscous oils, such as dimethylsilicone oil (viscosity 100 ± 8 mPa·s, 500 ± 8 mPa·s, 1000 ± 8 m mPa·s), soybean oil, lubricant oil, anti-wear hydraulic oil, and gasoline engine oil, with recovery efficiency above 94% as exhibited in Figure 6c.

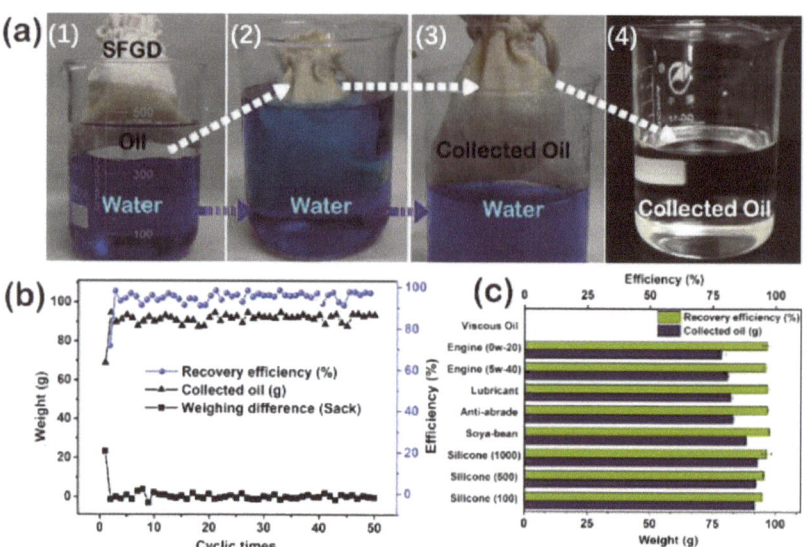

Figure 6. (a) Photographs for viscous oil removal from the water surface by SFGD. (b) The recovery of viscous silicone oil and the weighing difference of the sack before and after usage. (c) Recovery capacity and recovery efficiency of various kinds of viscous oils.

To evaluate the practical usage of the device (burlap sack: 15 × 20 cm, porous plastic ball diameter: 8 cm), 300 mL of viscous oil mixture containing dimethylsilicone oil (viscosity 100 ± 8 mPa·s), soybean oil, lubricant oil, anti-wear hydraulic oil, and gasoline engine oil at the ratio of 1:1:1:1:1 was added to in a 2000 mL beaker filled with 1500 mL of water (Figure 7a). The special wettability of the fabric provided the key foundation of the self-driven, gravity and liquid pressure aided, floating viscous oil collection device as illustrated. When dropped into the beaker, the enlarged SFGD with oil binding affinity was gradually wetted and adhered to by the oil mixture as shown in Figure 7b,c. The floating viscous oil mixture was driven to be filtered by the fabric surface and collected in the inner surface of the sack (Figure 7d,e) under the effectiveness of the surface tension, gravity, and liquid pressure and simultaneously. Though repelled by the hydrophobic fabric, the water seamlessly wrapped the device due to liquid pressure and indeed prevented the oil mixture from leaking from the sack. When taken out, the device could retain the oil mixture without any leakage, which mainly resulted from the adhesive properties and kinematic viscosity of the oil. Thus, the oil mixture was completely collected and removed from water (Figure 7f), exhibiting that this device with remarkable functionality could be easily scalable. Thus, the self-driven oil collection SFGD was successfully applied for the separation and removal of immiscible viscous oil/water mixtures with excellent recyclability and practicability.

Figure 7. Photographs for the collection process of an oil mixture from the water surface via SFGD. (**a**) SFGD, oil mixture, and water. (**b**–**e**) Oil collection process. (**f**) Cleaned water and recovered oil.

3. Conclusions

In summary, the burlap fabric coated by a layer of superhydrophobic/superoleophilic PDMS/SiO$_2$ composite aerogel coating were initially and creatively utilized for self-driven floating viscous oil collection. The SFGD could spontaneously attract, filter, and collect the floating oils under the synergetic effect of surface tension, gravity, and liquid pressure, requiring no extra operations such as pumping, oil/water mixture collecting, and squeezing. The superhydrophobic/superoleophilic burlap sack with excellent mechanical robustness and functionality could be easily scaled by a flexible dip-coating method and utilized for oil/water separation, displaying excellent oil recovery efficiency even after 50 cycles of usage. Furthermore, during long-time water resistance detection, the SFGD filled with the collected oil remained floating on water for 15 days with oil recovery efficiency no less than 95.8%, showing excellent endurance and practicability. It is worth stating that the SFGD could be scaled for the efficient elimination of large multiple oil mixtures on water, demonstrating the versatility for oil removal. We firmly believe that such a smartly designed oil-collection system could provide a unique perspective for dealing with floating viscous oil pollution.

4. Materials and Methods

4.1. Materials

Burlap sack (length and width: 10 cm × 12 cm and 15 cm × 20 cm) and porous suspended plastic ball (diameter: 6 mm and 8 mm) were supplied by Alibaba. PDMS was provided by Dow Corning Co., Ltd (Sylgard 184, Midland, MI, USA)). Tetraethoxysilane (TEOS, chemically pure), and NH$_3$·H$_2$O (28%) were obtained from Tianjin Kaitong Chemical Reagent Co, Ltd (Tianjin, China). Dimethylsilicone oil (viscosity 100 ± 8 mPa·s), dimethylsilicone oil (viscosity 500 ± 8 mPa·s), and dimethylsilicone oil (viscosity 1000 ± 8 mPa·s) were purchased from Shanghai Aladdin Biochemical Technology Co., Ltd (Shanghai, China). Methylene blue, ethanol, and n-hexane were provided by Sinopharm Chemical Reagents Co., Ltd. (Shanghai, China). Soybean oil was procured from a local market. Lubricant oil, anti-wear hydraulic oil, gasoline engine oil (0W-20), and gasoline engine oil (5W-40) were obtained from Petro-China Co., Ltd. (Beijing, China). All chemicals were used as received without further purification.

4.2. Synthesis of Superhydrophobic and Superoleophilic Burlap Sack

Firstly, the nano-silica aerogel particles were prepared with an alkali-base catalyzed sol–gel method by drying under ambient pressure. In brief, ethanol (90 mL), TEOS (10 mL) and deionized water (10 mL) were mixed and stirred to prepare a mixture solution, and then 2 mL NH$_3$·H$_2$O used as the catalyst was added dropwise into the mixture at room temperature under magnetic stirring (400 rpm) for 0.5 h. The gelation process usually takes around 12 h to complete, and the gels are subsequently allowed to age for 72 h in order to enhance the gel network. N-hexane was used to replace the solvent of wet gel. After rinsing with n-hexane 3–4 times, the wet gel was dried under ambient pressure to obtain the bulk silica aerogel. The bulk silica aerogel was ground into silica aerogel particles.

Secondly, a colloidal solution composed of PDMS and nano-silica aerogel particles was prepared. The colloidal solution was prepared by sonicating 90 mL of n-hexane, 3 g of PDMS prepolymer (Sylgard 184A), 0.3 g of PDMS prepolymer (Sylgard 184B) and 3.6 g of nano-silica aerogel particles for 20 min using an ultrasonic disruptor, followed by magnetic stirring for 10 min.

Finally, the burlap sack was totally washed with deionized water, n-hexane, and ethanol under ultrasonic treatment for 20 min, respectively. Then, the burlap fabric was immersed into the colloidal solution. Subsequently, the immersed fabric sack was taken out, rinsed with n-hexane, and dried at 60 °C for 3 h. The resulting burlap fabric was coated with a layer of superhydrophobic/superoleophilic PDMS/SiO_2 composite aerogel coating that adheres uniformly to its surface.

4.3. Characterization

The morphological surfaces of the samples were observed by a scanning electron microscope (SEM, FEI QUANTA 200, Hillsboro, OR, USA) operating at 15 kV. The surface chemistry composition was detected by Fourier transform infrared spectroscopy (FT-IR, Thermos Fisher Scientific, Nicolet 6700, Waltham, MA, USA) and X-ray photoelectron spectroscopy (XPS, ESCALAB 250Xi, Thermo Fischer Scientific, Waltham, MA, USA). The contact angles were measured with a 5 µL deionized water droplet or an oil droplet at room temperature using an optical contact angle meter (OCA20 system, DataPhysics Instruments GmbH, Filderstadt, Germany), and the final contact angle was determined by averaging the measurements taken from at least five different positions on samples which were adhered onto the glass slide by double-sided adhesive tape.

4.4. Stability Test

Mechanical stability superhydrophobic/superoleophilic PDMS/SiO_2 composite aerogel coating on surface of burlap fabric was assessed by an oscillating abrasion tester and recycling experiment for viscous oil/water separation. In order to be easily measured by a contact angle meter, the as-prepared waterproof fabric was cut into pieces (1 cm × 3 cm) before adhering one of them onto a glass slide (2.5 cm × 7.6 cm) that was sticky on the central area of a square board (10 cm × 10 cm) by double-sided adhesive tape as well. Then, the fabric side of the above sample was placed into the oscillating abrasion tester equipment with 800 mL of grit covering the whole sample. During the oscillation process, the rolling grit constantly abraded the textile surface at a speed of 130 cycles/min. The oscillating process was repeated for a certain number of times to detect the stability of superhydrophobicity. The structural stability of superhydrophobic/superoleophilic composite aerogel coating was assessed by cycling experiments of viscous oil/water mixture separation.

4.5. Separation of the Immiscible Viscous Oil/Water Mixture

Typically, an immiscible viscous oil/water mixture consisting of 300 mL water dyed by methylene blue and 100 mL of silicone oil (viscosity 100 ± 8 mPa·s) was prepared and served in a beaker container. Then, the SFGD was put into the beaker container for 3 h for viscous oil gathering and removal. The SFGD could be directly reused to collect the floating viscous oil without any clean-up process. Meanwhile, the fabric device could be applied to collect various kinds of daily used viscous oil such as soybean oil, vacuum pump oil, anti-wear hydraulic oil, and gasoline engine oil (5W-40) following the same separation process.

The oil recovery efficiency was defined as W (%) and calculated by Equation (2):

$$W = \frac{m_1}{m_0} \times 100\% \qquad (2)$$

where m_0 and m_1 were the mass of the oil before and after the separation process, respectively.

In this study, the average mass of the oil-coated SFGD was 37.61 g and the density of the silicone oil was 0.963 g/mL. The calculated theoretical value of the maximum volume of the collected oil under certain conditions was calculated as 1016.5 mL when reaching equilibrium (see the Supplementary Information, Figure S1), exhibiting that the collected volume totally covered the volume of the SFGD containing a porous plastic ball with a volume about 113 mL. Therefore, 100 mL of viscous oil was selected for the demonstrative experiments of oil collection and removal.

Supplementary Materials: The following supporting information can be downloaded at: https://www.mdpi.com/article/10.3390/gels9050405/s1. Figure S1: (**a**) Force equilibrium analysis diagram of the oil-filled and partially submerged SFGD and (**b**) the corresponding photograph. Figure S2: Photographs of an oil droplet quickly spread on the surface in the air once contacting the superoleophilic fabric surface. Figure S3: Underoil WCAs of various oils. Figure S4: EDX elemental analysis images of (**a**) PDMS coated fabric and (**b**) PDMS/SiO$_2$ layer coated fabric. Figure S5: Photo images of the blue-dyed water droplet rolling off the fabric surface after being treated by an oscillating abrasion tester for 1500 cycles. Figure S6: (**a**) Photographs of the SFGD dropped into a beaker containing oil/water mixture, (**b**) the oil-filled SFGD partially submerged in water for 15 days, (**c**) photographs of the oil-filled SFGD taken out of the container, and (**d**) the volume of the collected oil. Table S1: Surface relative composition of XPS analysis for pristine burlap fabric, burlap fabric coated with PDMS only and PDMS/SiO$_2$ layers.

Author Contributions: Conceptualization, F.L. and X.D.; methodology, F.L. and X.D.; software, X.S. and X.W.; validation, X.D. and X.W.; formal analysis, X.D. and M.W.; investigation, X.D. and T.Y.; resources, F.L.; data curation, F.L., C.W. and Y.L.; writing—original draft preparation, F.L. and X.D.; writing—review and editing, F.L. and X.D.; visualization, Y.L.; supervision, C.W. and Y.L.; project administration, J.L., C.W. and Y.L.; funding acquisition, F.L., C.W. and Y.L. All authors have read and agreed to the published version of the manuscript.

Funding: This research was funded by the Fundamental Research Funds for the Central Universities (Grant no. 2572017AB16), the Fundamental Research Funds for the Central Universities (Grant no. 2572021CG02), and the National Natural Science Foundation of China (Grant no. 32171693).

Institutional Review Board Statement: Not applicable.

Informed Consent Statement: Not applicable.

Data Availability Statement: Not applicable.

Conflicts of Interest: The authors declare no conflict of interest.

References

1. Cheng, X.; Ye, Y.; Li, Z.; Chen, X.; Bai, Q.; Wang, K.; Zhang, Y.; Drioli, E.; Ma, J. Constructing Environmental-Friendly "Oil-Diode" Janus Membrane for Oil/Water Separation. *ACS Nano* **2022**, *16*, 4684–4692. [CrossRef] [PubMed]
2. Dong, D.; Zhu, Y.; Fang, W.; Ji, M.; Wang, A.; Gao, S.; Lin, H.; Huang, R.; Jin, J. Double-Defense Design of Super-Anti-Fouling Membranes for Oil/Water Emulsion Separation. *Adv. Funct. Mater.* **2022**, *32*, 2113247. [CrossRef]
3. Song, J.; Huang, S.; Lu, Y.; Bu, X.; Mates, J.E.; Ghosh, A.; Ganguly, R.; Carmalt, C.J.; Parkin, I.P.; Xu, W.; et al. Self-driven one-step oil removal from oil spill on water via selective-wettability steel mesh. *ACS Appl. Mater. Interfaces* **2014**, *6*, 19858–19865. [CrossRef] [PubMed]
4. Wang, B.; Liang, W.; Guo, Z.; Liu, W. Biomimetic super-lyophobic and super-lyophilic materials applied for oil/water separation: A new strategy beyond nature. *Chem. Soc. Rev.* **2015**, *44*, 336–361. [CrossRef]
5. Zhu, Z.; Jiang, L.; Liu, J.; He, S.; Shao, W. Sustainable, Highly Efficient and Superhydrophobic Fluorinated Silica Functionalized Chitosan Aerogel for Gravity-Driven Oil/Water Separation. *Gels* **2021**, *7*, 66. [CrossRef]
6. Wang, X.; Liu, F.; Li, Y.; Zhang, W.; Bai, S.; Zheng, X.; Huan, J.; Cao, G.; Yang, T.; Wang, M.; et al. Development of a facile and bi-functional superhydrophobic suspension and its applications in superhydrophobic coatings and aerogels in high-efficiency oil–water separation. *Green Chem.* **2020**, *22*, 7424–7434. [CrossRef]

7. Yang, J.; Li, H.-N.; Chen, Z.-X.; He, A.; Zhong, Q.-Z.; Xu, Z.-K. Janus membranes with controllable asymmetric configurations for highly efficient separation of oil-in-water emulsions. *J. Mater. Chem. A* **2019**, *7*, 7907–7917. [CrossRef]
8. Kuang, Y.; Chen, C.; Chen, G.; Pei, Y.; Pastel, G.; Jia, C.; Song, J.; Mi, R.; Yang, B.; Das, S.; et al. Bioinspired Solar-Heated Carbon Absorbent for Efficient Cleanup of Highly Viscous Crude Oil. *Adv. Funct. Mater.* **2019**, *29*, 1900162. [CrossRef]
9. Ielo, I.; Giacobello, F.; Castellano, A.; Sfameni, S.; Rando, G.; Plutino, M.R. Development of Antibacterial and Antifouling Innovative and Eco-Sustainable Sol–Gel Based Materials: From Marine Areas Protection to Healthcare Applications. *Gels* **2022**, *8*, 26. [CrossRef]
10. Remuiñán-Pose, P.; López-Iglesias, C.; Iglesias-Mejuto, A.; Mano, J.F.; García-González, C.A.; Rial-Hermida, M.I. Preparation of Vancomycin-Loaded Aerogels Implementing Inkjet Printing and Superhydrophobic Surfaces. *Gels* **2022**, *8*, 417. [CrossRef]
11. Zhang, C.; Yang, Y.; Luo, S.; Cheng, C.; Wang, S.; Liu, B. Fabrication of Superhydrophobic Composite Membranes with Honeycomb Porous Structure for Oil/Water Separation. *Coatings* **2022**, *12*, 1698. [CrossRef]
12. Peng, K.; Wang, C.; Chang, C.; Peng, N. Phosphonium Modified Nanocellulose Membranes with High Permeate Flux and Antibacterial Property for Oily Wastewater Separation. *Coatings* **2022**, *12*, 1598. [CrossRef]
13. Sperling, M.; Gradzielski, M. Droplets, Evaporation and a Superhydrophobic Surface: Simple Tools for Guiding Colloidal Particles into Complex Materials. *Gels* **2017**, *3*, 15. [CrossRef] [PubMed]
14. Tan, J.; Sun, J.; Ma, C.; Luo, S.; Li, W.; Liu, S. pH-Responsive Carbon Foams with Switchable Wettability Made from Larch Sawdust for Oil Recovery. *Polymers* **2023**, *15*, 638. [CrossRef] [PubMed]
15. Tran, V.T.; Xu, X.; Mredha, M.T.I.; Cui, J.; Vlassak, J.J.; Jeon, I. Hydrogel bowls for cleaning oil spills on water. *Water Res.* **2018**, *145*, 640–649. [CrossRef] [PubMed]
16. Kim, T.; Lee, J.S.; Lee, G.; Seo, D.K.; Baek, Y.; Yoon, J.; Oh, S.M.; Kang, T.J.; Lee, H.H.; Kim, Y.H. Autonomous Graphene Vessel for Suctioning and Storing Liquid Body of Spilled Oil. *Sci. Rep.* **2016**, *6*, 22339. [CrossRef]
17. Zhang, W.; Shi, Z.; Zhang, F.; Liu, X.; Jin, J.; Jiang, L. Superhydrophobic and superoleophilic PVDF membranes for effective separation of water-in-oil emulsions with high flux. *Adv. Mater.* **2013**, *25*, 2071–2076. [CrossRef]
18. Yang, H.C.; Xie, Y.; Chan, H.; Narayanan, B.; Chen, L.; Waldman, R.Z.; Sankaranarayanan, S.; Elam, J.W.; Darling, S.B. Crude-Oil-Repellent Membranes by Atomic Layer Deposition: Oxide Interface Engineering. *ACS Nano* **2018**, *12*, 8678–8685. [CrossRef]
19. Wu, M.-B.; Hong, Y.-M.; Liu, C.; Yang, J.; Wang, X.-P.; Agarwal, S.; Greiner, A.; Xu, Z.-K. Delignified wood with unprecedented anti-oil properties for the highly efficient separation of crude oil/water mixtures. *J. Mater. Chem. A* **2019**, *7*, 16735–16741. [CrossRef]
20. Liu, Y.; Wang, X.; Fei, B.; Hu, H.; Lai, C.; Xin, J.H. Bioinspired, Stimuli-Responsive, Multifunctional Superhydrophobic Surface with Directional Wetting, Adhesion, and Transport of Water. *Adv. Funct. Mater.* **2015**, *25*, 5047–5056. [CrossRef]
21. Nam, C.; Li, H.; Zhang, G.; Lutz, L.R.; Nazari, B.; Colby, R.H.; Chung, T.C.M. Practical Oil Spill Recovery by a Combination of Polyolefin Absorbent and Mechanical Skimmer. *ACS Sustain. Chem. Eng.* **2018**, *6*, 12036–12045. [CrossRef]
22. Dong, C.; Hu, Y.; Zhu, Y.; Wang, J.; Jia, X.; Chen, J.; Li, J. Fabrication of Textile Waste Fibers Aerogels with Excellent Oil/Organic Solvent Adsorption and Thermal Properties. *Gels* **2022**, *8*, 684. [CrossRef] [PubMed]
23. Liu, F.; Jiang, Y.; Feng, J.; Li, L.; Feng, J. Bionic Aerogel with a Lotus Leaf-like Structure for Efficient Oil-Water Separation and Electromagnetic Interference Shielding. *Gels* **2023**, *9*, 214. [CrossRef] [PubMed]
24. Bayraktaroglu, S.; Kizil, S.; Bulbul Sonmez, H. A highly reusable polydimethylsiloxane sorbents for oil/organic solvent clean-up from water. *J. Environ. Chem. Eng.* **2021**, *9*, 106002. [CrossRef]
25. Pandey, K.; Bindra, H.S.; Jain, S.; Nayak, R. Sustainable lotus leaf wax nanocuticles integrated polydimethylsiloxane sorbent for instant removal of oily waste from water. *Colloids Surf. A Physicochem. Eng. Asp.* **2022**, *634*, 127937. [CrossRef]
26. Zhao, M.; Ma, X.; Chao, Y.; Chen, D.; Liao, Y. Super-Hydrophobic Magnetic Fly Ash Coated Polydimethylsiloxane (MFA@PDMS) Sponge as an Absorbent for Rapid and Efficient Oil/Water Separation. *Polymers* **2022**, *14*, 3726. [CrossRef]
27. Qiu, S.; Bi, H.; Hu, X.; Wu, M.; Li, Y.; Sun, L. Moldable clay-like unit for synthesis of highly elastic polydimethylsiloxane sponge with nanofiller modification. *RSC Adv.* **2017**, *7*, 10479–10486. [CrossRef]
28. Mo, S.; Mei, J.; Liang, Q.; Li, Z. Repeatable oil-water separation with a highly-elastic and tough amino-terminated polydimethylsiloxane-based sponge synthesized using a self-foaming method. *Chemosphere* **2021**, *271*, 129827. [CrossRef]
29. Prasanthi, I.; Raidongia, K.; Datta, K.K.R. Super-wetting properties of functionalized fluorinated graphene and its application in oil–water and emulsion separation. *Mater. Chem. Front.* **2021**, *5*, 6244–6255. [CrossRef]
30. Cao, C.; Ge, M.; Huang, J.; Li, S.; Deng, S.; Zhang, S.; Chen, Z.; Zhang, K.; Al-Deyab, S.S.; Lai, Y. Robust fluorine-free superhydrophobic PDMS–ormosil@fabrics for highly effective self-cleaning and efficient oil–water separation. *J. Mater. Chem. Front. A* **2016**, *4*, 12179–12187. [CrossRef]
31. Xu, L.; Wan, J.; Yuan, X.; Pan, H.; Wang, L.; Shen, Y.; Sheng, Y. Preparation of durable superamphiphobic cotton fabrics with self-cleaning and liquid repellency. *J. Adhes. Sci. Technol.* **2022**, *36*, 1–20. [CrossRef]
32. Su, X.; Li, H.; Lai, X.; Zhang, L.; Liang, T.; Feng, Y.; Zeng, X. Polydimethylsiloxane-Based Superhydrophobic Surfaces on Steel Substrate: Fabrication, Reversibly Extreme Wettability and Oil-Water Separation. *ACS Appl. Mater. Interfaces* **2017**, *9*, 3131–3141. [CrossRef]
33. Xue, Z.; Wang, S.; Lin, L.; Chen, L.; Liu, M.; Feng, L.; Jiang, L. A Novel Superhydrophilic and Underwater Superoleophobic Hydrogel-Coated Mesh for Oil/Water Separation. *Adv. Mater.* **2011**, *23*, 4270–4273. [CrossRef] [PubMed]

34. Xiang, B.; Sun, Q.; Zhong, Q.; Mu, P.; Li, J. Current research situation and future prospect of superwetting smart oil/water separation materials. *J. Mater. Chem. A* **2022**, *10*, 20190–20217. [CrossRef]
35. Liu, M.; Wang, S.; Wei, Z.; Song, Y.; Jiang, L. Bioinspired Design of a Superoleophobic and Low Adhesive Water/Solid Interface. *Adv. Mater.* **2009**, *21*, 665–669. [CrossRef]
36. Chen, C.; Weng, D.; Mahmood, A.; Chen, S.; Wang, J. Separation Mechanism and Construction of Surfaces with Special Wettability for Oil/Water Separation. *ACS Appl. Mater. Interfaces* **2019**, *11*, 11006–11027. [CrossRef] [PubMed]
37. Wei, C.; Zhang, G.; Zhang, Q.; Zhan, X.; Chen, F. Silicone Oil-Infused Slippery Surfaces Based on Sol-Gel Process-Induced Nanocomposite Coatings: A Facile Approach to Highly Stable Bioinspired Surface for Biofouling Resistance. *ACS Appl. Mater. Interfaces* **2016**, *8*, 34810–34819. [CrossRef] [PubMed]
38. Li, K.; Zeng, X.; Li, H.; Lai, X. Facile fabrication of a robust superhydrophobic/superoleophilic sponge for selective oil absorption from oily water. *RSC Adv.* **2014**, *4*, 23861. [CrossRef]

Disclaimer/Publisher's Note: The statements, opinions and data contained in all publications are solely those of the individual author(s) and contributor(s) and not of MDPI and/or the editor(s). MDPI and/or the editor(s) disclaim responsibility for any injury to people or property resulting from any ideas, methods, instructions or products referred to in the content.

Article

Biomass Chitosan-Based Tubular/Sheet Superhydrophobic Aerogels Enable Efficient Oil/Water Separation

Wenhui Wang [1,2,†], Jia-Horng Lin [1,2,3,4,†], Jiali Guo [1,2], Rui Sun [1,2], Guangting Han [5], Fudi Peng [6], Shan Chi [7] and Ting Dong [1,2,5,*]

1. College of Textile and Clothing, Qingdao University, 308, Ningxia Road, Qingdao 266071, China
2. Advanced Medical Care and Protection Technology Research Center, Qingdao University, 308 Ningxia Road, Qingdao 266071, China
3. Advanced Medical Care and Protection Technology Research Center, Department of Fiber and Composite Materials, Feng Chia University, Taichung City 407102, Taiwan
4. School of Chinese Medicine, China Medical University, Taichung City 404333, Taiwan
5. Key Laboratory of Bio-Fibers and Eco-Textiles, Qingdao University, 308 Ningxia Road, Qingdao 266071, China
6. Fujian Aton Advanced Materials Science and Technology Co., Ltd., Fujian 350304, China
7. Bestee Material Co., Ltd., Qingdao 266001, China
* Correspondence: tingdong09@qdu.edu.cn
† These authors contributed equally to this work.

Citation: Wang, W.; Lin, J.-H.; Guo, J.; Sun, R.; Han, G.; Peng, F.; Chi, S.; Dong, T. Biomass Chitosan-Based Tubular/Sheet Superhydrophobic Aerogels Enable Efficient Oil/Water Separation. Gels 2023, 9, 346. https://doi.org/10.3390/gels9040346

Academic Editor: Pavel Gurikov

Received: 17 March 2023
Revised: 12 April 2023
Accepted: 15 April 2023
Published: 18 April 2023

Copyright: © 2023 by the authors. Licensee MDPI, Basel, Switzerland. This article is an open access article distributed under the terms and conditions of the Creative Commons Attribution (CC BY) license (https://creativecommons.org/licenses/by/4.0/).

Abstract: Water pollution, which is caused by leakage of oily substances, has been recognized as one of the most serious global environmental pollutions endangering the ecosystem. High-quality porous materials with superwettability, which are typically constructed in the form of aerogels, hold huge potential in the field of adsorption and removal of oily substances form water. Herein, we developed a facile strategy to fabricate a novel biomass absorbent with a layered tubular/sheet structure for efficient oil/water separation. The aerogels were fabricated by assembling hollow poplar catkin fiber into chitosan sheets using a directional freeze-drying method. The obtained aerogels were further wrapped with -CH$_3$-ended siloxane structures using CH$_3$SiCl$_3$. This superhydrophobic aerogel (CA ≈ 154 ± 0.4°) could rapidly trap and remove oils from water with a large sorption range of 33.06–73.22 g/g. The aerogel facilitated stable oil recovery (90.07–92.34%) by squeezing after 10 sorption-desorption cycles because of its mechanical robustness (91.76% strain remaining after 50 compress-release cycles). The novel design, low cost, and sustainability of the aerogel provide an efficient and environmentally friendly solution for handling oil spills.

Keywords: chitosan; poplar catkin fiber; superhydrophobic aerogels; layered tubular/sheet structures; oil/water separation

1. Introduction

Gasoline is increasingly in demand because of recent industrial developments, yet leakage of oily substances during the processes of exploitation and transportation becomes a dire consequence [1,2]. For example, the outbreak of the Gulf War in 1991 resulted in a leak of oil of about 1.5 million tons, forming an oil band that was 16 km long and 3 km wide near Saudi Arabia at a spreading rate of 24 km/day. Unfortunately, any minor leakage of oil negatively affects the marine ecological system in various ways [3]. For example, the frequency of marine red tide is in direct proportion to the content and frequency of oil leakage [4,5]. Oil leakage over oceans may also adversely affect the health of infants and toddlers, potentially inflicting them with asthma and a higher death toll, of which the results may take shape over numerous years. As a result, the removal of leaked oils over the sea becomes a worldwide concern, and traditional management involves physical, chemical, and biological methods [6,7]. The physical methods, such as mechanical skimmers, have the disadvantages of costly oil/water separation machines and tremendous consumption of both manpower and material resources. Chemical methods include chemical dispersants and in situ burning, of which the former involves spraying a dispersion agent containing a toxic substance, while the latter generates tremendous toxic gases, causing secondary pollution and energy waste [8]. Biological methods rely on microorganisms to decompose leaked oils. However, these methods are only suitable for small-scale oil leakage. By contrast, the sorption-based method has been regarded as an energy-saving approach for oil contaminant disposal because of low production cost and low demand for manpower.

The oil/water separation efficiency via absorption is dependent on the hydrophilic/hydrophobic attributes and micro-pore structure of the absorbent. In particular, high-quality porous materials with superwettability, which are typically constructed in the forms of aerogels, hold huge potential in the field of adsorption and removal of oily substances form water because of their abundant and tunable porous structure, lightweight feature, and programmable surface groups. To date, a large number of aerogels have been developed, including the magnetic superhydrophobic melamine sponge (oil absorption ratio: 39.8–78.7 g/g) [9], polyurethane foam coated with polysiloxane-modified clay nanotubes (oil absorption ratio: 20–105 g/g) [10], graphene-coated carbon nanofiber (G-CNF) foam (oil absorption ratio: 86–153 g/g) [11], carbon nanotube sponges (oil absorption ratio: 80–180 g/g) [12], graphene/PDMS sponge (oil absorption ratio: 4.2–13.7 g/g) [13], and graphene/nanofiber aerogels (oil absorption ratio: 230–734 g/g) [14]. At present, the majority of aerogels are composed of synthetic materials that are favorable for mass production, and their waste either cannot be decomposed or has a very small decomposition rate, leading to severe secondary pollution. Carbon-based aerogels have the advantage of excellent oil absorption capacity, yet their fabrication demands tremendous energy and resources, which makes their mass production impossible. For example, carbon nanofiber (CNF) aerogels need to extract nano-cellulose from plants. However, the original plant fibers contain drastic hydrogen bonding between the fibrous groups, which means the extraction of nano-cellulose requires enzyme catalysis or chemical pre-treatment followed by mechanical decomposition. Compared to synthetic materials and carbon-based materials, such as natural fiber (kapok fiber [15,16], cotton fiber [17], populus seed fibers [18], kenaf core fiber, milkweed floss [19], etc.), they feature abundant natural resources, a low production cost, and biodegradation. Biomass aerogels are usually produced via freeze-drying in pace with superhydrophobic modification via the principle of constructing low surface energy and hierarchically rough surface [20]. At present, the majority of biomass-based materials lack sufficient mechanical endurance, and thus the oil absorption performance is compromised after multiple squeeze cycles. Moreover, due to the intrinsic fragility and internal chaotic porous structure, biomass materials are prone to have a collapse of internal structure and a swelling-then-dissolution feature when used in a water environment, which in turn severely restricts the biomass aerogel from the oil-water separation application. In nature, some natural materials like seaweed, lotus stems, and wood show an orderly, organized interior and a regular structure, hence demonstrating extraordinary mechanical properties [21,22]. Derived from conventional freeze-drying techniques, directional freeze-drying technology generates a temperature gradient in a single direction, which promotes the directional growth of ice crystals in the precursor to form an oriented porous structure along the temperature gradient after freeze-drying [23–25]. This method become an efficient method of improving the mechanical compression performance of aerogels, such as a high-strength (CNF)/polyvinyl alcohol (PVA)/graphene oxide (GO) aerogel showing an anisotropic porous structure [26] and a wood-inspired elastic biomass aerogel with special spring-like morphology [27].

On the other hand, environmental concerns regarding discarded solid debris containing oil residues due to the indiscriminate use of various sorption materials are often overlooked. These discarded oil-containing materials, which take up lots of land, have led to global environmental concerns regarding the release of microplastics due to the long-term environmental weathering [28,29]. For example, synthetic polymer absorbents, such as PU sponges, PS sponges, PP nonwovens, and PET nonwovens, which are non-degradable, have been proved to be the main contributor of detected microplastics in the terrestrial soil. The occurrence of microplastics in the terrestrial soil will cause hetero- or homo-aggregation with various microorganisms and macromolecules and disturb the vital aspects of soil like soil colloids and soil micro flora and fauna [30]. Biomass materials are biodegradable and environmentally friendly, but many biomass-based aerogels, which are generally fabricated by freeze-drying in pace with chemical cross-linkage to obtain the necessary strength, are less biodegradable [18]. Therefore, it is of global significance to develop sustainable alternatives combining high oil sorption performance, excellent biodegradation, and good reusability as oil absorbents.

Poplar is a common wood that grows worldwide. Its fruits grow ripe in the spring, and the cracked fruits produce numerous catkin fibers (PC) flying around in the air, which generally causes environmental pollution, possible fires, and allergic reactions in people. PC fibers show a unique hollow structure, and their composition enables good liquid adsorption capacity. Chitosan (CS), which is a linear polymer of β-(1 → 4)-linked 2-acetylamino-2-deoxy-β-D-glucopylanopylanoid and 2-amino-2-2 deoxy-β-d-glucopylanoid, is mainly extracted from crab/shrimp shells and obtained through the deacetylation of chitin. The degree of deacetylation of commercial CS is generally above 60%. As low-cost, biodegradable, non-toxic, and biocompatible, CS has been reported to be widely used in food additives, drug release, oil adsorption, heavy metal adsorption, and tissue engineering scaffolds [31–33]. Herein, we developed a facile strategy to fabricate a novel biomass absorbent with a layered tubular/sheet structure by a directional freeze-drying method, through which hollow PC fiber was assembled into chitosan sheets. The obtained aerogels were further wrapped with -CH_3-ended siloxane structures through a facile chemical vapor deposition (CVD) process using CH_3SiCl_3. This aerogel was used as an oil sorbent to efficiently trap and remove oils from water. The aerogels also showed mechanical robustness, which facilitated stable oil recovery for repeated oil/water separation by squeezing. The novel design, low cost, and sustainability of the sorbent reported here provides an efficient and environmentally friendly solution for the handling of oil spills.

2. Results and Discussion

The PC fibers were collected from the Qingdao University of Shandong province in China and were highly hollow, with a fiber wall thickness of 330 nm and an inner diameter of 6.63 μm, meaning the hollow part took up 90.7% of the total volume (Figure S1). On top of the hollow structure, the wax layer also provided PC fibers with hydrophobic features. In this study, to scatter the PC fibers in the water, the wax over the surface was first removed. Figure 1a–c compares the stereomicroscopic and SEM images of the PC fibers, and the treated PC fibers have a sleeker surface while retaining their hollow structure. The results suggest that pre-treatment does not change the intrinsic structure or features of PC fibers. The treated PC fibers and the CS (as a thickening agent) were mixed to form a stabilized suspension. The suspension was poured into a PTEF mold that was connected to a copper plate and placed in a freezer. As a result, ice crystals grew along a specified direction, and eventually the longer PC fibers became curly and entangled. Meanwhile, the CS became the connective points among the fibers, and after the crystals were removed via the freeze-drying process, there was an initiating configuration of PC/CS aerogel (Figure 2a). In the aerogel, PC fibers that had a hollow structure were assembled into chitosan sheets showing a layered tubular/sheet structure (Figure 1a–c). PC fibers that can be used as a second-pore capillary have a positive influence over the oil transport of aerogel, while the CS serves as bonding points that make the aerogel mechanically robust. The resulting tubular/sheet structure has a sheet structure (from freeze-drying the CS) as the first gradient of oil absorption as well as a concurrent hollow structure (from the PC fibers) as the second gradient of oil transport. As a result, the aerogel demonstrates a highly strengthened oil sorption capacity that is guaranteed by its super high porosity and lower volume density (0.011 g/cm^3).

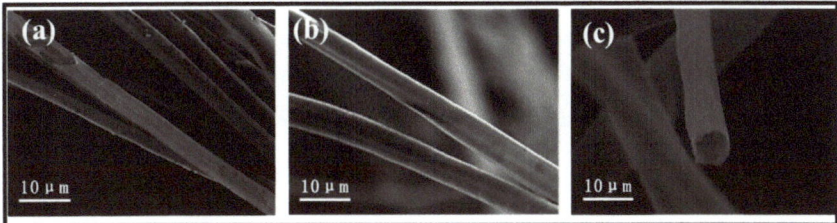

Figure 1. (a) SEM image of non-treated PC and (b,c) SEM images of treated PC.

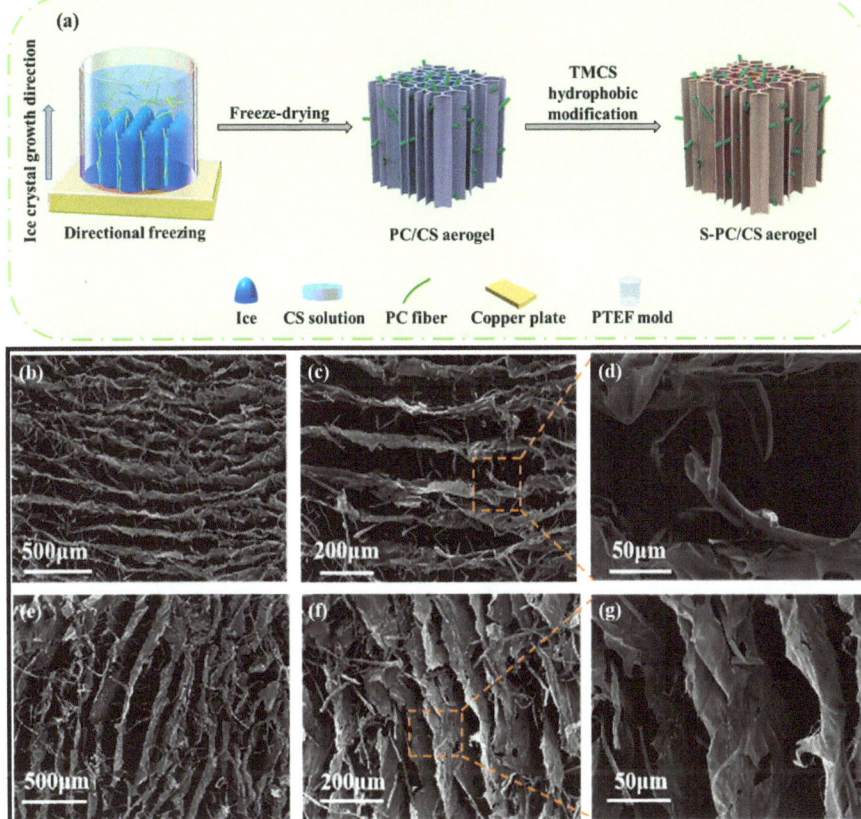

Figure 2. (a) Manufacturing process of S-PC/CS aerogel; (b–d) horizontally cutting section and (e–g) vertically cutting section of S-PC/CS aerogel.

As seen in Figure 3a,b, FTIR spectra is mainly used to study the differences in the surface functional group of non-treated PC fibers and the S-PC/CS aerogel. PC fibers, as cellulose fibers, contain many characteristic function groups, such as -OH (3340 cm^{-1}), C-H (2921 cm^{-1}), C-O (1737 cm^{-1} and 1237 cm^{-1}), C=C (1590 cm^{-1}), C-O-C (1106 cm^{-1}), and C-O (1037 cm^{-1}) [18]. By contrast, the treated PC fibers demonstrate significantly attenuated bands at 1598 cm^{-1}, 1242 cm^{-1}, and 1456 cm^{-1} that corresponded to the tensile vibration of the aromatic C skeleton, which confirms that the wax is removed from the PC fibers. Moreover, after the superhydrophobic treatment, S-PC/CS aerogel exhibits characteristic peaks at 773 cm^{-1} corresponding to the Si-O-Si bond and an asymmetric stretching band at 1271 cm^{-1} corresponding to the C-Si-O group. The Si-O bond indicates that the derived -OH from PC/CS has a drastic interaction with organosilane, which in turn forms a silicon-oxygen bond over the surface of the PC/CS aerogel, suggesting a chemical reaction between TMCS and the PC/CS aerogel. The peaks of the hydroxyl group in the S-PC/CS are still present. This is because the connection between MTCS and material is relatively complicated. During the process, MTCS undergoes self-polymerization with H$_2$O vapor to produce 3D methylsiloxanes with reactive trifunctional silanes that can bond with hydroxyl groups on the fiber surface via -Si(OH)$_3$ radicals. However, due to steric hindrance, these surfaces should contain "holes" between randomly attached disiloxane groups that are smaller than the disiloxane, contain surface hydroxyl groups, and cannot be filled by further reaction (Figure S2) [34].

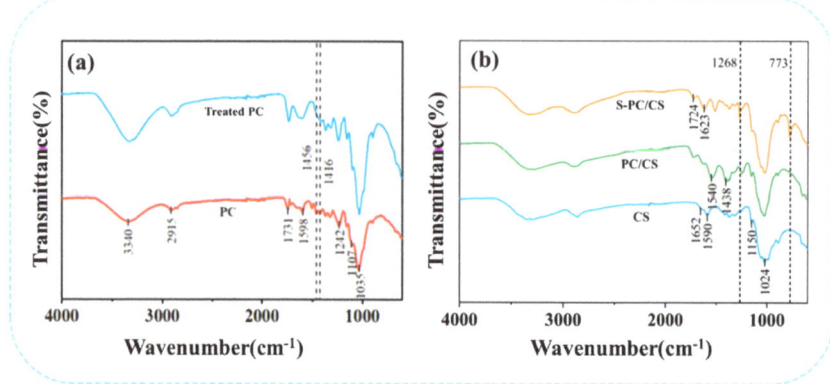

Figure 3. FTIR spectra of (**a**) treated/non-treated PC fibers as well as (**b**) CS, PC/CS, and S−PC/CS aerogel.

Figure 4a,b shows the cyclic compression performance of the S-PC/CS aerogel tested with a constant 20–60% strain. The deformation behavior of the S-PC/CS aerogel includes the linear elastic area when ε < 20%, the subsequent plateau stage when 20% < ε < 40%, and the densification stage when the stress accelerates. With a compression strain (ε = 40%), the S-PC/CS aerogel demonstrates marginal plastic deformation (6.81%) and a height recovery rate of 93.19%. Furthermore, when a greater compression stress (ε = 60%) is exerted, the S-PC/CS aerogel presents a plastic deformation of 13.23%. The test results also indicate that the aerogel exhibits excellent fatigue resistance. The produced plastic deformations are 4.83% after the first cycle and 8.57% after 20 cycles. Nonetheless, the effects of multiple plastic deformations are accumulated, and the aerogel at the 50th cycle of loading-unloading is inflicted with irreversible damage of 11.29%. To data, a majority of biomass aerogels still present crucial limitations on structural and mechanical stability. As illustrated in Table 1, the existing biomass aerogels have a maximum compressive stress of 3.5–55 kPa, and many of them have the problem of structural instability and plastic deformation exceeding 15%, such as dialdehyde carboxymethyl cellulose aerogels (<10 kPa, 15–20% plastic deformation after 50 cycles) [35], cellulose nanocrystals/PVA aerogels (<35 kPa, >15% plastic deformation after 50 cycles) [36], cellulose nanofibrils/N-alkylated chitosan/poly(vinyl alcohol) aerogels (<55 kPa, but the maximum stress will reduce to 17 kPa after 50 cycles, 18–20% plastic deformation) [37], seed hairs of typha orientalis aerogels (<25 kPa, 14.8% plastic deformation after 10 cycles) [38], and so on. The excellent rebound property of the S-PC/CS is primarily attributed to the unique sheet structure. When the aerogel is compressed, the sheet structure provides enough space for elastic deformation while saving energy. When the external force is withdrawn, the energy in need is released, which allows the aerogel to recover its original state. Meanwhile, the PC fibers among the sheets of aerogel also provide a proportion of support, which benefits the compression strength of the aerogel. The excellent mechanical properties and compression recovery prove that the materials can be repetitively used.

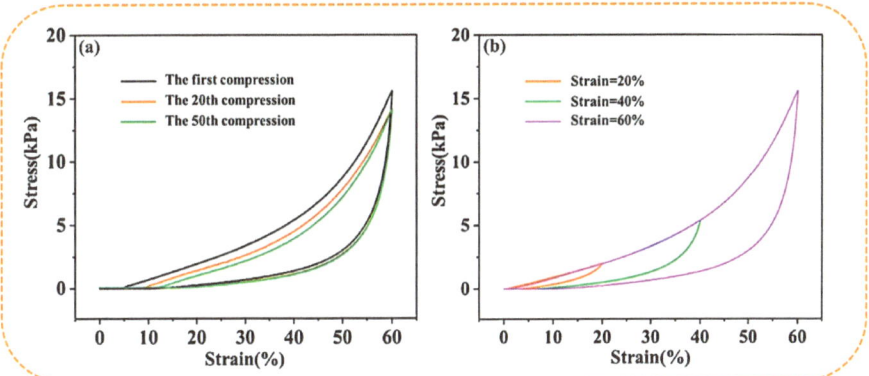

Figure 4. The stress-strain curves of S−PC/CS aerogel as related to (**a**) constant 60% strain and (**b**) various strains.

Figure 5a–c shows the effects of three types of oil on the S-PC/CS aerogel, and the difference in the speed of infiltrating the aerogel among the different oils is ascribed to the viscosity and mobility. It takes the aerogel only 3.1 s to absorb vegetable oil completely, 5.9 s for viscous motor oil [1#] 5W−40, and 3.4 s for viscous motor oil [2#] 0W−20, which substantiates that the modified aerogel exhibits excellent oil sorption rate regardless of the oil type. In addition, the original PC/CS is highly hydrophilic, and the modification of CH_3SiCl_3 results in the replacement of the hydroxyl groups of the materials by -CH_3-ended siloxane structures. This process effectively transfers the hydrophobic properties of the aerogels with an average WCA of about 154 ± 0.4° which is highly hydrophobic and satisfies the requirement of oil/water separation. In order to test the hydrophobicity and selective adsorption of the aerogel, the following tests were performed. As shown in Figure 5d, the aerogel was put into water stained with methylene blue for a period of time and taken out, and it was found that the aerogel was not dyed. As shown in Figure 5e, a few drops of soybean oil dyed with oil red O were dropped on the water surface, and then the aerogel was immersed in water. After a period of time, the soybean oil on the water surface was completely adsorbed by the aerogel, which proves that the aerogel has good hydrophobicity and selective adsorption of oil. Different liquids are employed to examine the oil sorption capacity of the aerogel. Figure 5f shows that the aerogel demonstrates the maximal and minimal sorption capacity for dichloromethane and hexane, respectively. Furthermore, the sorption capacity is ascending for diesel, soybean oil, motor oil [1#], motor oil [2#], motor oil[3#] 20W-50, and pump oil in a range of 33.06–73.22 g/g^{-1}, and the corresponding sorption capacity is dependent on the density of the liquids. In addition, the repetitive use of aerogel was tested, as seen in Figure 5g. An aerogel was immersed in the test oil for 10 minutes and then placed over a filter for another 1 min to remove redundant oil. After 10 cycles of sorption-desorption, the sorption capacity of the aerogel was decreased by 5.48 g/g for dichloromethane, 4.51 g/g for soybean, 4.10 g/g for diesel, and 2.69 g/g for hexane. During the 10 cycles of sorption-desorption, the aerogels retained 90.07–92.34% of their initial sorption capacity. The oil sorption tests of the aerogels under 50 sorption-desorption cycles were also carried out using soybean oil. As shown in Figure S3a, the sorption capacity of aerogel was decreased by 0.985 g/g after 20 cycles, 1.378 g/g after 30 cycles, 1.15 g/g after 40 cycles, and 0.81 g/g after 50 cycles. During the 50 cycles of sorption-desorption, the aerogels retained over 85.6% of their initial sorption capacity. In other words, the oil sorption capacity of the aerogel was not significantly compromised by the test, which suggests that the aerogel can be repetitively used and is an ideal oil/water separation material. In addition, the oil/water selectivity performance in Figure S3b shows that the aerogel retained a high oil sorption capacity of 47.48–44.49 g/g in 10 testing cycles, while the water sorption capacity was very low and in the range of 0.002–0.020 g/g, indicating a high oil/water selectivity of 222.45–19,869.46. The above characteristics feature S-PC/CS aerogel as a great application prospect in controlling oil spills. Table 1 compares the S-PC/CS with a wide range of current state-of-the-art biomass-based oil absorbents. Compared with biomass aerogels, which are generally derived from cellulose, chitosan, sodium alginate, lignin, etc., the S-PC/CS outperforms the majority of aerogels in terms of oil sorption capacity and hydrophobicity, such as dialdehyde carboxymethyl cellulose aerogels (sorption capacity: 20–30 g/g, WCA = 144.5°) [35], graphene ox-ide/halloysite nan-otubes (RGO/HNTs) membrane (WCA = 82.43°) [39], chitin/halloysite nanotubes sponges (sorption capacity: 11.23 g/g, WCA = 88–98°) [40], HNTC-FG-PU sponges (sorption capacity: 50.8 g/g, WCA = 145 ± 2°) [41], and alginate/oil gelator aerogels (sorption capacity: 32 g/g, WCA = 155 ± 5°) [42]. The results indicate that the tubular/sheet structure facilitates fast oil transport. The aerogel contains chitosan that forms narrow channels in tidy alignment, which provides sufficient space for oil transmission. Moreover, the tubular structure of PC fibers serves as a second channel that expedites the infusion of oil, and thus the aerogel demonstrates excellent oil sorption performance. At the same time, the WCA of the aerogel modified by super hydrophobicity is better than most aerogels. The excellent hydrophobicity endows the aerogel with the characteristic of selective adsorption. In addition, the raw material of aerogel comes from pure biomass material, which is low cost and easy to obtain, biodegradable, and environmentally friendly. These attractive advantages of the aerogel give it a broad application prospect, and it is expected to be used in industrial wastewater treatment and marine oil and water separation.

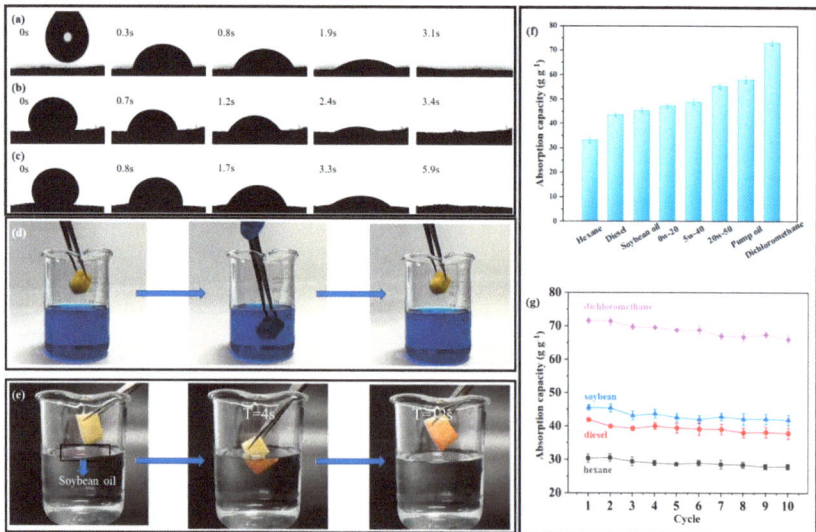

Figure 5. The images of S−PC/CS absorbing (**a**) soybean, (**b**) motor oil 0W−20, and (**c**) motor oil 5W−40. (**d,e**) Hydrophobic performance test. (**f**) Sorption capacity (as related to the liquid) and (**g**) Sorption capacity as related to the liquid and the multiple cycles.

Table 1. Comparison of properties of different sorption materials.

Sorbent Material	Maximum Compression Stress (kPa)	Number of Compression Cycles	Compressive Properties — Plastic Deformation	WCA	Sorption Capacity (g/g^{-1})	Preparation Method	Reference
Lignin/Agarose/PVA aerogels	<16	10	20%	150°	18	Indirectional freeze-drying	Jiang, J., et al., 2017 [43]
Dialdehyde carboxymethyl cellulose aerogels	<10	50	15–20%	144.5°	20–30	Indirectional freeze-drying	Zhang, F., et al., 2022 [35]
Cellulose nanocrystals/PVA aerogels	<35	50	>15%	136°	<35	Indirectional freeze-drying	Gong, X., et al., 2019 [36]
Carboxylated cellulose nanofibers/PEI aerogels	<9	1	20%	-	20–60	Indirectional freeze-drying	Tang, R., et al., 2023 [37]
Cellulose nanofibrils/N-alkylated chitosan/poly(vinyl alcohol) aerogels	<55 kPa	50	18–20%	147°	19–51	Indirectional freeze-drying	Li, M., et al., 2021 [44]
Seed hairs of typha orientalis aerogels	<25	10	14.8%	153°	42–160	Carbonized	Yang, J., et al., 2018 [38]
Bacterial cellulose aerogels	<3.5	100	5%	131 ± 3.5°	37–89	Pyrolysis	Ieamviteevanich, P., et al., 2020 [45]
Alginate/oil gelator aerogels	<9 kPa	-	-	155 ± 5°	32	Indirectional freeze-drying	Wang, Y., et al., 2022 [42]
graphene oxide/halloysite nanotubes (RGO/HNTs) membrane	-	-	-	82.43°	-	Hummers method	Liu Y., et al., 2018 [39]
chitin/halloysite nanotubes sponge	-	-	-	88–98°	11.23	Freeze-drying	Zhao X., et al., 2019 [40]
HNTC-FG-PU sponges	-	-	-	145 ± 2°	50.8	Dip-coating	Prasanthi, L., et al., 2022 [41]
S-PC/CS aerogels	16.5 (ε = 60%)	50	8.25%	154 ± 0.4°	33.06–73.22	YES	This work

3. Conclusions

In this study, chitosan is used as the basic material and is combined with PC fibers to form a CS-based aerogel with a unique tubular/sheet structure that mechanically improves the biomass aerogel. After 50 cycles of a compression resistance test, the aerogel only exhibits marginal irreversible deformation (8.24%). The aerogel exhibits an oil sorption range of 33.06–73.22 g g^{-1}, which outperforms the majority of recently reported sponges. The multiple sorption-desorption results indicate that the aerogel retains a stable oil sorption capacity that is over 90%, suggesting that the aerogel can be repetitively used. As the proposed aerogel is pure biomass material, it can be decomposed in nature and is eco-friendly. Also, the needed raw material for the chitosan-based pure biomass aerogel is easily accessible, and the aerogel can be expected to be used in oil/water separation on a large scale compared to the conventional aerogel.

4. Material and Methods

4.1. Materials

The PC fibers (hereafter referred to as PC fibers) were collected from poplar trees at Qingdao University, Shandong. Chitosan (CS) (95% degree of deacetylation, 100–200 mPa·s) was purchased from Aladdin, Industrial Co., Ltd., Shanghai, China. Hexane, absolute alcohol, acetic acid, dichloromethane, chloroform, and ethyl acetate were purchased commercially without further purification. Methyltrichlorosilane (MTCS, 98%) was obtained from Sigma, America. Methylene blue and oil red O were both from Hefei Sifu Biotechnology Co., Ltd., Hefei, China. Several oils were purchased commercially (see Table S1).

4.2. Pre-Treatment of PC Fibers

The PC fibers were trimmed to a length of 5–10 mm and rinsed with deionized water and ethanol several times in order to remove any impurities from the surface. Next, NaClO$_2$ (3 g) was dissolved in 300 mL of deionized water, after which 0.9 mL of acetic acid and 3.0 g of PC fibers were added to be heated in a water bath at 75 °C for 2.5 h. Afterwards, PC fibers were filtrated and once again rinsed with deionized water until the residual liquid becomes neutral. The fibers were removed and dried in an oven at 70 °C for 12 h.

4.3. Preparation of Aerogels

PC fibers (0.8 g) and CS (1.4 g) were added to deionized water (200 mL, 60 °C), after which 2 mL of acetic acid (1%, v/v) was added. The blends were mixed using a household blender to form suspensions with uniform quality. Next, the suspension was infused into a PTFE mold, with the bottom of the mold in contact with a copper plate, and was then frozen in liquid nitrogen. After being totally frozen, the suspension underwent the freeze-drying process for 48 h, resulting in the aerogel (hereafter referred to as PC/CS).

4.4. Modification of Aerogels

The PC/CS aerogel was placed in an environment at a humidity of 60–70% for 24 h. Next, the aerogel was placed in sealed glass, where 0.2 mL of TMCS was added and left for a 12-h reaction. The sample was removed and heated to 60 °C in an oven to remove the redundant TMCS, thereby obtaining a superhydrophobic aerogel (hereafter referred to as S-PC/CS).

4.5. Characterizations

A freezer dryer used (LGJ-18, Beijing Songyuan Huaxing Technology Development Co., LTD., Beijing, China) in the freeze drying process. A field emission scanning electron microscope (SEM, Zeiss Sigma500, Oberkochen, Germany) was used to observe the structure of the aerogel. The working distance and energy beam for the SEM were 3 and 5 mm, respectively, with voltage 10 keV and current 10 µA. The samples were metallized before analysis. Fourier-transform infrared spectroscopy (FTIR, Thermo Fisher, Waltham, MA, USA) was used to analyze the function groups of treated PC fibers, CS, PC/CS, and

S-PC/CS in the range of 500–4000 cm^{-1}. A universal material testing machine (Instron-3300, Norwood, MA, USA) was used for the compression stress-strain test with a strain rate of 20 mm/min. A drop-shaped analyzer (Theta, Biolin Corporation, Goteborg, Switzerland) was used to measure the water contact angle (WCA) of the aerogel with the specified volume of droplets being 5.0 µL. The porosity was measured by an automatic mercury intrusion porosimetry instrument (PoreMaster-33, Quantachrome, FL, USA). The volume density was calculated by the following formula:

$$p = \frac{v}{m} \quad (1)$$

where p is the volume density, v is the aerogel volume, and m is the aerogel mass.

4.6. Oil Sorption Capacity

Different types of oil were dripped over the aerogel, and the time that the aerogel required to absorb the whole droplet was measured. To evaluate the oil absorption capacity, S-PC/CS aerogel was immersed in different oils and organic solvents for 10 min, and the oil-loaded aerogel was placed over a filter for one minute to remove the excess oil. After ten cycles of the absorption-squeeze test, the oil absorption capacity and reusability of samples were recorded. The oil sorption capacity was calculated according to the following formula:

$$\text{Oil absorption rate}: (Q) = \frac{m - m_0}{m_0} \quad (2)$$

where the m_0 is for the quality of the aerogel oil absorption before, and m is for the quality of the aerogel after oil absorption. In addition, to measure the oil/water selectivity, the aerogel was completely immersed in a mixture of soybean oil/water (v/v = 1:1) for 1 h, and then the saturated aerogel was dried in the oven at 60 °C for 1 h to remove the absorbed water. The masses of absorbed water (m_w) and oil (m_o) were calculated using the following formulas:

$$m_w = m_1 - m_2. \quad (3)$$

$$m_o = m_2 - m_0. \quad (4)$$

where m_0 is the initial mass of the gel, m_1 is the mass of the adsorbed saturated sample, and m_2 is the mass of the sample after removing the water. The oil/water selectivity is calculated by the ratio of oil sorption mass to the water sorption mass.

Supplementary Materials: The following supporting information can be downloaded at https://www.mdpi.com/article/10.3390/gels9040346/s1: Figure S1: (a/b) Camera photos of PC trees/fibers; (c) microscopic image of PC; (d/e) microscopic image of treated PC; Figure S2: Reaction mechanism of MTCS with the material; Figure S3: (a) Adsorption capacity of soybean oil for 50 cycles, (b) Oil absorption and water absorption of aerogel in 10 cycles; Table S1: The oil used in experiments.

Author Contributions: Conceptualization, J.-H.L.; Software, R.S.; Validation, W.W.; Formal analysis, W.W.; Writing—original draft, W.W.; Writing—review & editing, J.-H.L., J.G., R.S. and G.H.; Supervision, J.-H.L. and G.H.; Project administration, T.D.; Funding acquisition, F.P., S.C. and T.D. All authors have read and agreed to the published version of the manuscript.

Funding: The research is financially supported by "National Natural Science Foundation of China (52203118)," "Natural Science Foundation of Shandong Province (ZR201911130239)," and "Qingdao Shinan District Science and technology planning project" (2022-1-016-ZH).

Institutional Review Board Statement: Not applicable.

Informed Consent Statement: Not applicable.

Data Availability Statement: Not applicable.

Conflicts of Interest: The authors declare no conflict of interest.

References

1. Hua, Y.; Dong, T. Multi-functional flame-retardant superhydrophobic ceramic fiber felt: Oil/Water mixture separation and oil mist interception. *Colloids Surf. A Physicochem. Eng. Asp.* **2021**, *629*, 127454. [CrossRef]
2. Dong, T.; Li, Q.; Tian, N.; Zhao, H.; Zhang, Y.; Han, G. Concus Finn Capillary driven fast viscous oil-spills removal by superhydrophobic cruciate polyester fibers. *J. Hazard. Mater.* **2021**, *417*, 126133. [CrossRef] [PubMed]
3. Dong, T.; Liu, Y.; Tian, N.; Zhang, Y.; Han, G.; Peng, F.; Lou, C.-W.; Chi, S.; Liu, Y.; Liu, C.; et al. Photothermal and Concus Finn capillary assisted superhydrophobic fibrous network enabling instant viscous oil transport for crude oil cleanup. *J. Hazard. Mater.* **2023**, *443*, 130193. [CrossRef]
4. Liu, X.; Zhang, C.; Geng, R.; Lv, X. Are oil spills enhancing outbreaks of red tides in the Chinese coastal waters from 1973 to 2017? *Environ. Sci. Pollut. Res.* **2021**, *28*, 56473–56479. [CrossRef] [PubMed]
5. Li, Y.; Fan, T.; Cui, W.; Wang, X.; Ramakrishna, S.; Long, Y. Harsh environment-tolerant and robust PTFE@ZIF-8 fibrous membrane for efficient photocatalytic organic pollutants degradation and oil/water separation. *Sep. Purif. Technol.* **2023**, *306*, 122586. [CrossRef]
6. Li, T.-T.; Li, S.; Sun, F.; Shiu, B.-C.; Ren, H.-T.; Lou, C.-W.; Lin, J.-H. pH-responsive nonwoven fabric with reversibly wettability for controllable oil-water separation and heavy metal removal. *Environ. Res.* **2022**, *215*, 114355. [CrossRef]
7. Li, K.; Yu, H.; Yan, J.; Liao, J. Analysis of Offshore Oil Spill Pollution Treatment Technology. *IOP Conf. Ser. Earth Environ. Sci.* **2020**, *510*, 042011. [CrossRef]
8. Abidli, A.; Huang, Y.; Cherukupally, P.; Bilton, A.M.; Park, C.B. Novel separator skimmer for oil spill cleanup and oily wastewater treatment: From conceptual system design to the first pilot-scale prototype development. *Environ. Technol. Innov.* **2020**, *18*, 100598. [CrossRef]
9. Li, Z.-T.; Wu, H.-T.; Chen, W.-Y.; He, F.-A.; Li, D.-H. Preparation of magnetic superhydrophobic melamine sponges for effective oil-water separation. *Sep. Purif. Technol.* **2019**, *212*, 40–50. [CrossRef]
10. Wu, F.; Pickett, K.; Panchal, A.; Liu, M.; Lvov, Y. Superhydrophobic Polyurethane Foam Coated with Polysiloxane-Modified Clay Nanotubes for Efficient and Recyclable Oil Absorption. *ACS Appl. Mater. Interfaces* **2019**, *11*, 25445–25456. [CrossRef]
11. Jin, X.; Al-Qatatsheh, A.; Subhani, K.; Salim, N.V. Biomimetic and flexible 3D carbon nanofiber networks with fire-resistant and high oil-sorption capabilities. *Chem. Eng. J.* **2021**, *412*, 128635. [CrossRef]
12. Gui, X.; Wei, J.; Wang, K.; Cao, A.; Zhu, H.; Jia, Y.; Shu, Q.; Wu, D. Carbon Nanotube Sponges. *Adv. Mater.* **2010**, *22*, 617–621. [CrossRef]
13. Pan, Z.; Guan, Y.; Liu, Y.; Cheng, F. Facile fabrication of hydrophobic and underwater superoleophilic elastic and mechanical robust graphene/PDMS sponge for oil/water separation. *Sep. Purif. Technol.* **2021**, *261*, 118273. [CrossRef]
14. Xiao, J.; Lv, W.; Song, Y.; Zheng, Q. Graphene/nanofiber aerogels: Performance regulation towards multiple applications in dye adsorption and oil/water separation. *Chem. Eng. J.* **2018**, *338*, 202–210. [CrossRef]
15. Dong, T.; Cao, S.; Xu, G. Highly efficient and recyclable depth filtrating system using structured kapok filters for oil removal and recovery from wastewater. *J. Hazard. Mater.* **2017**, *321*, 859–867. [CrossRef] [PubMed]
16. Dong, T.; Cao, S.; Xu, G. Highly porous oil sorbent based on hollow fibers as the interceptor for oil on static and running water. *J. Hazard. Mater.* **2016**, *305*, 1–7. [CrossRef] [PubMed]
17. Wang, W.; Dong, C.; Liu, S.; Zhang, Y.; Kong, X.; Wang, M.; Ding, C.; Liu, T.; Shen, H.; Bi, H. Super-hydrophobic cotton aerogel with ultra-high flux and high oil retention capability for efficient oil/water separation. *Colloids Surf. A Physicochem. Eng. Asp.* **2023**, *657*, 130572. [CrossRef]
18. Dong, T.; Tian, N.; Xu, B.; Huang, X.; Chi, S.; Liu, Y.; Lou, C.-W.; Lin, J.-H. Biomass poplar catkin fiber-based superhydrophobic aerogel with tubular-lamellar interweaved neurons-like structure. *J. Hazard. Mater.* **2022**, *429*, 128290. [CrossRef]
19. Panahi, S.; Moghaddam, M.K.; Moezzi, M. Assessment of milkweed floss as a natural hollow oleophilic fibrous sorbent for oil spill cleanup. *J. Environ. Manag.* **2020**, *268*, 110688. [CrossRef]
20. Tian, N.; Wu, S.; Han, G.; Zhang, Y.; Li, Q.; Dong, T. Biomass-derived oriented neurovascular network-like superhydrophobic aerogel as robust and recyclable oil droplets captor for versatile oil/water separation. *J. Hazard. Mater.* **2022**, *424*, 127393. [CrossRef]
21. Dong, T.; Li, Q.; Nie, K.; Jiang, W.; Li, S.; Hu, X.; Han, G. Facile Fabrication of Marine Algae-Based Robust Superhydrophobic Sponges for Efficient Oil Removal from Water. *ACS Omega* **2020**, *5*, 21745–21752. [CrossRef] [PubMed]
22. Liu, F.; Jiang, Y.; Feng, J.; Li, L.; Feng, J. Bionic Aerogel with a Lotus Leaf-like Structure for Efficient Oil-Water Separation and Electromagnetic Interference Shielding. *Gels* **2023**, *9*, 214. [CrossRef]
23. Ma, X.; Tian, N.; Wang, G.; Wang, W.; Miao, J.; Fan, T. Biomimetic vertically aligned aerogel with synergistic photothermal effect enables efficient solar-driven desalination. *Desalination* **2023**, *550*, 116397. [CrossRef]
24. Chatterjee, S.; Ke, W.-T.; Liao, Y.-C. Elastic nanocellulose/graphene aerogel with excellent shape retention and oil absorption selectivity. *J. Taiwan Inst. Chem. Eng.* **2020**, *111*, 261–269. [CrossRef]
25. Hu, Y.; Yang, B.; Hao, M.; Chen, Z.; Liu, Y.; Ramakrishna, S.; Wang, X.; Yao, J. Preparation of high elastic bacterial cellulose aerogel through thermochemical vapor deposition catalyzed by solid acid for oil-water separation. *Carbohydr. Polym.* **2023**, *305*, 120538. [CrossRef] [PubMed]
26. Zhou, L.; Zhai, S.; Chen, Y.; Xu, Z. Anisotropic Cellulose Nanofibers/Polyvinyl Alcohol/Graphene Aerogels Fabricated by Directional Freeze-drying as Effective Oil Adsorbents. *Polymers* **2019**, *11*, 712. [CrossRef]

27. Yi, L.; Yang, J.; Fang, X.; Xia, Y.; Zhao, L.; Wu, H.; Guo, S. Facile fabrication of wood-inspired aerogel from chitosan for efficient removal of oil from Water. *J. Hazard. Mater.* **2020**, *385*, 121507. [CrossRef] [PubMed]
28. Ye, S.; Cheng, M.; Zeng, G.; Tan, X.; Wu, H.; Liang, J.; Shen, M.; Song, B.; Liu, J.; Yang, H.; et al. Insights into catalytic removal and separation of attached metals from natural-aged microplastics by magnetic biochar activating oxidation process. *Water Res.* **2020**, *179*, 115876. [CrossRef]
29. He, S.; Jia, M.; Xiang, Y.; Song, B.; Xiong, W.; Cao, J.; Peng, H.; Yang, Y.; Wang, W.; Yang, Z.; et al. Biofilm on microplastics in aqueous environment: Physicochemical properties and environmental implications. *J. Hazard. Mater.* **2022**, *424*, 127286. [CrossRef]
30. Ganie, Z.A.; Khandelwal, N.; Tiwari, E.; Singh, N.; Darbha, G.K. Biochar-facilitated remediation of nanoplastic contaminated water: Effect of pyrolysis temperature induced surface modifications. *J. Hazard. Mater.* **2021**, *417*, 126096. [CrossRef]
31. Rocha-Pimienta, J.; Navajas-Preciado, B.; Barraso-Gil, C.; Martillanes, S.; Delgado-Adámez, J. Optimization of the Extraction of Chitosan and Fish Gelatin from Fishery Waste and Their Antimicrobial Potential as Active Biopolymers. *Gels* **2023**, *9*, 254. [CrossRef] [PubMed]
32. Lisuzzo, L.; Cavallaro, G.; Milioto, S.; Lazzara, G. Layered composite based on halloysite and natural polymers: A carrier for the pH controlled release of drugs. *New J. Chem.* **2019**, *43*, 10887–10893. [CrossRef]
33. Chijcheapaza-Flores, H.; Tabary, N.; Chai, F.; Maton, M.; Staelens, J.-N.; Cazaux, F.; Neut, C.; Martel, B.; Blanchemain, N.; Garcia-Fernandez, M.J. Injectable Chitosan-Based Hydrogels for Trans-Cinnamaldehyde Delivery in the Treatment of Diabetic Foot Ulcer Infections. *Gels* **2023**, *9*, 262. [CrossRef] [PubMed]
34. Zhang, W.; Gu, J.; Tu, D.; Guan, L.; Hu, C. Efficient Hydrophobic Modification of Old Newspaper and Its Application in Paper Fiber Reinforced Composites. *Polymers* **2019**, *11*, 842. [CrossRef]
35. Zhang, F.; Wang, C.; Mu, C.; Lin, W. A novel hydrophobic all-biomass aerogel reinforced by dialdehyde carboxymethyl cellulose for oil/organic solvent-water separation. *Polymer* **2022**, *238*, 124402. [CrossRef]
36. Gong, X.; Wang, Y.; Zeng, H.; Betti, M.; Chen, L. Highly Porous, Hydrophobic, and Compressible Cellulose Nanocrystals/Poly(vinyl alcohol) Aerogels as Recyclable Absorbents for Oil–Water Separation. *ACS Sustain. Chem. Eng.* **2019**, *7*, 11118–11128. [CrossRef]
37. Tang, R.; Xu, S.; Hu, Y.; Wang, C.; Lu, C.; Wang, L.; Zhou, Z.; Liao, D.; Zhang, H.; Tong, Z. Multifunctional nano-cellulose aerogel for efficient oil–water separation: Vital roles of magnetic exfoliated bentonite and polyethyleneimine. *Sep. Purif. Technol.* **2023**, *314*, 123557. [CrossRef]
38. Yang, J.; Xu, P.; Xia, Y.; Chen, B. Multifunctional carbon aerogels from typha orientalis for oil/water separation and simultaneous removal of oil-soluble pollutants. *Cellulose* **2018**, *25*, 5863–5875. [CrossRef]
39. Liu, Y.; Tu, W.; Chen, M.; Ma, L.; Yang, B.; Liang, Q.; Chen, Y. A mussel-induced method to fabricate reduced graphene oxide/halloysite nanotubes membranes for multifunctional applications in water purification and oil/water separation. *Chem. Eng. J.* **2018**, *336*, 263–277. [CrossRef]
40. Zhao, X.; Luo, Y.; Tan, P.; Liu, Y. Hydrophobically modified chitin/halloysite nanotubes composite sponges for high efficiency oil-water separation. *Int. J. Biol. Macromol.* **2019**, *132*, 406–415. [CrossRef]
41. Prasanthi, I.; Rani Bora, B.; Raidongia, K.; Datta, K.K.R. Fluorinated graphene nanosheet supported halloysite nanoarchitectonics: Super-wetting coatings for efficient and recyclable oil sorption. *Sep. Purif. Technol.* **2022**, *301*, 122049. [CrossRef]
42. Wang, Y.; Yu, X.; Fan, W.; Liu, R.; Liu, Y. Alginate-oil gelator composite foam for effective oil spill treatment. *Carbohydr. Polym.* **2022**, *294*, 119755. [CrossRef] [PubMed]
43. Jiang, J.; Zhang, Q.; Zhan, X.; Chen, F. Renewable, Biomass-Derived, Honeycomblike Aerogel As a Robust Oil Absorbent with Two-Way Reusability. *ACS Sustain. Chem. Eng.* **2017**, *5*, 10307–10316. [CrossRef]
44. Li, M.; Liu, H.; Liu, J.; Pei, Y.; Zheng, X.; Tang, K.; Wang, F. Hydrophobic and self-recoverable cellulose nanofibrils/N-alkylated chitosan/poly(vinyl alcohol) sponge for selective and versatile oil/water separation. *Int. J. Biol. Macromol.* **2021**, *192*, 169–179. [CrossRef]
45. Ieamviteevanich, P.; Palaporn, D.; Chanlek, N.; Poo-arporn, Y.; Mongkolthanaruk, W.; Eichhorn, S.J.; Pinitsoontorn, S. Carbon Nanofiber Aerogel/Magnetic Core–Shell Nanoparticle Composites as Recyclable Oil Sorbents. *ACS Appl. Nano Mater.* **2020**, *3*, 3939–3950. [CrossRef]

Disclaimer/Publisher's Note: The statements, opinions and data contained in all publications are solely those of the individual author(s) and contributor(s) and not of MDPI and/or the editor(s). MDPI and/or the editor(s) disclaim responsibility for any injury to people or property resulting from any ideas, methods, instructions or products referred to in the content.

Article

Alleviating Effect of a Magnetite (Fe₃O₄) Nanogel against Waterborne-Lead-Induced Physiological Disturbances, Histopathological Changes, and Lead Bioaccumulation in African Catfish

Afaf N. Abdel Rahman [1,*], Basma Ahmed Elkhadrawy [2], Abdallah Tageldein Mansour [3,4,*], Heba M. Abdel-Ghany [5], Engy Mohamed Mohamed Yassin [6], Asmaa Elsayyad [7], Khairiah Mubarak Alwutayd [8], Sameh H. Ismail [9] and Heba H. Mahboub [1,*]

Citation: Rahman, A.N.A.; Elkhadrawy, B.A.; Mansour, A.T.; Abdel-Ghany, H.M.; Yassin, E.M.M.; Elsayyad, A.; Alwutayd, K.M.; Ismail, S.H.; Mahboub, H.H. Alleviating Effect of a Magnetite (Fe₃O₄) Nanogel against Waterborne-Lead-Induced Physiological Disturbances, Histopathological Changes, and Lead Bioaccumulation in African Catfish. *Gels* **2023**, *9*, 641. https://doi.org/10.3390/gels9080641

Academic Editors: Daxin Liang, Ting Dong, Yudong Li and Caichao Wan

Received: 7 July 2023
Revised: 2 August 2023
Accepted: 3 August 2023
Published: 8 August 2023

Copyright: © 2023 by the authors. Licensee MDPI, Basel, Switzerland. This article is an open access article distributed under the terms and conditions of the Creative Commons Attribution (CC BY) license (https://creativecommons.org/licenses/by/4.0/).

[1] Department of Aquatic Animal Medicine, Faculty of Veterinary Medicine, Zagazig University, Zagazig 44519, Egypt
[2] Department of Forensic Medicine and Toxicology, Faculty of Veterinary Medicine, University of Sadat City, Sadat City 32897, Egypt; basma.elkhadrawy@vet.usc.edu.eg
[3] Animal and Fish Production Department, College of Agricultural and Food Sciences, King Faisal University, P.O. Box 420, Hofuf 31982, Saudi Arabia
[4] Fish and Animal Production Department, Faculty of Agriculture (Saba Basha), Alexandria University, Alexandria 21531, Egypt
[5] Department of Pathology, Faculty of Veterinary Medicine, Zagazig University, Zagazig 44519, Egypt; heba.vet@yahoo.com
[6] Department of Biochemistry, Faculty of Veterinary Medicine, Zagazig University, Zagazig 44519, Egypt; drengyyassin7@gmail.com
[7] Department of Pharmacology, Faculty of Veterinary Medicine, Mansoura University, Mansoura 35516, Egypt; asmaa_ezat@mans.edu.eg
[8] Department of Biology, College of Science, Princess Nourah bint Abdulrahman University, P.O. Box 84428, Riyadh 11671, Saudi Arabia; kmalwateed@pnu.edu.sa
[9] Faculty of Nanotechnology for Postgraduate Studies, Cairo University, Sheikh Zayed Branch Campus, Giza 12588, Egypt; drsameheltayer@yahoo.com
* Correspondence: afne56@gmail.com (A.N.A.R.); amansour@kfu.edu.sa (A.T.M.); hhhmb@yahoo.com (H.H.M.)

Abstract: Heavy metal toxicity is an important issue owing to its harmful influence on fish. Hence, this study is a pioneer attempt to verify the in vitro and in vivo efficacy of a magnetite (Fe₃O₄) nanogel (MNG) in mitigating waterborne lead (Pb) toxicity in African catfish. Fish ($n = 160$) were assigned into four groups for 45 days. The first (control) and second (MNG) groups were exposed to 0 and 1.2 mg L^{-1} of MNG in water. The third (Pb) and fourth (MNG + Pb) groups were exposed to 0 and 1.2 mg L^{-1} of MNG in water and 69.30 mg L^{-1} of Pb. In vitro, the MNG caused a dramatic drop in the Pb level within 120 h. The Pb-exposed group showed the lowest survival (57.5%) among the groups, with substantial elevations in hepato-renal function and lipid peroxide (MDA). Moreover, Pb exposure caused a remarkable decline in the protein-immune parameters and hepatic antioxidants, along with higher Pb residual deposition in muscles and obvious histopathological changes in the liver and kidney. Interestingly, adding aqueous MNG to Pb-exposed fish relieved these alterations and increased survivability. Thus, MNG is a novel antitoxic agent against Pb toxicity to maintain the health of *C. gariepinus*.

Keywords: *Clarias gariepinus*; health status; lead toxicity; magnetite nanogel; nanotechnology; tissue architecture

1. Introduction

With the introduction of harmful compounds into the aquatic environment, public health issues connected to environmental pollution are receiving much attention. Heavy metal (HM) pollution is considered one of the most disastrous problems threatening aquatic and human life [1]. Fish are considered a pivotal indicator of aquatic environments for the assessment of the severity of HM toxicity, which constitutes a major hazard for all fish consumers [2,3].

African catfish (*Clarias gariepinus*) has been used to assess HM toxicity. The recent literature reveals the susceptibility of *C. gariepinus* to various HMs and verifies the deleterious impacts of HMs by inducing behavioral changes, immune–antioxidant impairments, and bioaccumulation [4–6]. Lead (Pb) is among the most hazardous HMs and is toxic even in low amounts for aquatic animals and humans, resulting in toxic impacts and accelerating different diseases [7]. In aquaculture, exposure to Pb induces oxidative stress, bioaccumulation, neurotoxicity, and immune dysfunction [8]. In *C. gariepinus* and Nile tilapia (*Oreochromis niloticus*), Pb toxicity causes several issues, including hepato-renal toxicity, oxidative damage, histopathological changes, and higher mortality rates [9,10].

Currently, the application of nanomaterials has been proven to have great success in drug delivery, antimicrobial uses, and remediating toxicity caused by either chemical toxicants or HMs in freshwater fishes [11–14]. Regarding the removal of HMs, engineered nanomaterials represent novel and successful approaches compared to traditional methods. Among the recently formulated nanoparticles, magnetite (Fe_3O_4) nanoparticles have interesting electric and magnetic properties and unlimited physical and chemical characteristics at the nanoscale [15,16]. The nano-magnetite form of iron has wide applications in the industry (magnetic recording media, soft magnetic materials, and coloring) and medical sectors (drug delivery, in vivo therapeutic technology, cell separation, and imaging) [17,18]. The magnetite nanocomposites prepared by the sol–gel method have several advantages, including low-cost preparation, toxicity-free iron salts, small particle size, and good dispersion in the solvent [16]. Magnetite nanoparticles (Fe_3O_4) succeeded in removing 66% of copper from a solution after 15 min [19]. Magnetite nanoparticles (Fe_3O_4 NPs) have been used in *O. niloticus* to chelate mercury (Hg) in vitro, in addition to boosting the immune–antioxidant status and liver and kidney function in vivo [14]. Nanogels (NGs) refer to small, aqueous, swollen nanoparticles composed of nano-scaled polymeric chains [20]. Recently, NGs have emerged as very promising and flexible biomaterials utilized in several applications, such as catalysts, sensing materials, or environmental adsorbents. Their characteristics (such as their wide surface area, flexibility in size, ability to carry molecules, and encapsulation of a high percentage of water when suspended in the fluid) enable their use for drug delivery [21,22]. NGs have been reported in novel environmental fields to eliminate organic toxicants and agrochemicals [23,24]. These contaminants can be trapped inside the NGs, and then removed from the environment [25,26]. In addition, a magnetic nanocomposite sol–gel of iron oxide nanoparticles coated with titanium dioxide efficiently removed aluminum and iron ions from contaminated water [27].

Therefore, this novel study is carried out to investigate the potent magnetic power of a magnetite nanogel (MNG) to mitigate the waterborne toxicity induced by Pb ions via testing their adsorption capacity and, accordingly, testing their magnetic effect to prevent Pb bioaccumulation in muscles. In addition, this study provides an assessment of the promising role of MNG on the protein profile, hepato-renal function, immune responses, tissue antioxidants, and the histological picture of African catfish.

2. Results

2.1. MNG Characterization

Figures 1–3 display various types of MNG characterization findings. X-ray diffraction (XRD) analysis demonstrated the fingerprint curve and data for magnetite according to the Brucker Database library, which conformed to validate our synthesis method without any secondary phases (Figure 1A). Dynamic light scattering (DLS) and zeta potential data showed a homogenous size (one peak) of 60 nm (Figure 1B). Due to a substantial degree of zeta potential (-35 mV), the results demonstrated a superior colloidal structure in aqueous solution (Figure 1C).

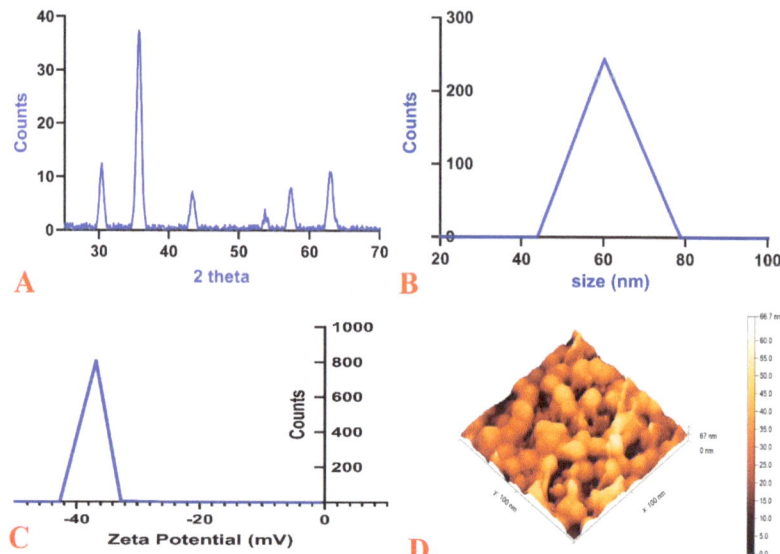

Figure 1. Characterization patterns of magnetite nanogel: (**A**) XRD, (**B**) DLS, (**C**) Zeta potential, and (**D**) AFM.

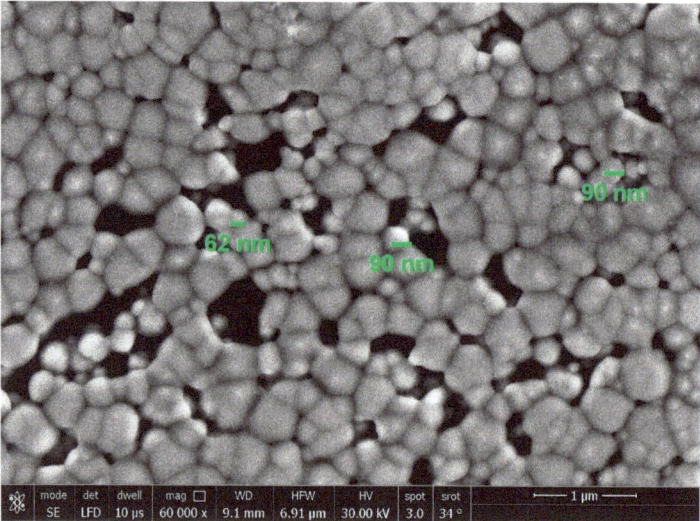

Figure 2. SEM image (1 µm) of magnetite nanogel.

Meanwhile, gel formation had no characteristic peaks due to its amorphous nature. The morphology illustrated by atomic force microscopy (AFM), scanning electron microscopy (SEM), and transmission electron microscopy (TEM) showed the spherical shape of MNG (Figure 1D, Figure 2, and Figure 3).

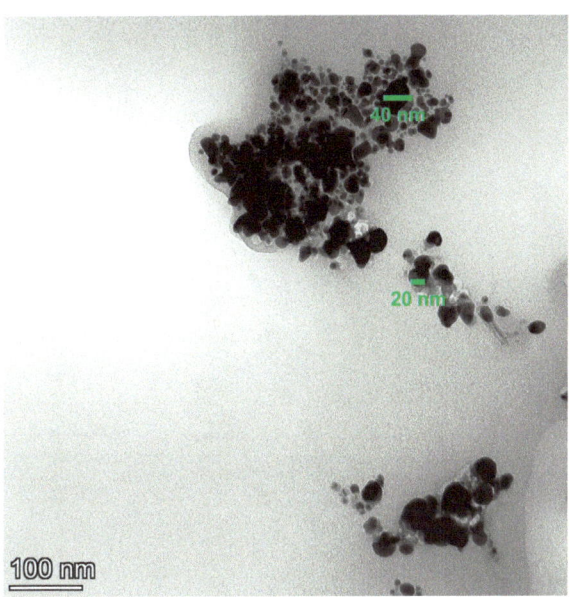

Figure 3. TEM image (100 nm) of magnetite nanogel.

2.2. Absorption of Pb Ions by MNG

Figure 4A shows that MNG caused a dramatic drop in the concentration of Pb ions throughout all sampling points. The concentration decreased from 169.53 mg L^{-1} at the beginning of the experiment to 82.87 mg L^{-1} after 120 h.

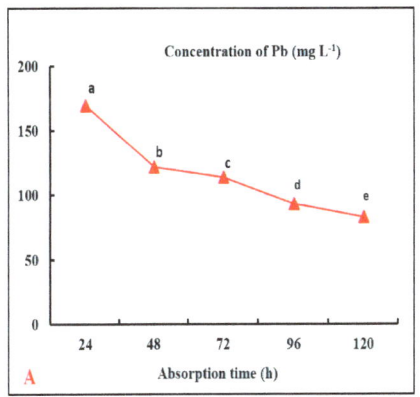
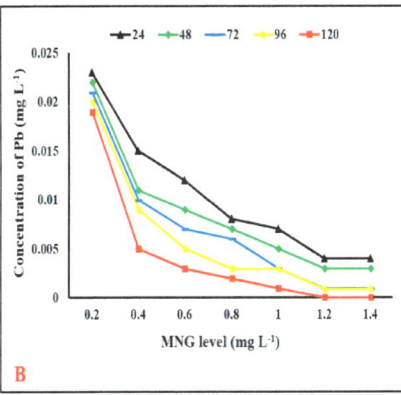

Figure 4. (**A**) Absorption of lead (Pb) by magnetite nanogel (MNG) across 24, 48, 72, 96, and 120 h. (**B**) Impact of MNG level on the concentration of Pb ions across 24, 48, 72, 96, and 120 h. Values that did not have the same superscripts differ significantly (one-way ANOVA; $p < 0.05$).

The various MNG concentrations affected the elimination of the Pb ions, as seen in Figure 4B. The findings showed that raising the MNG level lowered the amount of Pb ions in the aquarium water and allowed for the removal of reduced Pb metal. The outcomes also showed that 1.2 and 1.4 mg/L of MNG were the ideal doses that produced the greatest Pb ion adsorption loading.

2.3. Mortality and Clinical Observations

Based on Kaplan–Meier curves (Figure 5A), the survival rate was 100% in the control and MNG groups during the experimental period (45 days). The lowest survival rate was recorded in the Pb group (57.5%). There was a marked elevation in the survival rate in the MNG + Pb group (82.5%) compared with the Pb group.

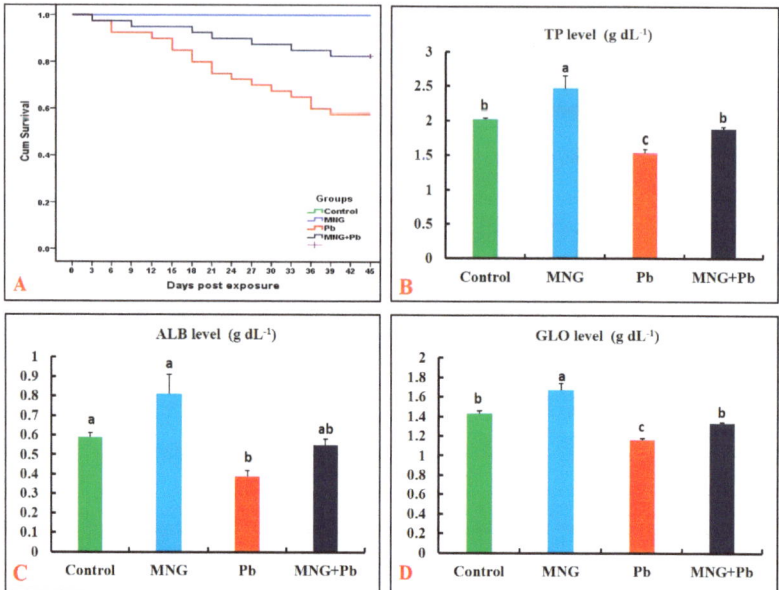

Figure 5. Cumulative survival ($n = 40$/group) and protein profile parameters ($n = 12$/group) of *C. gariepinus* exposed to magnetite nanogel (MNG) and/or lead (Pb) as a water exposure for 45 days. (**A**) Survival curves (Kaplan–Meier). (**B**) Total proteins (TP). (**C**) Albumin (ALB). (**D**) Globulins (GLO). Bars (means ± SE) that did not have the same superscripts differ significantly (one-way ANOVA; $p < 0.05$).

The clinical examination of the treated fish showed that neither the control nor the MNG groups exhibited any atypical behaviors or disease symptoms during the 45 days of exposure. On the contrary, the fish of the Pb group featured symptoms of respiratory distress manifested by rapidly moving the operculum and air gasping from the surface. Fish also developed a slimy appearance, severe skin rot, darkening, and erosions with hemorrhages. Internally, the gills were pale with congestion of internal organs. The Pb and MNG (MNG + Pb)-exposed group showed a remarkable return to the typical appearance with minimal fin rot and a mildly congested liver.

2.4. Hepato-Renal Function Biomarkers

Table 1 demonstrates no discernible variations in the values of hepato-renal biomarkers (ALT, AST, ALP, creatinine, and urea) between the MNG and control groups. These biomarkers displayed the highest values ($p < 0.05$) in the Pb group compared to the control. In contrast, treatment of Pb-exposed fish with MNG resulted in a significant decrease ($p < 0.05$) in these variables compared to Pb exposure alone.

Table 1. Liver and kidney function biomarkers of *C. gariepinus* exposed to magnetite nanogel (MNG) and/or lead (Pb) as a water exposure for 45 days (n = 12/group).

Parameters	Control	MNG	Pb	MNG + Pb
ALT (U L^{-1})	16.33 ± 0.93 [c]	17.75 ± 1.91 [c]	25.08 ± 1.17 [a]	20.25 ± 1.23 [b]
AST (U L^{-1})	44.95 ± 1.22 [c]	46.70 ± 0.85 [c]	94.33 ± 2.20 [a]	82.58 ± 1.68 [b]
ALP (U L^{-1})	34.24 ± 1.08 [c]	34.88 ± 1.35 [c]	50.20 ± 1.53 [a]	41.12 ± 0.62 [b]
Urea (mg dL^{-1})	1.44 ± 0.05 [c]	1.56 ± 0.04 [c]	2.75 ± 0.10 [a]	2.21 ± 0.05 [b]
Creatinine (mg dL^{-1})	0.27 ± 0.02 [b]	0.30 ± 0.03 [b]	0.49 ± 0.50 [a]	0.34 ± 0.01 [b]

ALT, alanine aminotransferase; AST, aspartate aminotransferase; ALP, alkaline phosphatase. Values (means ± SE) in the same row that did not have the same superscripts differ significantly (one-way ANOVA; p < 0.05).

2.5. Protein Profile and Immune Status

Figure 5B–D and Figure 6A–D demonstrate substantial augmentations (p < 0.05) in the protein profile (TP, ALB, and GLO) and immune (LYZ, C3, NO, and IgM) parameters in the MNG group related to the control. Meanwhile, the lowest concentrations of these biomarkers were observed in the Pb-exposed fish, followed by the MNG + Pb fish.

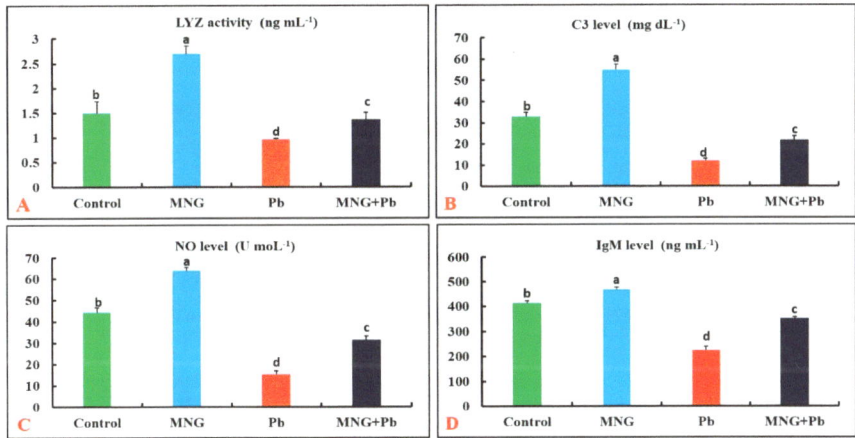

Figure 6. Immune parameters of *C. gariepinus* exposed to magnetite nanogel (MNG) and/or lead (Pb) as a water exposure for 45 days (n = 12/group). (**A**) Lysozyme activity (LYZ). (**B**) Complement 3 (C3). (**C**) Nitric oxide (NO). (**D**) Immunoglobulin M (IgM). Bars (means ± SE) that did not have the same superscripts differ significantly (one-way ANOVA; p < 0.05).

2.6. Hepatic Oxidant/Antioxidant Status

Table 2 shows the levels of MDA and antioxidants in the liver (GSH, SOD, and CAT) of *C. gariepinus* after the exposure period (45 days). There was no noticeable variation in the MNG group's MDA level compared with the control one; however, a significant elevation (p < 0.05) in the GSH, SOD, and CAT values was noticed. Pb exposure induced a profound elevation in the MDA level and lessened the antioxidant values relative to the control. The values of these variables showed more improvement in the MNG + Pb group than in the Pb group.

Table 2. Hepatic oxidant/antioxidant biomarkers of *C. gariepinus* exposed to magnetite nanogel (MNG) and/or lead (Pb) as a water exposure for 45 days ($n = 12$/group).

Parameters	Control	MNG	Pb	MNG + Pb
MDA (nmol mg^{-1})	0.64 ± 0.15 [c]	0.99 ± 0.05 [c]	11.55 ± 0.58 [a]	3.06 ± 0.32 [b]
GSH (ng mg^{-1})	113.57 ± 1.84 [b]	143.76 ± 2.42 [a]	41.21 ± 0.43 [d]	71.69 ± 1.19 [c]
SOD (U mg^{-1})	88.23 ± 1.79 [b]	157.67 ± 3.55 [a]	12.73 ± 0.49 [d]	62.82 ± 1.31 [c]
CAT (ng mg^{-1})	22.20 ± 0.57 [b]	47.30 ± 1.65 [a]	4.91 ± 0.19 [d]	8.79 ± 0.15 [c]

MDA, malondialdehyde; GSH, reduced glutathione content; SOD, superoxide dismutase; CAT, catalase. Values (means ± SE) in the same row that did not have the same superscripts differ significantly (one-way ANOVA; $p < 0.05$).

2.7. Histopathological Findings

According to the histopathological investigations, the livers of the control and MNG fish both displayed normal histological structures of hepatic acini and vasculatures (Figures 7A and 7B, respectively). On the contrary, the Pb exposure caused areas of fatty changes, congested hepatic blood vessels, and perivascular inflammatory cell infiltrates (Figure 7C). The livers of the MNG + Pb group exhibited an improvement of lesions as depicted by the appearance of microvacuoles within a small number of hepatocytes, congested hepatic blood vessels, inflammatory cells aggregated within the portal area, and perivascular aggregation of melanomacrophage (Figure 7D).

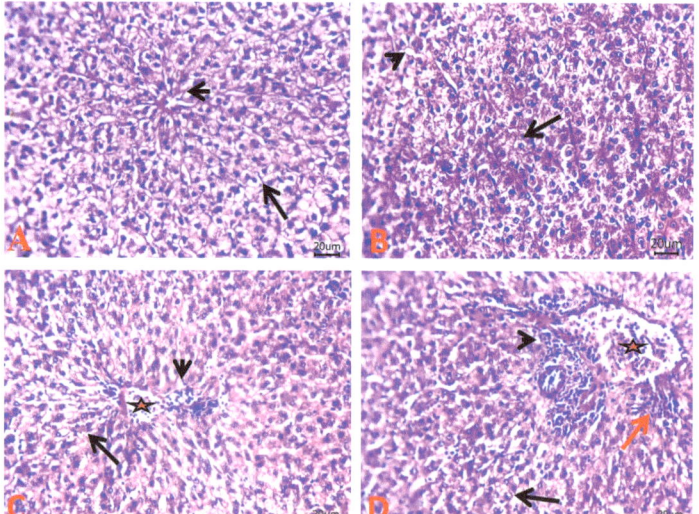

Figure 7. Photomicrograph of H&E-stained liver sections of *C. gariepinus* exposed to magnetite nanogel (MNG) and/or lead (Pb) as a water exposure for 45 days. (**A**,**B**) Liver of the control and MNG groups, respectively, showing normal histological structures of hepatic acini (arrow) and vasculatures (arrowheads). (**C**) Liver of the Pb group showing a focal area of fatty change (arrow), congested hepatic blood vessel (star), and perivascular inflammatory cell infiltrates (arrowhead). (**D**) Liver of the MNG +Pb group showing microvacuoles within a few numbers of hepatocytes (arrow), congested hepatic blood vessels (star), inflammatory cell aggregate within the portal area (arrowhead), and perivascular aggregation of melanomacrophage (red arrow). Scale Bar: 20 μm.

Moreover, normal renal structures with preserved glomerular capillary tufts, renal tubular epithelium, and hemopoietic cells were clear in the fish kidney of the control and MNG groups (Figures 8A and 8B, respectively). However, Pb exposure induced histopathological alterations in the kidney, which appeared as marked necrotic changes in tubular epithelium and maintained glomerular architectures. Further, a depletion of the hemopoietic center replaced by a pale eosinophilic substance was obvious (Figure 8C). Treatment of Pb-exposed fish with MNG markedly improved these alterations and revealed normal histopathological structures of renal tubules and glomerular corpuscles (Figure 8D).

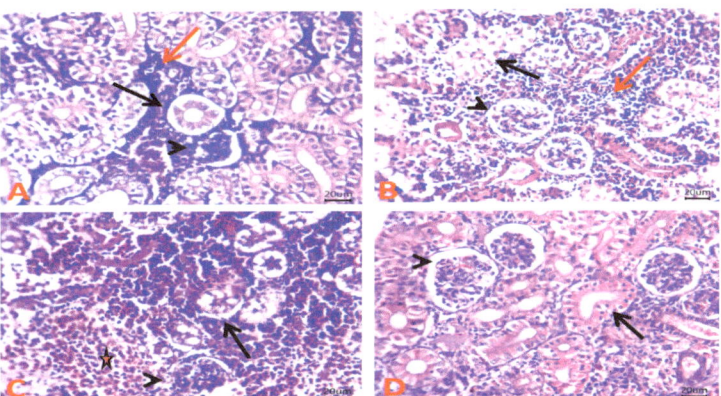

Figure 8. Photomicrograph of H&E-stained kidney sections of *C. gariepinus* exposed to magnetite nanogel (MNG) and/or lead (Pb) as a water exposure for 45 days. (**A,B**) Kidney of the control and MNG groups, respectively, showing normal renal structures with preserved glomerular capillary tufts (arrowheads), renal tubular epithelium (arrows), and the presence of hemopoietic cells (red arrows). (**C**) Kidney of the Pb group showing marked necrotic changes in the tubular epithelium (arrow), maintained glomerular architectures (arrowhead), and depletion of hemopoietic center replaced by pale eosinophilic substance (star). (**D**) Kidney of the MNG + Pb group showing normal histomorphological structures of the renal tubule (arrow) and glomerular corpuscle (arrowhead). Scale Bar: 20 μm.

2.8. Bioaccumulation of Pb^{2+} in Fish Muscles

The concentration of Pb ions in the muscles of the MNG and control groups did not alter significantly ($p > 0.05$), as presented in Figure 9. The muscles of the Pb group had the highest levels of Pb ions. Still, the MNG + Pb group had considerably lower levels of Pb residues.

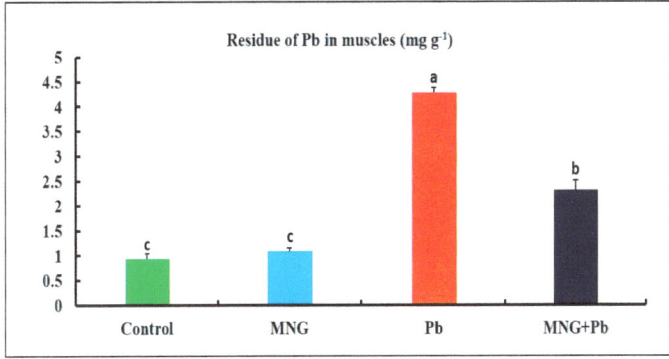

Figure 9. Residues of lead (Pb) in muscles of *C. gariepinus* exposed to magnetite nanogel (MNG) and/or Pb as a water exposure for 45 days ($n = 12$/group). Bars (means ± SE) that did not have the same superscripts differ significantly (one-way ANOVA; $p < 0.05$).

3. Discussion

The toxicity of Pb has a slow-acting cumulative impact that results in major health problems for aquatic animals and humans because of its use in various industrial processes that contaminate water [28]. The current report is an innovative trial to underpin the effectiveness of MNG to alleviate the toxicity of Pb in vitro and in vivo via assessing its magnetic power to protect fish muscles from Pb bioaccumulation and studying its potential role on protein picture, hepato-renal function, immune-antioxidant response, and tissue architecture in African catfish.

Among heavily studied nanoparticles, the magnetite nanoscale has attracted more interest owing to the potent power of magnetite to adsorb heavy metal ions in solutions. The nanosized Fe_3O_4 particles remove heavy metals via their magnetic properties, high surface area, chemical stability, easy synthesis, and low toxicity [29]. Pb was chosen for testing the adsorption capacity of MNG because it is one of the most predominant dangerous HMs in aquaculture practice [30]. Our findings showed that MNG had a potent adsorbing power of Pb ions that caused a clear reduction in Pb ions with time (120 h). Also, the best adsorption loading was provided by 1.2 and 1.4 mg L^{-1} MNG concentrations relative to the other concentrations. The adsorption power was raised at the start of the experiment and then declined by progressing the adsorption time. These findings could be related to the adsorption and the decrease of the Pb^{2+} ions to Pb metal on the surface of the MNG until the saturation of the MNG surfaces by the Pb ions. The magnetite has a specific crystal structure consisting of free electrons (Fe^{2+} and Fe^{3+}) conjugated with oxygen. In the crystal structure, the localization of free electrons [31] is responsible for the magnetic properties, which elevate the surface activities and the adsorption power of the magnetite to lessen the Pb ions on the surface of the MNG. The sorption reaction between magnetite and Pb is chemical adsorption [32]. In the same manner, Hong et al. [33] verified the efficacy of MNG in getting rid of more than 80% of Pb, chromium (Cr), and cadmium (Cd) from contaminated water at 1 mg L^{-1} because of the emerging electrostatic attraction between the positive metal ions and the negatively charged ions of iron oxide.

Considering the clinical picture and post mortem examination, exposure to Pb alters the general health of *C. gariepinus*. Additionally, fish suffered anorexia, major signs of respiratory manifestation, profound mucous secretion production, fin rot, severe erythema, erosion in the skin, and the lowest survival rate (57.5%). It is assumed that the Pb ions irritate the skin and gills because of their direct contact with fish in the aquatic environment, inducing respiratory distress and erythema with more mucous production as a defense reaction against the toxic Pb ions. Our findings were concurrent with those of Alfakheri et al. [34] and Abdel Rahman et al. [10], who noted that the exposure of *C. gariepinus* and *O. niloticus* to Pb toxicity induces respiratory problems and mortalities. On the other hand, exposure to MNG improved the clinical picture and reduced the mortalities in Pb-exposed fish. It is assumed that there are two reasons: the first is the potent magnetism of the magnetite, which enables MNG to adsorb the Pb ions, resulting in a decrease in its level. The other reason is the verified potent antioxidant activity of the NGs, which counteracts the oxidative damage produced by Pb ions. Likewise, Mahboub et al. [14] recorded no mortalities in the mercury-exposed *O. niloticus* with Fe_3O_4 NPs.

LYZ, complement activity, NO, and IgM are non-specific and important components that mainly indicate innate immunity in fish [35,36]. TP indicates activated humeral immunity in aquatic organisms [37]. Herein, we reveal the occurrence of immune suppression upon exposure to Pb reflected by a clear reduction in immune parameters, including lysozymes, C3, NO, IgM, and TP. Concurrent with an earlier study, Shah [38] recorded that lethal and even sub-lethal exposure to Pb alters the immunological biomarkers in tench (*Tinca tinca*). Likely, Alandiyjany et al. [12] found a clear depression in the level of TP following the exposure of *O. niloticus* to Pb.

On the other hand, immunomodulation has been reported upon the exposure of *C. gariepinus* to aqueous MNG, which is indicated by a noticeable increase in the immunological biomarkers.

The mechanism of action of Fe_3O_4 NPs on the immune system was recently documented by Huang et al. [39], who revealed that the degradation products of the magnetic nanoparticles improve immune stimulation via the interferon gene activating protein (STING) pathway, which, in turn, enhances cellular immune response. A similar report found that Fe_3O_4 NPs had an immunological influence by augmenting the activity of LYZ in *O. niloticus* exposed to mercury toxicity [14].

Detoxification of HMs is mainly carried out in the hepatic tissue, followed by filtration and excretion in the renal tissue. Hence, elevating the concentration of HMs induces an increased rate of filtration and detoxification in the fish body, which in turn causes hepato-renal dysfunction [40]. In the present investigation, the biomarkers of renal functions (creatinine and urea) and hepatic enzymes (ALT, ALP, and AST) exhibited an elevation in their levels upon exposure to Pb. It is assumed that Pb induces necrosis in the liver, and accordingly, this damage leads to the leakage of hepatic enzymes into the bloodstream, producing an elevation. Furthermore, Pb toxicity impairs renal function by minimizing its ability to excrete urine, urea and by impairing the glomerular filtration rate, as Akturk et al. [41] described. These attributions were confirmed by the histopathological alteration that was observed in the liver and kidney in our study. In line with the present findings, Abdel-Tawwab et al. [42] revealed that a noticeable increase in the values of urea and creatinine was recorded in *O. niloticus* after intoxication with a mixture of HMs, including Pb. Histological alterations of the liver and gill tissue of *C. gariepinus* were reported post-exposure to Pb, including fibrosis of hepatic cords and necrosis of parenchyma cells besides collapsing blood vessels [43].

On the other side, a restoration of hepato-renal biomarkers in the aqueous MNG + Pb-treated group and a clear regeneration of histological changes indicated the protective effect of MNG against Pb-induced hepato-renal damages. It is suggested that aqueous MNG can mitigate the hazardous effects of Pb toxicity by lessening the Pb-generated reactive oxygen species (ROS) on hepatic cells. Similarly, Mahboub et al. [14] reported that Fe_3O_4 NPs had a promising effect on improving hepato-renal functions of *O. niloticus* and could enhance the levels of liver enzymes and renal parameters upon exposure to mercury toxicity. A recent study conducted by Alandiyjany et al. [12] reported severe histopathological changes in the liver and gills of *O. niloticus* following exposure to Pb, and a noticeable improvement was detected in the magnetized silica-received group.

HMs induce oxidative damage by generating ROS. The antioxidant defense mechanism involves various enzymes, such as CAT, SOD, and GSH, which protect cells from oxidative stress by detoxifying ROS [44]. The current work showed that oxidative damage in the Pb-exposed group reflected a clear elevation in MDA level and a reduction in GSH, CAT, and SOD. It is opined that Pb causes excess production of ROS, resulting in oxidative damage. In line with recent work, Alandiyjany et al. [12] detected decreased serum CAT, SOD, and GSH activity levels in Pb-exposed *O. niloticus*.

Contrarily, the exposure of fish to MNG in the Pb-exposed group has an antioxidant-protecting effect indicated by a clear modulation in the antioxidant biomarkers (elevated SOD and CAT activities) resulting in protection from oxidative damage. In line with a recent finding, Răcuciu et al. [45] confirmed that Fe_3O_4 NPs have potent antioxidant enzymatic activity via modulating the levels of CAT and SOD and aid in plant development. Moreover, Fe_3O_4 NPs can enhance the antioxidant status and reduce the oxidative stress of *O. niloticus* and Indian major carp (*Labeo rohita*) [14,46,47].

HM toxicity produces variable immunological and physiological responses, allowing for the bioaccumulation of metals in different fish tissues [7]. Here, we find that the Pb-exposed group's muscles have a greater level of Pb. In line with a recent report, Alandiyjany et al. [12] detected bioaccumulation of Pb in the muscles of *O. niloticus* following exposure to Pb.

In contrast, the MNG + Pb group reflected the least accumulation of Pb, indicating its efficacy in removing Pb. It is assumed that the magnetic power of magnetite found in MNG, plus the formulation of NGs, enables it to absorb Pb strongly. Previous studies supported our outcomes and documented that the structure of NGs causes them to be easily biocompatible and biodegradable and can absorb and release molecules for decontaminating water, catalysis, and sensors [48,49]. Furthermore, Neamtu et al [50] added that NGs can absorb active materials through chemical interactions such as hydrogen or hydrophobic bonding and salt formation. Similar outcomes were observed by Alandiyjany et al. [12] in the muscles of *O. niloticus*.

4. Conclusions

The present study demonstrates that Pb is a hazardous heavy metal that causes a decline in the survival rate, suppresses immune-antioxidant status, and deteriorates hepatorenal functions and histopathological structure of the liver and kidney tissues. Also, Pb exposure results in high bioaccumulation in the muscles of the treated African catfish. The basic attention is directed to the magnetic antitoxic power of MNG to adsorb Pb ions and protect fish from bioaccumulation in muscles. Additionally, MNG enhances the immune-antioxidant profile, improves the hepato-renal function, and regenerates the histopathological picture. Further studies are mandatory to assess other applications of MNG in various fish species and to assess the safe use on a large scale for sustaining aquaculture and maintaining human health.

5. Materials and Methods

5.1. Synthesis and Characterization of MNG

Firstly, Fe_3O_4 NPs were synthesized following the protocol of Hamdy et al. [51]. About 0.4 g of the hematite ore (Fe_3O_4) was added drop by drop to 40 mL of H_2O_2. At the same time, the mixture was subjected to ultrasound at 60 kHz for 2.5 h in an ultrasonic device (Sonica 4200 EPS3, Milano, Italy) until the black particles of Fe_3O_4 were obtained. After 1.5 h, the Fe_3O_4 NPs (black color) precipitated from the supernatant (reddish color). The Fe_3O_4 NPs were separated from the solution by centrifugation at 4000 rpm, and, finally, the Fe_3O_4 NPs were washed four times using methanol.

For the synthesis of Fe_3O_4 NPs/carbopol hybrid nanogel, 0.2 g of Fe_3O_4 NP desperation in 25 mL of ethanol was added to a solution of 0.25 g carbopol dissolved in 25 mL of ethanol and the mixture was stirred using a mechanical stirrer for 50 min. Then, 0.75 mL of trimethylamine was added drop by drop and stirred for another 40 min until obtaining a black gel. The Fe_3O_4 NPs/carbopol hybrid nanogel was prepared in high and low viscosities.

Characterization protocols were categorized into three groups: morphology, identification, and index class, according to the Hassan et al. [52] approach.

5.2. Preparation of Pb Ion Solution

In this experiment, lead chloride ($PbCl_2$; purity 98%) of Merck, Darmstadt, Germany was utilized as a source of Pb ions. To reach the proper concentrations, $PbCl_2$ was primarily dissolved in de-ionized water to create a stock solution (1000 mg L^{-1}) which was then diluted to the necessary concentration before being used in aquarium water. According to Alfakheri et al. [34], the 96 h median lethal concentration (LC_{50}) for Pb was 231 mg L^{-1} and 30% of 96 h LC_{50} (69.30 mg L^{-1}) was used.

5.3. Adsorption Capacity of MNG

In two different studies, the capacity of MNG to adsorb the Pb ions was evaluated. In the first experiment, at 24 °C and pH 6.0, an exact amount of $PbCl_2$ (20 mg) was mixed with 100 mL of ultrapure water. In a glass vial, 20 mL of prepared $PbCl_2$ and 20 mg of MNG were mixed and vortexed for 10 min to assess the adsorption kinetics per the Kôsak et al. [53] technique. Daily, for five days (24, 48, 72, 96, and 120 h), and using an atomic absorption spectrophotometer (Buck Scientific, Norwalk, CT, USA), the concentration of Pb^{+2} ions was calculated. Three copies of each sample were tested.

The second experiment examined how varied MNG concentrations (0.2, 0.4, 0.6, 0.8, 1, 1.2, and 1.4 mg L^{-1}) affected the adsorption ability of Pb ions. It involved setting up seven aquariums with water in them, adding 0.025 mg L^{-1} of Pb ions to each aquarium at pH 6.0, and then adding the seven concentrations of MNG directly to each Pb-exposed aquarium [4]. Then, using the atomic absorption spectrophotometry technique, the level of Pb ions was assessed after 24, 48, 72, 96, and 120 h. The safe recommended level of iron in fish, which varies between 0.35 and 1.7 mg/L [54,55], was considered when choosing the concentrations of MNG.

5.4. Ethical Agreement and Fish Acclimation

The Institutional Animal Care and Use Committee of Zagazig University in Egypt (ZU-IACUC/2/F/309/2022) approved the experimental strategy. Two hundred and forty African catfish (100 ± 7.39 g) were selected from the Al-Abbassa private fish farm in Sharkia Governorate, Egypt. The fish were kept for ten days in 100 L of well-aerated aquaria for acclimation. Part of the water was partially exchanged (25%). The fish were supplemented with a basal diet at 3% of their body weight twice daily during acclimation and experimental trial. Assessment of physio-chemical parameters of the rearing water was carried out daily, including temperature, dissolved oxygen, ammonia, and pH, and recorded as 24 ± 2 °C, 6 ± 0.26 mg L^{-1}, 0.01 ± 0.04 mg L^{-1}, and 7 ± 0.13, respectively.

5.5. Assessing the Initial Concentration of MNG

Fish (n = 80) were exposed to 8 various concentrations of MNG for 15 days (Table 3) to determine the starting concentration for the treatment experiment. These concentrations were 0, 0.2, 0.4, 0.6, 0.8, 1, 1.2, and 1.4 mg L^{-1} of MNG. The clinical observations were kept track of every day during the preliminary trial. MNG concentrations were safe in the 0.2 to 1.4 mg L^{-1} range, and 1.2 mg L^{-1} was determined to be the dose used for treatment.

Table 3. Mortality and clinical observations of *C. gariepinus* exposed to different concentrations of magnetite nanogel (MNG) for 15 days.

Conc. (mg L^{-1})	Mortality (n = 10)	Clinical Observations		
		Erratic Swimming	Loss of Escape Reflex	External Symptoms (Hemorrhages, Darkness, Fin Rot, and Ulcerations)
0.0	0/10	-	-	-
0.2	0/10	-	-	-
0.4	0/10	-	-	-
0.6	0/10	-	-	-
0.8	0/10	-	-	-
1	0/10	-	-	-
1.2	0/10	-	-	-
1.4	0/10	-	-	-

(-) No abnormal observations.

5.6. Experimental Design

For 45 days, fish (n = 160) were randomly assigned into four groups (10 fish/replicate; 40/group). The first and second (MNG) groups were exposed to 0 and 1.2 mg/L MNG in water, where the control was the first group. The third (Pb) and fourth (MNG + Pb) groups were exposed to 0 and 1.2 mg L^{-1} MNG in water, respectively, and 69.30 mg L^{-1} of lead chloride. Fish were moved to freshly produced solutions with the same concentrations daily for 45 days during the experiment. Every day, about 25% of the aquarium's contents were replenished. Clinical observation and mortalities were kept track of throughout the trial.

5.7. Sampling

Fish were randomly selected (12 fish per group) at the end of the experiment (45 days) to collect samples. According to Neiffer and Stamper's [56] approach, fish were anesthetized with a benzocaine solution (100 mg L^{-1}), and blood was then drained from the caudal blood vessels using tubes devoid of the anticoagulant. Samples were centrifuged at 1750× g for 10 min after being incubated at room temperature (21 ± 3 °C) for 5 h. Clear serum was then kept at 20 °C until biochemical and immunological assays. Liver tissues (12 fish/group) were gathered and kept in liquid nitrogen for the oxidant/antioxidant assay. Additionally, liver and kidney samples (12 fish/group) were used for histopathology analysis, and samples of muscles (12 fish/group) were picked for determining Pb residues.

5.8. Evaluation of Hepato-Renal Function Biomarkers

The activity of hepatic function biomarkers, including aspartate aminotransferase (AST, Catalog No.; EK12276) (Biotrend Co., Laurel, MD, USA), alanine aminotransferase (ALT, Catalog No.; MBS038444) (MyBioSource Co., CA, USA), and alkaline phosphatase (ALP, Catalog No.; TR11320) (Thermo Fisher Scientific, Swindon, UK) were computed. Also, the total protein (TP, Catalog No.; MBS9917835), albumin (ALB, Catalog No.; MBS019237), and urea (Catalog No.; MBS9374784) (MyBioSource Co., CA, USA) were measured. All the biomarkers mentioned above were computed using a spectrophotometer (Lambda EZ201; Perkin Elm, Beaconsfield, UK). The globulin (GLO) level was determined by subtracting ALB from TP. The creatinine (Catalog No.; MAC080) level was estimated at a wavelength of 340 nm using a spectrophotometric protocol (Centromic Gmbit kit manual, Wartenberg, Germany).

5.9. Immune Assays

The immune parameters, including lysozyme activity (LYZ), were estimated using the inhibition zone method in agarose gel plates, depending on the protocol of Lee and Yang [57]. The level of complement 3 (C3) was measured by immunoturbidimetry using the method of Abdollahi et al. [58] with separated Eastbiopharm ELISA kits (Hangzhou Eastbiopharm CO., LTD., Torrance, CA, USA).

To quantify the serum nitric oxide (NO), about 100 mL of each serum sample was added to the Griess reagent, which was then incubated for 10 min at 27 °C [59]. Immunoglobulin M (IgM) was quantified in serum spectrophotometrically using ELISA kits (Cusabio Biotech Co., Ltd., Wuhan, China) as directed by the manufacturer, following Schultz's [60] approach.

5.10. Hepatic Oxidant/Antioxidant Assays

According to the Siroka et al. [61] assay, the liver samples were prepared to estimate the levels of oxidant/antioxidant biomarkers (malondialdehyde (MDA), reduced glutathione content (GSH), catalase (CAT), and superoxide dismutase (SOD)). Liver samples were subjected to homogenization in a buffer with a pH of 7.5 to obtain the supernatant, which was then obtained by centrifuging them at 4 °C for 15 min at 10,000 × g for 1 h to recover the final supernatant.

The level of MDA was assessed using the Sigma assay kit (MAK085) according to the protocol of Ohkawa et al. [62]. The content of GSH and SOD activity was computed depending on the assays of Beutler et al. [63] and Velkova-Jordanoska et al. [64]. The GSH was estimated at 412 nm using 5,5'-dithio-bis-2-nitrobenzoic acid in the supernatant fraction. The level of SOD was calculated using the xanthine oxidase–cytochrome protocol using a spectrophotometer at 505 nm. Xanthine interacted with 2-[4-iodophenyl]-3-[4-nitrophenyl]-5-phenyl-tetrazolium chloride (INT) to compose superoxide radicals producing red-colored formazan. This product was utilized to measure the activity of SOD as SOD conjugate with superoxide radicals and consequently controls the formazan synthesis.

The activity of CAT was monitored depending on the decrease in hydrogen peroxide (H_2O_2) at 240 nm using a light plate and a spectrophotometer with 1.0 mL quartz cuvettes according to the method of Aksenes and Njaa [65].

5.11. Histopathological Investigation

Samples from the liver and kidneys were gathered from all investigated groups, fixed using 10% buffered neutral formalin, then exposed to dehydration in ascending degrees of alcohol, cleared using xylene, and soaked in paraffin. Paraffin sections of about 5 μm in thickness were arranged and stained using hematoxylin and eosin (H&E) and then inspected by an optical microscope, depending on the protocol of Suvarna et al. [66].

5.12. Determination of Pb Residues in Fish Muscles

After terminating the experiment (45 days), representative samples were dissected from the dorsal muscle of each group and dried in an oven at 85 °C until they reached a stabilized weight. The prepared samples were weighed (1 g dry weight) and placed in a muffle furnace (Shelton, CT, USA) for 6 h of ashing. After the procedure outlined by Golberg et al. [67], the samples were digested using 5 mL of freshly made perchloric acid ($HCLO_4$; 70%) and nitric acid (HNO_3; 65% v/v) to Teflon beakers and heating at 50 °C for approximately 5 h to completely break down the organic matter. The digested solution was chilled at ~21 ± 2 °C and diluted using deionized water to reach a final volume of 50 mL. The atomic absorption spectrophotometer was used to analyze each sample separately for calculating the Pb ion residues [68].

5.13. Data Analysis

The Shapiro–Wilk test was first conducted to evaluate whether all the data were normal. To determine whether there was a statistically significant difference between treatments, a one-way analysis of variance (ANOVA) was performed with Tukey's post hoc analysis (SPSS version 18; IBM Corp., Armonk, NY, USA). The Kaplan–Meier protocol was used to analyze survival according to Kaplan and Meier [69]. A p-value of less than 0.05 represents statistical variance, including all tests.

Author Contributions: Conceptualization, A.N.A.R., A.T.M. and H.H.M.; Methodology, A.N.A.R., B.A.E., A.T.M., H.M.A.-G., E.M.M.Y., A.E., K.M.A., S.H.I. and H.H.M.; Software and data curation: A.N.A.R., S.H.I. and H.H.M.; Writing—Original draft preparation: A.N.A.R. and H.H.M.; Writing—Reviewing and Editing: A.N.A.R., A.T.M. and H.H.M.; Funding acquisition, K.M.A. All authors have read and agreed to the published version of the manuscript.

Funding: The APC was funded by Princess Nourah bint Abdulrahman University Researchers Supporting Project number (PNURSP2023R402), Princess Nourah bint Abdulrahman University, Riyadh, Saudi Arabia. This work was supported by the Deanship of Scientific Research, Vice Presidency for Graduate Studies and Scientific Research, King Faisal University, Saudi Arabia (Grant No. GRANT 3867).

Institutional Review Board Statement: The experimentation was performed in the Aquatic Animal Medicine Department, Faculty of Veterinary Medicine, Zagazig University, and supervised by the Animal Use in Research Committee with ethical approval code ZU-IACUC/2/F/309/2022. All experimental procedures were conducted in compliance with the ethical guidelines approved by the National Institutes of Health for the Use and Treatment of Laboratory Animals.

Informed Consent Statement: Not applicable.

Data Availability Statement: The datasets generated or analyzed during the current study are not publicly available but are available from the corresponding author upon reasonable request.

Acknowledgments: The authors appreciate Princess Nourah bint Abdulrahman University Researchers Supporting Project number (PNURSP2023R402), Princess Nourah bint Abdulrahman University, Riyadh, Saudi Arabia. This work was supported by the Deanship of Scientific Research, Vice Presidency for Graduate Studies and Scientific Research, King Faisal University, Saudi Arabia (Grant No. GRANT 3867). We also would like to thank all staff of Aquatic Animal Medicine, Faculty of Veterinary Medicine, Zagazig University, for their kind support during the study.

Conflicts of Interest: The authors declare no conflict of interest.

References

1. Yap, C.K.; Al-Mutairi, K.A. Ecological-health risk assessments of heavy metals (Cu, Pb, and Zn) in aquatic sediments from the ASEAN-5 emerging developing countries: A review and synthesis. *Biology* **2022**, *11*, 7. [CrossRef] [PubMed]
2. Feng, W.; Wang, Z.; Xu, H. Species-specific bioaccumulation of trace metals among fish species from Xincun Lagoon, South China Sea. *Sci. Rep.* **2020**, *10*, 21800. [CrossRef] [PubMed]
3. Bella, C.D.; Calagna, A.; Cammilleri, G.; Chembri, P.; Monaco, D.L.; Ciprì, V.; Battaglia, L.; Barbera, G.; Ferrantelli, V.; Sadok, S.; et al. Risk assessment of cadmium, lead, and mercury on human health in relation to the consumption of farmed sea bass in Italy: A meta-analytical approach. *Front. Mar. Sci.* **2021**, *8*, 616488. [CrossRef]
4. El-Bouhy, Z.M.; Reda, R.M.; Mahboub, H.H.; Gomaa, F.N. Bioremediation effect of pomegranate peel on subchronic mercury immunotoxicity on African catfish, *Clarias gariepinus*. *Environ. Sci. Pollut. Res.* **2021**, *28*, 2219–2235. [CrossRef]
5. El-Bouhy, Z.M.; Reda, R.M.; Mahboub, H.H.; Gomaa, F.N. Chelation of mercury intoxication and testing different protective aspects of *Lactococcus lactis* probiotic in African catfish. *Aquaculture Res.* **2021**, *52*, 3815–3828. [CrossRef]
6. Apiamu, A.; Osawaru, S.U.; Asagba, S.O.; Evuen, U.F.; Achuba, F.I. Exposure of African catfish (*Clarias gariepinus*) to lead and zinc modulates membrane-bound transport protein: A plausible effect on Na+/K+-ATPase activity. *Biol. Trace Elem. Res.* **2022**, *200*, 4160–4170. [CrossRef]
7. Garai, P.; Banerjee, P.; Mondal, P.; Saha, N.C. Effect of heavy metals on fishes: Toxicity and bioaccumulation. *J. Clin. Toxicol.* **2021**, *18*, S18:001. [CrossRef]
8. Lee, J.W.; Choi, H.; Hwang, U.K.; Kang, J.C.; Kang, Y.J.; Kim, K.I. Toxic effects of lead exposure on bioaccumulation, oxidative stress, neurotoxicity, and immune responses in fish: A review. *Environ. Toxicol. Pharmacol.* **2019**, *68*, 101–108. [CrossRef]
9. Abdel-Warith, A.A.; Younis, E.M.I.; Al-Asgah, N.A.; Rady, A.M.; Allam, H.Y. Bioaccumulation of lead nitrate in tissues and its effects on hematological and biochemical parameters of *Clarias gariepinus*. *Saudi J. Biol. Sci.* **2020**, *27*, 840–845. [CrossRef]
10. Abdel Rahman, A.N.; ElHady, M.; Hassanin, M.E.; Mohamed, A.A.R. Alleviative effects of dietary Indian lotus leaves on heavy metals-induced hepato-renal toxicity, oxidative stress, and histopathological alterations in Nile tilapia, *Oreochromis niloticus* (L.). *Aquaculture* **2019**, *509*, 198–208. [CrossRef]
11. Mahboub, H.H.; Shahin, K.; Mahmoud, S.M.; Altohamy, D.E.; Husseiny, W.A.; Mansour, D.A.; Shalaby, S.I.; Gaballa, M.M.S.; Shaalan, M.; Alkafafy, M.; et al. Silica nanoparticles are novel aqueous additive mitigating heavy metals toxicity and improving the health of African catfish, *Clarias gariepinus*. *Aquat. Toxicol.* **2022**, *249*, 106238. [CrossRef]
12. Alandiyjany, M.N.; Kishawy, A.T.Y.; Hassan, A.A.; Eldoumani, H.; Elazab, S.T.; El-Mandrawy, S.A.M.; Saleh, A.A.; El Sawy, N.A.; Attia, Y.A.; Arisha, A.H.; et al. Nano-silica and magnetized-silica mitigated lead toxicity: Their efficacy on bioaccumulation risk, performance, and apoptotic targeted genes in Nile tilapia (*Oreochromis niloticus*). *Aquat. Toxicol.* **2022**, *242*, 106054. [CrossRef]
13. Abdel Rahman, A.N.; Ismail, S.H.; Fouda, M.M.S.; Abdelwarith, A.A.; Younis, E.M.; Khalil, S.S.; El-Saber, M.M.; Abdelhamid, A.E.; Davies, S.J.; Ibrahim, R.E. Impact of *Streptococcus agalactiae* challenge on immune response, antioxidant status and hepatorenal indices of Nile tilapia: The palliative role of chitosan white poplar nanocapsule. *Fishes* **2023**, *8*, 199. [CrossRef]
14. Mahboub, H.H.; Beheiry, R.R.; Shahin, S.E.; Behairy, A.; Khedr, M.H.E.; Ibrahim, S.M.; Elshopakey, G.E.; Daoush, W.M.; Altohamy, D.E.; Ismail, T.A.; et al. Adsorptivity of mercury on magnetite nano-particles and their influences on growth, economical, hemato-biochemical, histological parameters and bioaccumulation in Nile tilapia (*Oreochromis niloticus*). *Aquat. Toxicol.* **2021**, *235*, 105828. [CrossRef]
15. Kavas, H.; Günay, M.; Baykal, A.; Toprak, M.S.; Sozeri, H.; Aktaş, B. Negative Permittivity of Polyaniline-Fe$_3$O$_4$ Nanocomposite. *J. Inorg. Organomet. Polym. Mater.* **2013**, *23*, 306–314. [CrossRef]
16. Shaker, S.; Zafarian, S.; Chakra, C.S.; Rao, K.V. Preparation and characterization of magnetic nanoparticles by sol-gel method for water treatment. *Int. J. Innov. Res. Technol. Sci. Eng.* **2013**, *2*, 2969–2973.
17. Lemine, O.M.; Omri, K.; Zhang, B.; El Mir, L.; Sajieddine, M.; Alyamani, A.; Bououdina, M. Sol–gel synthesis of 8nm magnetite (Fe$_3$O$_4$) nanoparticles and their magnetic properties. *Superlattices Microstruct.* **2012**, *52*, 793–799. [CrossRef]
18. Cui, H.; Liu, Y.; Ren, W. Structure switch between α-Fe$_2$O$_3$, γ-Fe$_2$O$_3$ and Fe$_3$O$_4$ during the large scale and low temperature sol—Gel synthesis of nearly monodispersed iron oxide nanoparticles. *Adv. Powder Technol.* **2013**, *2*, 93–97. [CrossRef]
19. Neyaz, N.; Zarger, M.S.; Siddiq, W.A. Synthesis and characterisation of modified magnetite super paramagnetic nano composite for removal of toxic metals from ground water. *Int. J. Environ. Sci.* **2014**, *5*, 260–269. Available online: https://www.cabdirect.org/cabdirect/abstract/20153136507 (accessed on 6 July 2023).
20. Zhang, H.; Zhai, Y.; Wang, J. New progress and prospects: The application of nanogel in drug delivery. *Mater. Sci. Eng. C* **2016**, *60*, 560–568. [CrossRef]
21. Kaoud, R.M.; Heikal, E.J.; Jaafar, L.M. Novel nanogel applications: A review. *WJAHR* **2022**, *6*, 11–15.
22. Pinellia, F.; Saadatia, M.; Zareb, E.N.; Makvandic, P.; Masia, M.; Filippo, A.S. A perspective on the applications of functionalized nanogels: Promises and challenges. *Int. Mater. Rev.* **2023**, *68*, 1–25. [CrossRef]
23. Anandharamakrishnan, C. Trends and impact of Nanotechnology in agro-Food sector. In *Innovative Food Processing Technologies*; Knoerzer, K., Muthukumarappan, K., Eds.; Elsevier: Oxford, UK, 2021; pp. 523–531. [CrossRef]
24. Yiamsawas, D.; Kangwansupamonkon, W.; Kiatkamjornwong, S. Lignin-based nanogels for therelease of payloads in alkaline conditions. *Eur. Polym. J.* **2021**, *145*, 110241. [CrossRef]
25. Shoueir, K.R.; Sarhan, A.A.; Atta, A.M. Macrogel andnanogel networks based on crosslinked poly (vinylalcohol) for adsorption of methylene blue from aquasystem. *Environ. Nanotechnol. Monit. Manag.* **2016**, *5*, 62–73. [CrossRef]

26. Shah, M.T.; Alveroglu, E. Facile synthesis of nanogelsmodified Fe_3O_4@Ag NPs for the efficient adsorption of bovine & human serum albumin. *Mater. Sci. Eng. C* **2021**, *118*, 111390. [CrossRef]
27. Abdulhady, Y.A.; El-Shazly, M.M. Removal of some heavy metals and polluted antibacterial activities via synthesized magnetic nano-composite of iron oxide and derivatives: Chemical and microbial treatment case study: Al tard-bilraha drain Ismailia, EGYPT. *Egypt J. Desert Res.* **2018**, *68*, 15–36. [CrossRef]
28. Raj, L.; Das, A.P. Lead pollution: Impact on environment and human health and approach for a sustainable solution. *J. Environ. Chem. Ecotoxicol.* **2023**, *5*, 79–85. [CrossRef]
29. Sarma, G.K.; Sen Gupta, S.; Bhattacharyya, K.G. Nanomaterials as versatile adsorbents for heavy metal ions in water: A review. *Environ. Sci. Poll. Res.* **2019**, *26*, 6245–6278. [CrossRef]
30. Emenike, E.C.; Iwuozor, K.O.; Anidiobi, S.U. Heavy metal pollution in aquaculture: Sources, impacts and mitigation techniques. *Biol. Trace Elem. Res.* **2022**, *200*, 4476–4492. [CrossRef]
31. Daoush, W.M. Co-precipitation and magnetic properties of magnetite nanoparticles for potential biomedical applications. *J. Nanomed. Res.* **2017**, *5*, 118–123. [CrossRef]
32. Rajput, S.; Pittman, C.U.; Mohan, D. Magnetic magnetite (Fe3O4) nanoparticle synthesis and applications for lead (Pb^{2+}) and chromium (Cr6+) removal from water. *J. Colloid Interface Sci.* **2016**, *468*, 334–346. [CrossRef]
33. Hong, J.; Xie, J.; Mirshahghassemi, S.; Lead, J. Metal (Cd, Cr, Ni, Pb) removal from environmentally relevant waters using polyvinylpyrrolidone-coated magnetite nanoparticles. *RSC Adv.* **2020**, *10*, 3266–3276. [CrossRef]
34. Alfakheri, M.; Elarabany, N.; Bahnasawy, M. Effects of lead on some oxidative stress of the African catfish, *Clarias gariepinus*. *J. Egypt. Acad. Soc. Environ. Dev.* **2018**, *19*, 171–175. [CrossRef]
35. Ibrahim, R.E.; Elshopakey, G.E.; Abd El-Rahman, G.I.; Ahmed, A.I.; Altohamy, D.E.; Zagloul, A.W.; Younis, E.M.; Abdelwarith, A.A.; Davies, S.J.; Al-Harthi, H.F. Palliative role of colloidal silver nanoparticles synthetized by moringa against *Saprolegnia* spp. infection in Nile Tilapia: Biochemical, immuno-antioxidant response, gene expression, and histopathological investigation. *Aquac. Rep.* **2022**, *26*, 101318. [CrossRef]
36. Abdel Rahman, A.N.; Elshopakey, G.E.; Behairy, A.; Altohamy, D.E.; Ahmed, A.I.; Farroh, K.Y.; Alkafafy, M.; Shahin, S.A.; Ibrahim, R.E. Chitosan-*Ocimum basilicum* nanocomposite as a dietary additive in *Oreochromis niloticus*: Effects on immune-antioxidant response, head kidney gene expression, intestinal architecture, and growth. *Fish Shellfish Immunol.* **2022**, *128*, 425–435. [CrossRef]
37. Alexander, C.; Sahu, N.; Pal, A.; Akhtar, M. Haemato-immunological and stress responses of *Labeo rohita* (Hamilton) fingerlings: Effect of rearing temperature and dietary gelatinized carbohydrate. *J. Anim. Physiol. Anim. Nutr.* **2011**, *95*, 653–663. [CrossRef]
38. Shah, S.L. Alterations in the immunological parameters of tench (*Tinca tinca* L. 1758) after acute and chronic exposure to lethal and sublethal treatments with mercury, cadmium and lead. *Turkish. J. Vet. Anim. Sci.* **2005**, *29*, 1163–1168.
39. Huang, L.; Liu, Z.; Wu, C.; Lin, J.; Liu, N. Magnetic nanoparticles enhance the cellular immune response of dendritic cell tumor vaccines by realizing the cytoplasmic delivery of tumor antigens. *Bioeng. Transl. Med.* **2023**, *8*, e10400. [CrossRef]
40. Soto, M.; Marigomez, I.; Cancio, I. *Biological Aspects of Metal Accumulation and Storage*; University of the Basque Country: Leioa, Spain, 2003.
41. Akturk, O.; Demirin, H.; Sutcu, R.; Yilmaz, N.; Koylu, H.; Altuntas, I. The effects of diazinon on lipid peroxidation and antioxidant enzymes in rat heart and ameliorating role of vitamin E and vitamin C. *Cell Biol. Toxicol.* **2006**, *22*, 455–461. [CrossRef]
42. Abdel-Tawwab, M.; El-Sayed, G.O.; Monier, M.N.; Shady, S.H. Dietary EDTA supplementation improved growth performance, biochemical variables, antioxidant response, and resistance of Nile tilapia, *Oreochromis niloticus* (L.) to environmental heavy metals exposure. *Aquaculture* **2017**, *473*, 478–486. [CrossRef]
43. Olojo, E.A.A.; Olurin, K.B.; Mbaka, G.; Oluwemimo, A.D. Histopathology of the gill and liver tissues of the African catfish *Clarias gariepinus* exposed to lead. *African J. Biotechnol.* **2005**, *4*, 117–122.
44. Mondal, P.; Garai, P.; Chatterjee, A.; Saha, N.C. Toxicological and therapeutic effects of neem (*Azadirachta indica*) leaf powder in hole-in-the-head (HITH) disease of fish *Anabas testudineus*. *Aquac Res.* **2020**, *52*, 715–723. [CrossRef]
45. Răcuciu, M.; Tecucianu, A.; Oancea, S. Impact of magnetite nanoparticles coated with aspartic acid on the growth, antioxidant enzymes activity and chlorophyll content of maize. *Antioxidants* **2022**, *11*, 1193. [CrossRef]
46. Behera, T.; Swain, P.; Rangacharulu, P.V.; Samanta, M. Nano-Fe as feed additive improves the hematological and immunological parameters of fish, *Labeo rohita* H. *Appl. Nanosci.* **2014**, *4*, 687–694. [CrossRef]
47. Ates, M.; Demir, V.; Arslan, Z.; Kaya, H.; Yılmaz, S.; Camas, M. Chronic exposure of tilapia (*Oreochromis niloticus*) to iron oxide nanoparticles: Effects of particle morphology on accumulation, elimination, hematology and immune responses. *Aquat. Toxicol.* **2016**, *177*, 22–32. [CrossRef]
48. Yin, Y.; Hu, B.; Yuan, X. Nanogel: A versatile nano-delivery system for biomedical applications. *Pharmaceutics* **2018**, *12*, 290. [CrossRef]
49. El-Naggar, M.E.; Radwan, E.K.; El-Wakeel, S.T.; Kafafy, H.; Gad-Allah, T.A.; El-Kalliny, A.S.; Shaheen, T.I. Synthesis, characterization and adsorption properties of microcrystalline cellulose based nanogel for dyes and heavy metals removal. *Int. J. Biol. Macromol.* **2018**, *113*, 248–258. [CrossRef]
50. Neamtu, I.; Rusu, A.G.; Diaconu, A.; Nita, L.E.; Chiriac, A.P. Basic concepts and recent advances in nanogels as carriers for medical applications. *Drug Deliv.* **2017**, *24*, 539–557. [CrossRef]

51. Hamdy, A.; Ismail, S.H.; Ebnalwaled, A.A.; Mohamed, G.G. Characterization of superparamagnetic/monodisperse PEG-coated magnetite nanoparticles sonochemically prepared from the hematite ore for Cd (II) removal from aqueous solutions. *J. Inorg. Organomet. Polym. Mater.* **2021**, *31*, 397–414. [CrossRef]
52. Hassan, G.K.; Abdel-Karim, A.; Al-Shemy, M.T.; Rojas, P.; Sanz, J.L.; Ismail, S.H.; Mohamed, G.G.; El-gohary, F.A.; Al-Sayed, A. Harnessing Cu@Fe$_3$O$_4$ core shell nanostructure for biogas production from sewage sludge: Experimental study and microbial community shift. *Renew. Energy* **2022**, *188*, 1059–1071. [CrossRef]
53. Košak, A.; Lobnik, A.; Bauman, M. Adsorption of mercury (II), lead (II), cadmium (II) and zinc (II) from aqueous solutions using mercapto-modified silica particles. *Int. J. Appl. Ceram.* **2015**, *12*, 461–472. [CrossRef]
54. Phippen, B.; Horvath, C.; Nordin, R.; Nagpal, N. *Ambient Water Quality Guidelines for Iron, Prepared for Science and Information Branch, Water Stewardship Division*; Ministry of Environment: Madrid, Spain, 2008; p. 45.
55. Kasozi, N.; Tandlich, R.; Fick, M.; Kaiser, H.; Wilhelmi, B. Iron supplementation and management in aquaponic systems: A review. *Aquac. Rep.* **2019**, *15*, 100221. [CrossRef]
56. Neiffer, D.L.; Stamper, M.A. Fish sedation, anesthesia, analgesia, and euthanasia: Considerations, methods, and types of drugs. *ILAR J.* **2009**, *50*, 343–360. [CrossRef]
57. Lee, Y.C.; Yang, D. Determination of lysozyme activities in a microplate format. *Anal. Biochem.* **2002**, *310*, 223. [CrossRef]
58. Abdollahi, R.; Heidari, B.; Aghamaali, M. Evaluation of lysozyme, complement C3, and total protein in different developmental stages of Caspian kutum (*Rutilus frisii kutum* K.). *Arch. Pol. Fish.* **2016**, *24*, 15–22. [CrossRef]
59. Bai, F.; Ni, B.; Liu, M.; Feng, Z.; Xiong, Q.; Xiao, S.; Shao, G. Mycoplasma hyopneumoniae-derived lipid-associated membrane proteins induce apoptosis in porcine alveolar macrophage via increasing nitric oxide production, oxidative stress, and caspase-3 activation. *Vet. Immunol. Immunopathol.* **2013**, *155*, 155–161. [CrossRef]
60. Schultz, L. *Methods in Clinical Chemistry*; CV Mosby Company: St. Louis, MO, USA, 1987; pp. 742–746.
61. Siroka, Z.; Krijt, J.; Randak, T.; Svobodova, Z.; Peskova, G.; Fuksa, J.; Hajslova, J.; Jarkovsky, J.; Janska, M. Organic pollutant contamination of river Elbe assessed by biochemical markers. *Acta Vet.* **2005**, *74*, 293–303. [CrossRef]
62. Ohkawa, H.; Ohishi, N.; Yagi, K. Assay for lipid peroxides in animal tissues by thiobarbituric acid reaction. *Anal. Biochem.* **1979**, *95*, 351–358. [CrossRef]
63. Beutler, E.; Duron, O.; Kelly, B.M. Improved method for the determination of blood glutathione. *J. Lab. Clin. Med.* **1963**, *61*, 882–888. Available online: https://pubmed.ncbi.nlm.nih.gov/13967893/ (accessed on 6 July 2023).
64. Velkova-Jordanoska, L.; Kostoski, G.; Jordanoska, B. Antioxidative enzymes in fish as biochemical indicators of aquatic pollution. *Bulg. J. Agric. Sci.* **2008**, *14*, 235–237.
65. Aksnes, A.; Njaa, L.R. Catalase, glutathione peroxidase and superoxide dismutase in different fish species. *Camp. Biochem. Physiol.* **1981**, *69B*, 893–896. [CrossRef]
66. Suvarna, K.S.; Layton, C.; Bancroft, J.D. *Bancroft's Theory and Practice of Histological Techniques E-Book*; Elsevier Health Sciences: Amsterdam, The Netherlands, 2018; p. 672.
67. Goldberg, E.D.; Koide, M.; Hodge, V.; Flegel, A.R.; Martin, J. US mussel watch: 1977–1978 results on trace metals and radionuclides. *Estuar. Coast. Shelf Sci.* **1993**, *16*, 69–93. [CrossRef]
68. Yacoub, A.M.; Gad, N.S. Accumulation of some heavy metals and biochemical alterations in muscles of *Oreochromis niloticus* from the River Nile in Upper Egypt. *Int. J. Environ. Sci. Eng.* **2012**, *3*, 1–10.
69. Kaplan, E.L.; Meier, P. Nonparametric-estimation from incomplete observations. *J. Am. Stat. Assoc.* **1958**, *53*, 457–481. [CrossRef]

Disclaimer/Publisher's Note: The statements, opinions and data contained in all publications are solely those of the individual author(s) and contributor(s) and not of MDPI and/or the editor(s). MDPI and/or the editor(s) disclaim responsibility for any injury to people or property resulting from any ideas, methods, instructions or products referred to in the content.

Article

The Development of Fe₃O₄-Monolithic Resorcinol-Formaldehyde Carbon Xerogels Using Ultrasonic-Assisted Synthesis for Arsenic Removal of Drinking Water

Sasirot Khamkure [1,*], Prócoro Gamero-Melo [2], Sofía Esperanza Garrido-Hoyos [3], Audberto Reyes-Rosas [4], Daniella-Esperanza Pacheco-Catalán [5] and Arely Monserrat López-Martínez [2]

1. Postgraduate Department, CONAHCYT-Mexican Institute of Water Technology, Jiutepec 62550, Mexico
2. Sustainability of Natural Resources and Energy, Cinvestav Saltillo, Ramos Arizpe 25900, Mexico; procoro.gamero@cinvestav.edu.mx (P.G.-M.); arely.lopez@cinvestav.edu.mx (A.M.L.-M.)
3. Postgraduate Department, Mexican Institute of Water Technology, Jiutepec 62550, Mexico; sgarrido@tlaloc.imta.mx
4. Department of Bioscience and Agrotechnology, Research Center of Applied Chemistry, Saltillo 25294, Mexico; audberto.reyes@ciqa.edu.mx
5. Renewable Energy Unit, Yucatan Scientific Research Center, Merida 97302, Mexico; dpacheco@cicy.mx
* Correspondence: skhamkure@conahcyt.mx

Abstract: Inorganic arsenic in drinking water from groundwater sources is one of the potential causes of arsenic-contaminated environments, and it is highly toxic to human health even at low concentrations. The purpose of this study was to develop a magnetic adsorbent capable of removing arsenic from water. Fe₃O₄-monolithic resorcinol-formaldehyde carbon xerogels are a type of porous material that forms when resorcinol and formaldehyde (RF) react to form a polymer network, which is then cross-linked with magnetite. Sonication-assisted direct and indirect methods were investigated for loading Fe₃O₄ and achieving optimal mixing and dispersion of Fe₃O₄ in the RF solution. Variations of the molar ratios of the catalyst (R/C = 50, 100, 150, and 200), water (R/W = 0.04 and 0.05), and Fe₃O₄ (M/R = 0.01, 0.03, 0.05, 0.1, 0.15, and 0.2), and thermal treatment were applied to evaluate their textural properties and adsorption capacities. Magnetic carbon xerogel monoliths (MXRF600) using indirect sonication were pyrolyzed at 600 °C for 6 h with a nitrogen gas flow in the tube furnace. Nanoporous carbon xerogels with a high surface area (292 m²/g) and magnetic properties were obtained. The maximum monolayer adsorption capacity of As(III) and As(V) was 694.3 µg/g and 1720.3 µg/g, respectively. The incorporation of magnetite in the xerogel structure was physical, without participation in the polycondensation reaction, as confirmed by XRD, FTIR, and SEM analysis. Therefore, Fe₃O₄-monolithic resorcinol-formaldehyde carbon xerogels were developed as a potential adsorbent for the effective removal of arsenic with low and high ranges of As(III) and As(V) concentrations from groundwater.

Keywords: adsorption; arsenate and arsenite; carbon xerogels; resorcinol-formaldehyde; sonication

1. Introduction

A current environmental and human health problem is the availability of water due to the increasing demand and contamination of drinking water sources. Most of the accessible drinking water is found in aquifers, which are underground reservoirs of water. However, the presence of contaminants such as arsenic, which naturally occur in the environment due to geological factors, can migrate into groundwater through weathering processes.

Inorganic arsenic (As) is a well-known carcinogenic element and one of the most significant chemical pollutants worldwide, found in several countries across the globe. The Agency for Toxic Substances and Disease Registry (ATSDR) has ranked arsenic as the top substance with potential risks to public health on a global scale [1]. The World Health

Organization (WHO) recommends a guideline value of 10 µg/L for arsenic concentrations in drinking water, which is also considered acceptable by the United States Environmental Protection Agency (EPA) [2,3]. In Mexico, the NOM-127-SSA1-2021, "Environmental Health, Water for Human Use and Consumption—Permissible Limits of Quality and Treatments for Water Purification", sets an allowable limit of 10 µg/L for arsenic in drinking water [4]. Groundwater contamination by arsenic affects millions of people in various countries, including the United States, Argentina, Australia, Bangladesh, Cambodia, Chile, China, India, Laos, Myanmar, Mexico, Pakistan, Taiwan, Thailand, and Vietnam [5–7].

Arsenic contamination in drinking water is a serious problem in Mexico. The levels of arsenic in drinking water in some regions of the country exceed the recommended limit of 10 µg/L. Many people in Mexico are at risk due to consuming water with elevated arsenic levels. children are at risk of developing serious health problems as a result of their exposure to arsenic [8]. The concentrations of arsenic in water samples from Chihuahua ranged from 0.1 to 419.8 µg/L, which is associated with adverse health effects [9]. Groundwater background values in Guanajuato State were evaluated, and the arsenic values from sample wells were in the range of 0.068–0.777 mg/L. These values are due to geogenic sources containing volcanic rocks, specifically rhyolites, which have a presence of As and F in the hot deep flow [10]. Three Yaqui villages in southern Sonora, Mexico, have been studied for arsenic exposure through drinking water. The range of arsenic concentration in these villages was 11.8–70.01 µg/L, and it has been associated with lung function and inflammation, as well as respiratory infections in children [11].

Arsenic can exist in various oxidation states, but in natural water sources, it is predominantly found in its inorganic forms as trivalent arsenite (As(III)) or pentavalent arsenate (As(V)) oxyanions. The presence of arsenic-contaminated water that is used for drinking, food preparation, and agricultural irrigation poses a significant threat to public health. Prolonged exposure to arsenic through the ingestion of contaminated food and water can lead to the development of cancer and skin lesions [12]. Arsenic has been linked to various diseases affecting the cardiovascular, liver, neurological, immune, endocrine systems, as well as the skin. It has also been associated with diabetes and various types of cancer, such as skin, liver, lung, and bladder cancer, due to its absorption through the gastrointestinal tract, skin, and respiratory system [13]. Furthermore, elevated levels of arsenic in drinking water have been associated with an increased risk of myocardial infarction [14], adverse effects on fertility in women [15,16], and negative consequences for fetal development during pregnancy [17].

The processes and technologies for the removal of arsenic from water that are known currently are oxidation, precipitation, coagulation and softening with lime, reverse osmosis, microfiltration, nanofiltration, adsorption, biological treatments, phytoremediation, electrodialysis, and electrokinetics, among others [18–20]. Among these methods, the application of adsorption is a promising technique and has been widely extended in the treatment of water and wastewater due to its high efficiency, affordability, ease of design, operation, handling, and maintenance, and the variety of adsorbent materials that can be regenerated and reused. Furthermore, no additional chemicals are needed in the operation, and there is no production of sludge or generation of toxic byproducts [7,21].

Some emerging arsenic adsorbents are chemically modified zeolites [22], zeolitic imidazole frameworks [23], lanthanum hydroxide–doped graphene oxide biopolymer foam [24], metal–organic framework-based composite materials [25], and jarosites [26]. The vast majority of reported adsorbents are micro or nanometer-size powders; although some of these materials have a high adsorption capacity for micropollutants from water, their application at the pilot plant or industrial level is limited by the difficulty of separating the adsorbent from the treated water.

Iron-based adsorbents are excellent adsorbents for removing arsenic from water. Magnetite (Fe_3O_4) is one of the most well-known iron oxides/hydroxides due to its strong affinity for arsenic and ease of accessibility [27]. Iron-based adsorbents are non-toxic, low cost, and easily accessible in large quantities and offer promising results for arsenic removal from water [28,29].

The gels are mesoporous materials with texture, mechanical resistance, and chemical stability. They can be controlled and designed according to the variation of the synthesis and processing conditions. Gels are formed by the addition and/or polycondensation of a low molecular weight oligomer in an aqueous or alcoholic solution. First, a "sol" is formed, a colloidal solution of solid particles that grow and coalesce to the gel point, making the sol-gel transition at which the viscosity of the medium changes. The formed wet gel behaves like a giant molecule of equal size to the container where it is prepared. The gel progressively strengthens as residual unreacted oligomers bind to the developing network. This phenomenon is called aging or curing, and it allows favorable conditions for drying the gel with the least number of breaks in the structure formed. In other words, the gel is composed of a continuous solid skeleton formed by chains of monomer particles arranged in a pearl necklace that is immersed in a continuous liquid phase. By removing the liquid from the wet gel, a large pore volume can be obtained [30]. The solvent that saturates the pores can be evacuated by three following drying methods: subcritical, supercritical, and cryogenic. Subcritical or conventional drying under atmospheric conditions can generate drastic changes in the surface tension of the solvent once the vapor–liquid interface is formed, this difference between the surface tension of the coexisting vapor and liquid phases produces collapses in the pore structure of the gels. The result is a dense polymer called xerogel. A specific application of resorcinol formaldehyde (RF) gel, including the doping of RF gels with metals or metal oxides, is found in the removal of contaminants from drinking water and wastewater [31–34].

Ultrasound technology has been utilized for the synthesis of various materials, including nanoparticles, and has found numerous applications such as homogenizing, emulsifying, dispersing, deagglomeration, sonochemistry, and sono-catalysis. The effects of sonication on agglomeration, metal release, zeta potential, and the administered dose were evaluated using probe sonication for the synthesis of non-functionalized nanoparticles such as copper, aluminum, manganese, and zinc oxide [35]. The results showed that sonication can be used to control the size and morphology of nanoparticles, as well as to improve their dispersibility and zeta potential. Iron(III) trimesate xerogel was prepared using ultrasonic irradiation within a short time of 10 to 20 min and a low pH solution [36]. This method produced a product with a high specific surface area of 1042 m^2/g. The high specific surface area of the xerogel was attributed to the formation of a porous structure during the ultrasonic irradiation process. Furthermore, silica xerogels were prepared using the sol-gel method with ultrasonic treatment to accelerate aging and hydrophobic treatment [37]. The effect of ultrasonic frequency, specifically 100 kHz and 500 kHz, on the structure was investigated. It was found that 500 kHz accelerated the aging reaction, facilitated hydrophobization, and rapidly suppressed gel shrinkage. These studies demonstrate the potential of ultrasound technology for the synthesis of various materials with desired properties. Ultrasound can be used to control the size, morphology, dispersibility, zeta potential, and aging of nanoparticles. It can also be used to accelerate the formation of porous structures and to improve the hydrophobicity of materials. Ultrasonic technology can be considered an environmentally friendly application because it reduces processing time, increases cost efficiency, simplifies manipulation, enhances the purity of the final product, and lowers energy consumption [32,34].

Regarding the theoretical knowledge of resorcinol-formaldehyde xerogels and the applications of iron-based adsorbents such as Fe_3O_4 for arsenic removal, this study applied the ultrasonic-assisted synthesis of carbon xerogels to evaluate the effect of Fe_3O_4 loading through both direct and indirect methods on the removal of arsenic species in groundwater. The novelty of this work is a Fe_3O_4-monolithic resorcinol-formaldehyde carbon xerogel that, due to its chemical composition and ordered porous structure, is capable of removing arsenite and arsenate ions present in groundwater. Moreover, due to its magnetic properties, it is possible to easily recover it from the treated water.

This research focuses on generating magnetic carbon xerogels with morphological, magnetic, textural, and physical–chemical properties in the form of pellets, which makes them reusable. These materials have the capability to adsorb arsenates and arsenites. The synthesis procedure was developed considering the effect of Fe_3O_4 loading via ultrasonic methods, both direct and indirect, while varying the molar ratios of Fe, catalyst, and water. The arsenic adsorption test was conducted in batch, and the synthesized materials were characterized using various analytical techniques before and after the adsorption of arsenic. The intended purpose of this work is for the adsorbent materials produced to serve as viable alternatives within the technological advancements for water remediation. The magnetic properties of carbon xerogels facilitate the separation, reuse, regeneration, and recycling of the adsorbents so that their useful life is extended. The efficient separation of aged adsorbents also facilitates the recovery and final disposal of contaminants and strengthens the environmental sustainability of the water purification process.

The experimental reproducibility of Fe_3O_4-monolithic resorcinol-formaldehyde carbon xerogels involves several challenges, including the composition of the starting materials (variation in molar ratios), the homogeneity of dispersion (direct and indirect ultrasonication methods), the sonication conditions (power output, duration, and frequency), the gelation and curing conditions, and the post-synthesis treatments (pyrolysis). Controlling these parameters consistently across different experiments can be challenging, and this can affect the properties of the resulting monolithic carbon xerogels. Subsequently, the adsorption capacity of the final material was improved with increasing the initial concentration and the adsorption affinity for arsenic species.

The environmental sustainability of the arsenic adsorption process using Fe_3O_4-monolithic resorcinol-formaldehyde carbon xerogels can be evaluated through methodologies such as the lifecycle, planetary boundaries, and sustainable development goals [38]. However, applying these methodologies and their indicators is beyond the scope of this article.

2. Results and Discussion

2.1. Fe_3O_4-Monolithic Resorcinol-Formaldehyde Xerogels: Effect of Loading of Magnetite with Indirect and Direct Sonication, and Modification of Catalyst

This study focuses on the development and initial preparation of monolithic xerogels with magnetic properties using magnetite (Fe_3O_4, Lanxess) as an adsorbent material for water treatment. The magnetic xerogel monoliths (MCs and MXs) were synthesized through the sol-gel polymerization of resorcinol and formaldehyde (RF) with sodium carbonate (C) as a catalyst, employing indirect and direct sonication, respectively. The effect of varying the molar ratios of resorcinol/catalyst was evaluated to obtain the high adsorption capacity in the arsenate adsorption.

2.1.1. Characterization of MCs and MXs

To identify the phases in the xerogels, XRD analysis was carried out. XRD pattern of RFX revealed the presence of both crystalline and amorphous phases as shown in Figure 1, which is similar to the pattern reported by [39,40].

The XRD patterns of magnetic xerogels prepared using direct and indirect sonication methods and different R/C ratios (Figure 1) were found to be similar, with diffraction peaks at 2 θ values of 18°, 30°, 35.5°, 43°, 57°, and 62° corresponding to the crystallographic planes of magnetite 111, 220, 311, 400, 511, and 440, as reported in the ICCD card number 00-01900629. These findings align with the research of [41]. The percentage of crystalline phase for RFX, MX1, and MX2 was 10.54%, 12.45%, and 10.51%, respectively. Meanwhile, the percentage of crystalline phase for MC1-MC4 ranged from 10.51% to 12.64%. The percentage of crystalline phase for magnetic xerogels prepared using direct and indirect methods (MX1 and MC1) at the same molar ratios and gelation process was approximately 12%.

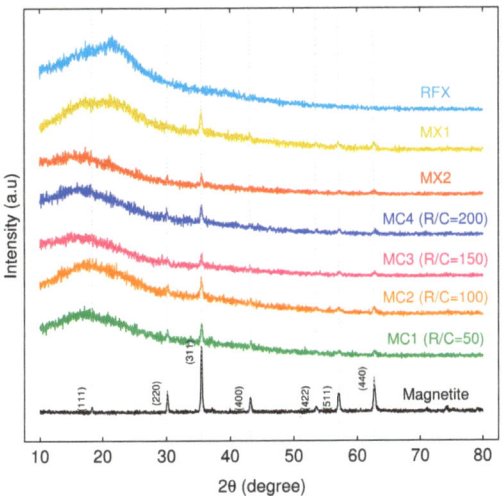

Figure 1. Powder X-ray diffraction patterns of the xerogels (RFX), Fe_3O_4-monolithic resorcinol-formaldehyde xerogels synthesized through direct sonication (MX1 and MX2), and indirect sonication with different R/C (MC1-MC4), along with their corresponding JCPDS card assignments.

The crystal size of the magnetic xerogels was determined by calculating the X-ray diffraction peak widths using Bragg's law and Debye Scherrer equation, as described by [42].

$$D = K\lambda / \beta \cos\theta, \qquad (1)$$

where D is the crystalline size, K denotes represents the Scherrer constant (0.98), λ represents the X-rays wavelength (1.54178 Å), β denotes the full width at half maximum (FWHM) and θ is the Bragg diffraction angle (radians).

Table 1 shows the average crystalline sizes of MC50, MC100, and MC200 were in the range of 22.94–25.88 nm. Additionally, their values of β (0.32–0.36 radians) and θ (35.52–35.53 radians) were similar. However, the crystallinity of particle of MC200 (R/C = 200) was higher than MC50 (R/C = 50) and MC100 (R/C = 100). This indicates that increasing the R/C ratio can result in increased crystallinity, which is similar to the results obtained by [43].

Table 1. The average crystal size of magnetic xerogels varying R/C Molar ratios.

Sample Name	R/C Molar Ratios	Crystal Size Average (nm)	FWHM (β) (Radian)	θ (Radian)
MC1	50	24.52	0.34	35.54
MC2	100	22.94	0.36	35.52
MC4	200	25.88	0.32	35.53

The morphology of RF xerogels (RFX) was observed using scanning electron microscopy (SEM). Figure 2a shows that RFX is composed of a large number of microclusters that are uniformly distributed. These microclusters contain the resorcinol-formaldehyde polymer. The RF gel network is formed with nearly spherical particles, showing similar results to those obtained by [44]. Furthermore, the interconnects between the microclusters were observed to form porous materials. This porosity is likely due to the gelation process used in the synthesis of RFX, which involves the formation of a three-dimensional network of interconnected polymer chains [39,45,46]. Therefore, RFX is a highly porous material with a complex microstructure.

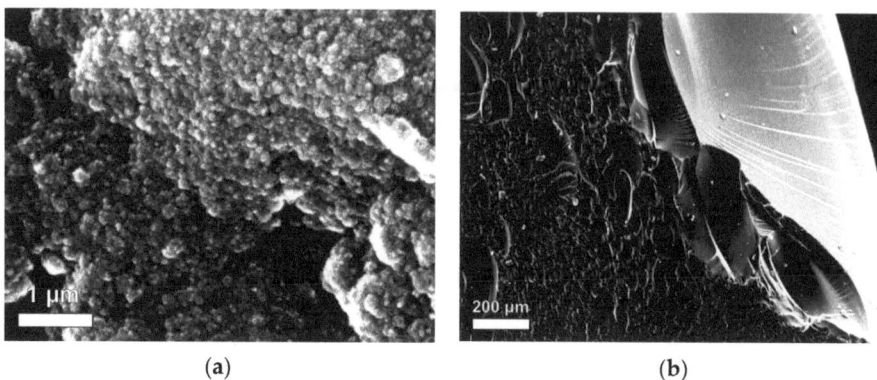

Figure 2. SEM images of RF gels between (**a**) inside and (**b**) outside.

Figure 2a shows the difference between the outside and inside of the RF gel. The microclusters in the outer region appear more compact than those in the inner region. This difference in microstructure is likely due to the contact of the outer region with the glass tube during the gelation process. During gelation, the RF solution is typically poured into a mold or tube and allowed to solidify. The contact of the outer region with the glass surface may have caused the microclusters to pack more tightly together, resulting in a more compact microstructure.

In this study, the SEM analysis revealed the effect of direct and indirect ultrasonication on the preparation of magnetic xerogels. Figure 3 depicts the morphology of magnetic gels, namely MX1 and MC4, synthesized with the same molar ratios. It can be observed that the morphology of MC4 (Figure 3b,d) characterizes nearly spherical particles that partially overlap, resulting in the formation of large pores. This morphology is likely attributed to the incorporation of magnetite particles into the RF gel during synthesis. Comparing MX1 and MC4 at the same magnification range of 15,000 and 25,000, it is evident that MX1 has smaller particle and pore sizes compared to MC4. Additionally, MX1 shows a more compact RF gel structure than MC4. Both techniques involve delivering energy to the RF solution with magnetite particles through probe sonication. However, the resulting particle sizes and mesoporosity of RF gels differ between the two methods. Indirect sonication generates cavitation in the water bath using high-intensity ultrasound through a water bath, while direct sonication involves the probe causing cavitation during sample processing. It can be explained that the particle size of MX1 decreases after ultrasonication, as observed by [37,47,48].

Energy-dispersive X-ray spectroscopy (EDX) is a technique used to determine the elemental composition of a material. In this study, EDX analysis was used to determine the Fe content, confirming the incorporation of Fe content in the structure of magnetic RF gels. Both magnetic gels demonstrate the physical incorporation of magnetite into the structure of RF gel without participating in the polycondensation reaction of RF, as stated by [47]. MC4 shows a more uniform distribution of magnetite contents in the structure of RF gel than MX1.

During the synthesis of MX1 with direct sonication, the RF solution was mixed, and the temperature increased dramatically from 45 °C to 79 °C within 5 min, resulting in the formation of a black gel. On the other hand, MC4 was prepared using indirect sonication. The mixed solution of MC1-MC4 allowed the dispersion of magnetite into the RF gel, and the temperature of the solution continuously increased from 33 °C to 85 °C, and finally leading to the formation of a black gel within the water bath for 60 min.

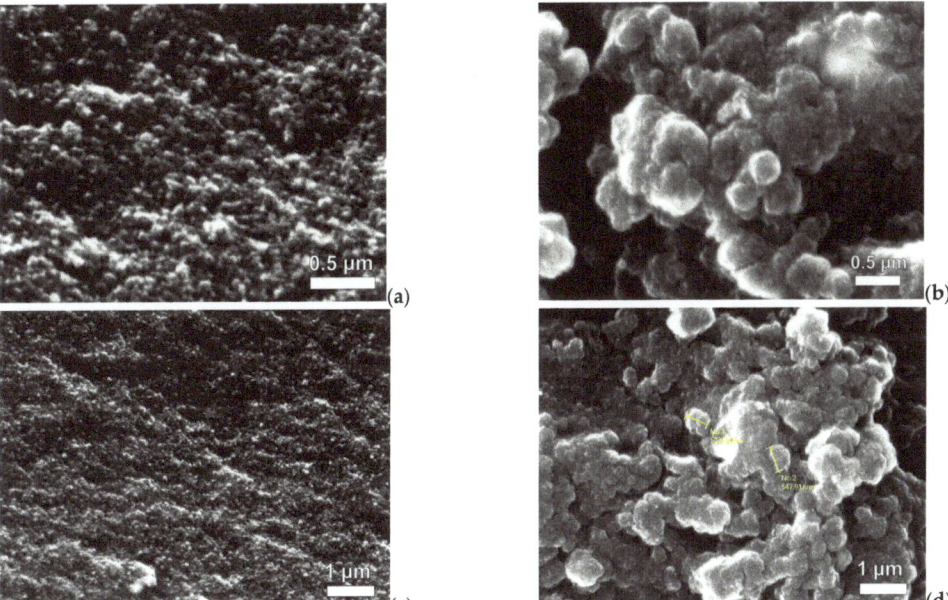

Figure 3. SEM images of Fe$_3$O$_4$-monolithic resorcinol-formaldehyde xerogels prepared by (**a**,**c**) direct (MX1) and (**b**,**d**) indirect (MC4) ultrasonication with magnifications.

It can be explained that direct sonication involves the use of a sonication probe directly immersed in the reaction mixture. The probe emits ultrasonic waves that directly interact with the sample, resulting in more localized and precise energy transfer. However, direct sonication generally requires shorter processing times compared to indirect sonication, as the energy efficiently reaches the desired regions, accelerating the required reactions. Consequently, the mixed solution of MX1 with ultrasonication experienced a significant increase in temperature, leading to reduced gelation times [49]. In this study, magnetite was added to the RF solution, and due to its natural behavior, magnetite tends to agglomerate within a short mixing time. Considering the variables involved in the solution, gelation, and curing processes, high temperatures during synthesis lead to porosity shrinkage [49].

After ultrasonication of magnetite into an aqueous RF solution, the particle size of magnetite was decreased, which can be clearly observed with MC4. The EDX analysis revealed that MC4 incorporated 1.19% Fe content (Figure 4d). The M/R ratio used in the synthesis was 0.01, indicating a low concentration of magnetite compared to the RF polymer. This suggests that even with a low M/R ratio, the incorporation of Fe in the RF gel was successful, due to the use of magnetite particles in the synthesis process.

The morphology of the MCs was studied by SEM as shown in Figure 5. A three-dimensional RF gel network was formed with nearly spherical particles [46,50]. MC1 and MC4 prepared with different Na$_2$CO$_3$ concentrations, the morphology and pore size distribution can be observed that MC1 with lower R/C molar and high initial pH solution exhibit smaller particles and pore sizes than other materials. pH variations can alter the nucleation and growth of the gel network, leading to changes in the average pore size, pore connectivity, and surface area of the xerogel. Higher pH values can promote the formation of smaller pores, while lower pH values may result in larger pores [50].

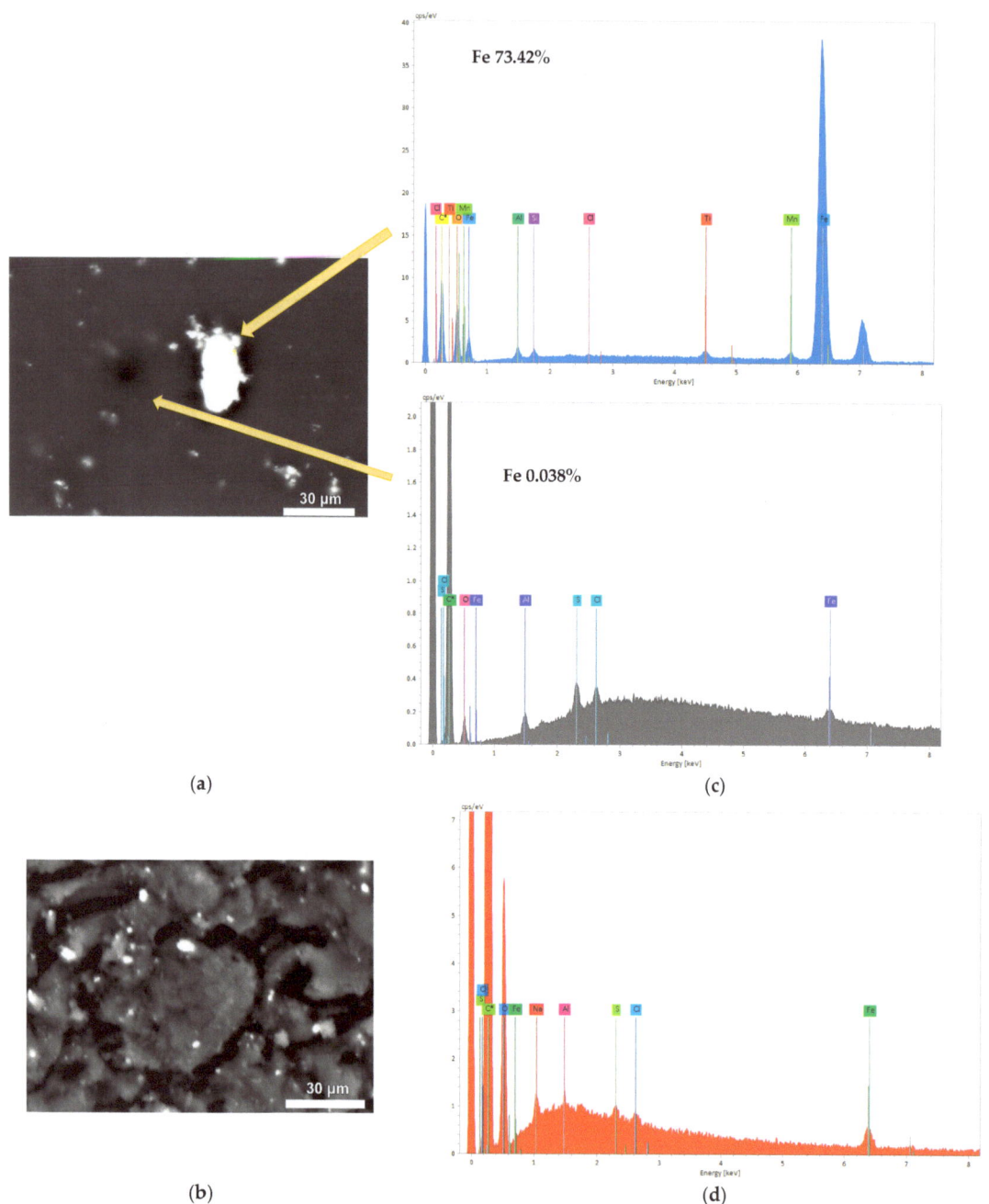

Figure 4. SEM and EDX images of Fe$_3$O$_4$-monolithic resorcinol-formaldehyde xerogels prepared by (**a**,**c**) direct (MX1) and (**b**,**d**) indirect (MC4) ultrasonication.

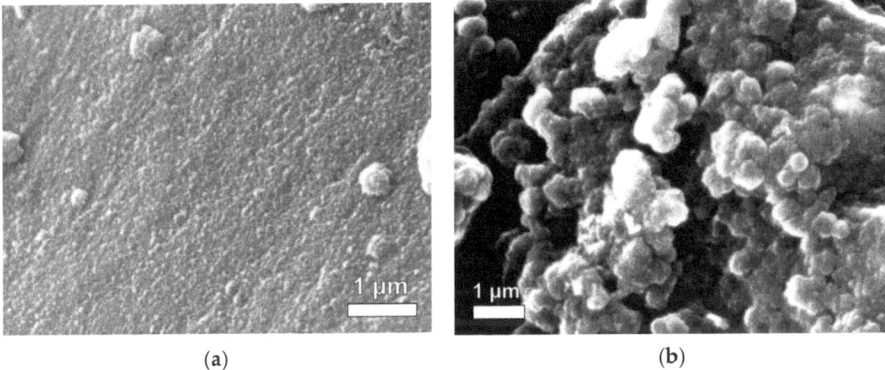

Figure 5. SEM Images of Fe_3O_4-monolithic resorcinol-formaldehyde xerogels with R/C ratios of (**a**) 50 (MC1) and (**b**) 200 (MC4).

The mesoporosity of RF gels increases with an increase in the R/C ratio, as reported in previous studies by [39,45,51]. This indicates that the porosity of RF gels can be controlled by adjusting the R/C ratio in the synthesis process. Mesopores are pores with diameters between 2 and 50 nm and are desirable for various applications such as adsorption.

Table 2 shows the textural properties of magnetic xerogels, the effect of direct, and indirect sonication on the textural properties of materials, specifically MX1 and MC4, respectively. The surface area of the magnetic xerogels for both MX1 and MC4 increased significantly compared to the xerogel. MC4 exhibited a higher surface area of 529.47 m^2/g, whereas MX1 had a surface area of 472.41 m^2/g. Additionally, the total pore volume and average pore diameter of MC4 were lower than those of MX1. This can be explained by the fact that MX1, prepared through direct sonication with a shorter sonication time for gelation, resulted in a lower surface area but higher total pore volume and larger average pore size.

Table 2. Textural parameters, pH_{pzc}, and IEP of xerogel and Fe_3O_4-monolithic resorcinol-formaldehyde xerogels.

Samples	Molar Ratio of R/C	Area BET (m^2/g)	Total Pore Volume (cm^3/g)	Average Pore Diameter (nm)	pH_{pzc}	IEP
RFX	200	399.19	0.517	5.23	2.99	2.74
MX1	200	472.41	0.842	7.57	4.54	3.09
MC1	50	365.93	0.255	2.79	6.63	3.40
MC2	100	545.09	0.549	4.03	6.12	3.59
MC4	200	529.47	0.683	5.16	4.35	3.70

The effect of pore structure in magnetic xerogels using the ultrasound-assisted sol-gel method was investigated through N_2 physisorption analysis. It can be explained that direct sonication promotes the formation of smaller and more uniform pores within the xerogel structure, as the energy can be precisely targeted to specific regions. On the other hand, indirect sonication may result in the generation of larger or more irregularly shaped pores in the xerogel due to less controlled and localized energy transfer. This observation is consistent with the findings in Figure 3 of SEM images and Figure 4 depicting particle distributions, which demonstrate that MX1 has a smaller particle size compared to MC4.

Moreover, Table 2 shows the effect of catalyst contents on the surface area and pore volume of the magnetic xerogels. The results of the RF gels using sodium carbonate as a catalyst show that MC200 had a higher average pore diameter (5.16 nm) than MC100 (4.03 nm), but MC200 had a lower surface area (529.47 m^2/g) than MC100 (545.09 m^2/g). However, MC200 (529.47 m^2/g) exhibited a surface area lower than MC100 (545.09 m^2/g).

These findings are consistent with those of [52], reported that increasing the molar ratios of R/C in gels prepared with Na_2CO_3 leads to an increase in average pore width. When lower molar ratios of R/C are used for RF gel preparation, a higher concentration of Na_2CO_3 results in the formation of smaller clusters with smaller average-sized pores. Therefore, MC100, with its lower R/C ratios, has a greater number of smaller pore diameters, and a higher surface area, making it more suitable for use in water treatment adsorption.

In this study, the obtained results of the mesoporous nature of magnetic xerogels with varying R/C demonstrated the effect on the surface area and pore volume of the RF polymer in magnetic xerogels. It can be explained that the pH of the RF solution is associated with the quantity of catalyst utilized during the synthesis. When the pH was decreased, both the surface area and pore volume of the RF polymer in xerogels increased. This indicates that lower pH values result in the formation of a greater number of pores and increased surface area within the RF polymer retained in the xerogels [53]. Moreover, it can be explained that the larger carbonate ions have a trigonal planar molecular geometry, which may cause steric hindrance. Consequently, the condensation of the intermediates leads to the generation of larger pores of the samples [54]. The use of a higher amount of catalyst leads to more rapid gelation, resulting in a less uniform structure with fewer and larger pores. Alternatively, the catalyst itself may interfere with the formation of crosslinks within the RF polymer, leading to a less porous structure. Therefore, higher amounts of catalyst used during the synthesis have a similar effect on the surface area and porosity of the resulting material, as observed in the results obtained by [55].

The determination of the isoelectric point (IEP) and point of zero charges (pH_{pzc}) of xerogels and magnetic xerogels was carried out by measuring the zeta potential and pH, as shown in Table 2. The IEP and pH_{pzc} of MX1 and MC4 prepared by direct and indirect sonication, respectively, with R/C 200 are in a similar range of values. However, the RF xerogel exhibits lower IEP and pH_{pzc} values compared to the other materials. These findings are similar to the results reported by [56], where organic xerogels demonstrated a pH_{pzc} value of 3.

Figure 6 shows the particle distribution of xerogel and magnetic xerogels prepared using the sol-gel method under ultrasonic irradiation. The particle size distribution in the obtained xerogels may vary because of sonication-assisted synthesis and variations in the R/C ratios. RFX exhibits a broader particle size distribution with larger particles compared to MX1 and MC4, which were prepared with the same molar ratios and drying process. RFX, prepared without sonication, showed a larger particle size, which is consistent with the findings of [57].

It can be observed that the average particle diameter of MX1 (28.05 nm) was lower than that of MC4 (32.65 nm), which is similar to the results obtained from SEM analysis. The use of direct sonication in the preparation of MX1 resulted in a narrower particle distribution due to localized energy transfer, leading to more consistent particle sizes in the obtained xerogel. On the other hand, MC4 exhibited a wider range of particle sizes due to less precise control of sonication energy distribution. Therefore, the direct method of sonication generally leads to a lower particle size distribution compared to the indirect method, due to the more localized and intense energy transfer that promotes effective fragmentation and reduction in particle size. Similar findings of the study of [58].

The initial pH of the solution is a factor influencing the polymerization of xerogels, especially when varying the molar ratio of the catalyst. The pH values of the RF solutions for MC1, MC2, MC3, and MC4 were 7.26, 7.05, 6.92, and 6.82, respectively, within the similar range of the study of [52]. It can be observed that higher catalyst concentrations with lower R/C molar ratios result in smaller particles and pore sizes, as reported by [53].

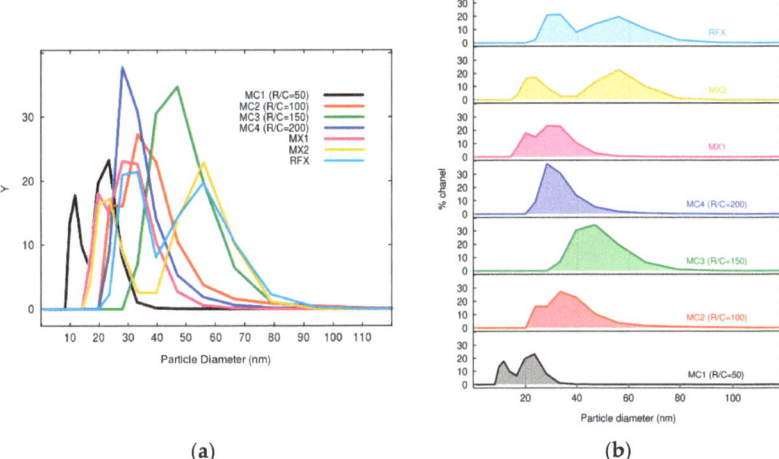

Figure 6. Particle size distribution of RF xerogel and Fe_3O_4-monolithic resorcinol-formaldehyde xerogels prepared by the sol-gel method under ultrasonic irradiation and presented in (**a**) grouping and (**b**) separation graphs.

The pH of the precursor solution plays a crucial role in determining the final structure of the obtained xerogel. It affects the kinetics of polymerization and crosslinking reactions, as well as the condensation and gelation processes. The mechanism of polymerization in RF gels involves two steps: the addition reaction to form hydroxymethyl derivatives of resorcinol and the condensation of these derivatives to form methylene or methylene ether bridged compounds [49]. In a high pH solution, the first addition reaction is favored. This leads to a higher rate of polymerization and crosslinking, resulting in a more extensively crosslinked network structure and a relatively quick process. This process often yields small nodules and narrow mesopores. Gelation kinetics, which refers to the rate of transition from a liquid precursor solution to a gel network, is strongly influenced by pH. Higher pH values generally promote faster gelation, while lower pH values slow down the process. The gelation kinetics can significantly impact the overall pore structure and porosity of the xerogel. When the condensation reaction occurs in the presence of small particles resulting from the high pH conditions, it produces materials with smaller pores, leading to a higher density or more compact RF gel structure [59]. On the other hand, lower pH values may result in a less densely crosslinked structure.

Figure 7a depicts the FTIR spectra of RFX and magnetic gels prepared using ultrasonication with direct and indirect techniques, covering a wavelength range of 4000–400 cm^{-1}. The characteristic FTIR bands of RFX, MX, and MC are similar. However, MC4, MX1, and MX2 exhibit an FTIR band at 478 cm^{-1} attributed to Fe-O stretching vibration [60,61]. The profiles of RFX, MC4, and MXs show the presence of six absorption bands: (i) O-H stretching at 3300 cm^{-1}, (ii) C-H stretching at 2900 cm^{-1}, (iii) C = C stretching in the aromatic ring at 1600 cm^{-1}, (iv) C-H bending vibration at 1400 cm^{-1}, (v) C-O stretching at 1200 cm^{-1}, and (vi) methylene ether C-O-C linkage stretching between two resorcinol molecules at 1000 cm^{-1} [62]. The FTIR spectra of RF gel and MC1-MC4 can be observed in Figure 7b, and all of them exhibit bands that are correlated with the bands described above.

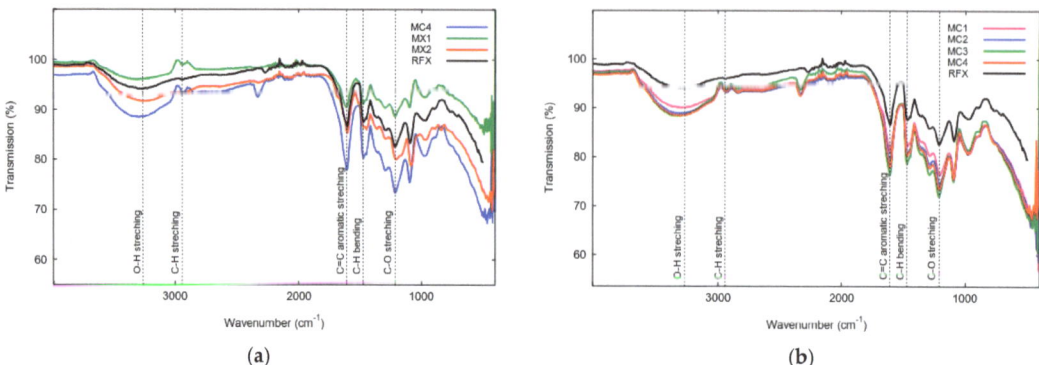

Figure 7. FTIR spectra of RF xerogel and Fe$_3$O$_4$-monolithic resorcinol-formaldehyde xerogels prepared by (**a**) direct and indirect sonication ultrasonication-assisted and (**b**) varying of R/C molar ratios.

Regarding the characterization of MCs and MXs, it can be observed that the preparation of monolithic resorcinol-formaldehyde xerogels involved different methods of sonication, utilizing low and high intensity, respectively. However, the results of their XRD and FTIR analyses show significant similarities. This is in contrast to the study conducted by [48], where the preparation of ZnO nanoparticles using direct and indirect sonication had an impact on the crystalline structure (XRD analysis) and resulted in different IR spectra of the samples. Due to the probable growth mechanisms of ZnO nanoparticles, various crystallization mechanisms were proposed. However, in the case of xerogels, ultrasonic irradiation aids in promoting aging and hydrophobization reactions. Additionally, [37] discovered that the preparation of silica xerogels can be accomplished in less than 1/5 of the time required by conventional methods.

2.1.2. Performance of Adsorption of Arsenic Using MCs and MXs

In the batch adsorption experiment of As(V) using MCs and MXs, the effect of pH in the range of 2 to 7 was used to evaluate their adsorption capacities, as shown in Figure 8. MC1 and MC2 demonstrated high adsorption capacities, q_e were more in the range of 63.26–73.47 µg/g and 59.18–61.22 µg/g, respectively, than other materials. MX1 and MX2 showed higher adsorption capacity in the acidic solution. Due to the pH$_{pzc}$ being the zero net charge on the surface of the adsorbent, the adsorbent surfaces are charged positively or negatively, depending on whether the pH of the solution is lower or higher than the pH$_{pzc}$ values, respectively [55]. The analysis result of pH$_{pzc}$ of MX1 was 4.54, meaning that MX1 adsorbed As(V) at pH values lower than this value. The same can be described for the adsorption of MC1 and MC2, whose pH$_{pzc}$ values were 6.63 and 6.12, respectively.

2.2. Fe$_3$O$_4$-Monolithic Resorcinol-Formaldehyde Xerogels with Direct and Indirect Sonications: Effects of Power Output of Ultrasonic Processor, Varying the Molar Ratios of M/R and R/W

Table 3 presents the molar ratios used in the synthesis of five magnetic xerogels, specifically MX3-MX7. These xerogels were prepared with molar ratios of M/R of 0.03, 0.05, 0.1, 0.15, and 0.2, respectively. The xerogels were synthesized using direct ultrasonic-assisted synthesis with an ultrasonic VCX130 operating at 130 watts and a 1/4" diameter probe. The M/R ratios increased with the increasing Fe contents, as determined by chemical composition analysis using ICP-Optical Emission Spectroscopy. However, MX6 and MX7 exhibited similar Fe content values. In this case, it can be explained that the high quantity of magnetite may not have fully incorporated into the gel matrix and some of it may have washed out during the solvent exchange, as evidenced by the observation of a brown solution after changing the acetone solution.

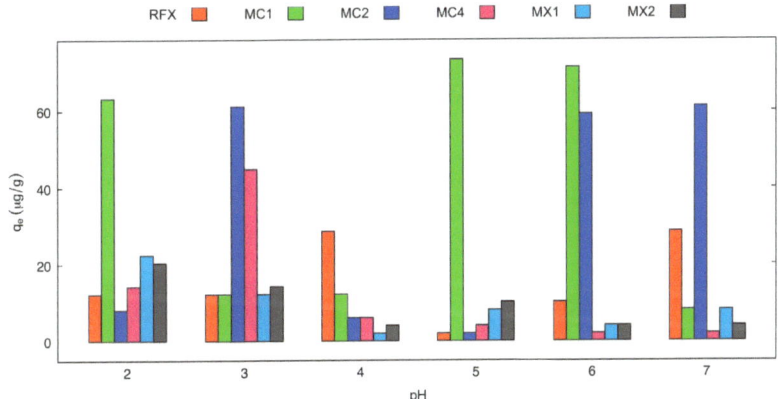

Figure 8. Effect of pH on the adsorption of As(V) using Fe_3O_4-monolithic resorcinol-formaldehyde xerogels (Condition: initial concentration 100 µg/L, dose 1 g/L, 6 h, and temperature 25 °C).

Table 3. Fe_3O_4-Monolithic resorcinol-formaldehyde xerogels prepared with direct sonication with low power output varying M/R ratios ranging from 0.03 to 0.2, resulting in different Fe content.

Molar Ratios	Direct Ultrasonic-Assisted Synthesis				
	MX3	MX4	MX5	MX6	MX7
R/W			0.04		
R/C			200		
R/F			0.5		
M/R	0.03	0.05	0.1	0.15	0.2
Fe content (w%)	3.48	5.62	9.29	13.39	13.13
Solids content (w/v%)	19.59	20.37	22.29	24.23	26.16

MX8-MX11 were synthesized using indirect sonication via the Q700 sonicator, which has a power output of 700 watts and a 1/2″ diameter probe. The molar ratios used in the synthesis, along with the corresponding Fe content, are shown in Table 4.

Table 4. Fe_3O_4-Monolithic resorcinol-formaldehyde xerogels prepared with indirect sonication with high power output varying M/R ratios ranging from 0.03 to 0.15.

Molar Ratios	Indirect Ultrasonic-Assisted Synthesis			
	MX8	MX9	MX10	MX11
R/W		0.05		
R/C		200		
R/F		0.5		
M/R	0.03	0.05	0.1	0.15
Fe content (w%)	3.41	5.82	11.59	16.09
Solids content (w/v%)	23.04	23.95	26.22	28.48

At the same molar ratios of M/R at 0.15 for direct (MX7) and indirect (MX11) sonication, MX11 demonstrated a higher Fe content than MX7. The theoretical calculations of Fe content for MX8, MX9, MX10, and MX11 are 4.27%, 6.85%, 12.52%, and 17.29%, respectively. These values are similar to the results obtained from ICP-OES analysis. This can be explained that increasing the power output to 700 watts makes the system more homogenous.

2.2.1. Characterization of Fe$_3$O$_4$-Monolithic Resorcinol-Formaldehyde Xerogels MX3-MX7 and MX8-MX11

SEM images and mapping analysis of MX3-MX7 are shown in Figure 9, with an increasing amount of magnetite through direct sonification. Figure 9a–e shows the SEM images of the surface morphology of magnetic xerogels composed of large numbers of microclusters with a three-dimensional network. However, some parts of them are agglomerated, and some bright particles can be observed. The elemental distribution of these particles can be confirmed with the corresponding EDX spectra, which demonstrate the existence of iron (Fe), oxygen (O), carbon (C), aluminum (Al), and sodium (Na).

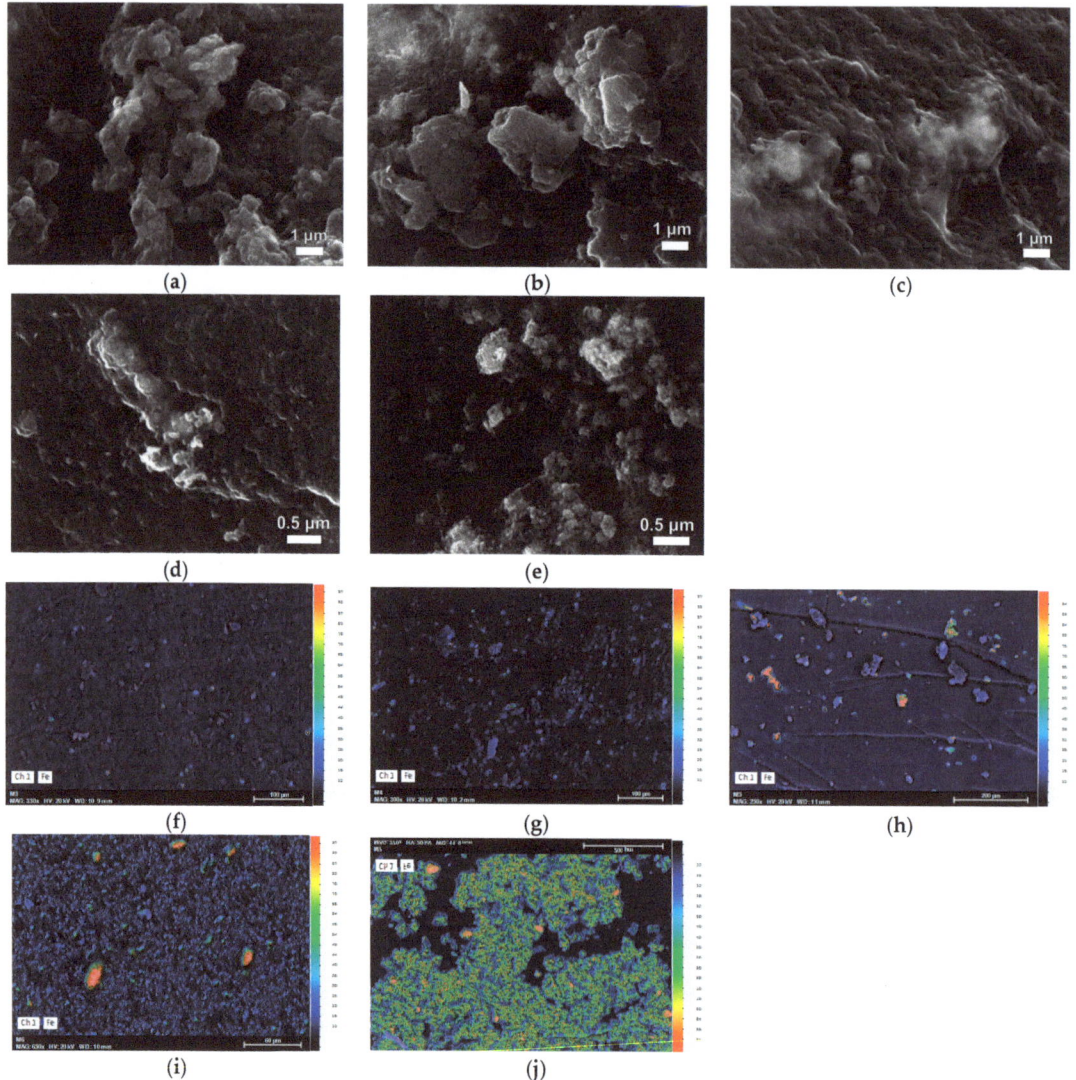

Figure 9. SEM images and EDX mapping analysis of magnetic xerogels prepared with R/C = 200, R/W = 0.04, and varying M/R of MX3 = 0.03 (**a,f**), MX4 = 0.05 (**b,g**), MX5 = 0.1 (**c,h**), MX6 = 0.15 (**d,i**), and MX7 = 0.2 (**e,j**), respectively.

Elemental mapping was analyzed to observe the distribution of synthesized magnetite on RF matrix gels, as shown in Figure 9f–j. The individual EDX mapping of Fe element distributions: blue = low, green = medium, and red = high. Magnetite particles were observed in a blue color and were evenly distributed on the RF surface, with a higher quantity corresponding to the increasing M/R ratios. Similar results were obtained from Table 3. The results of the mapping analysis show that the incorporation of Fe into the structure of the samples is homogeneously distributed, similar to the results obtained from activated carbon xerogels doped with iron (II) phthalocyanine by ultrasonication [63]. However, some of them had some accumulation of Fe due to the increase of high concentration of magnetite in the RF solution. It can be observed in Figure 9h,i, where EDX mappings for Fe display a red color in several regions, indicating a high concentration of Fe within the RF gels. The agglomeration of the microclusters and the presence of bright particles on the surface of the MX5-MX7 xerogels suggest that the synthesis process could be improved. Further studies are needed to optimize the synthesis conditions in order to produce magnetic xerogels with improved properties.

The XRD patterns of MX3-MX7, prepared by direct sonication, and MX8-MX11, prepared by indirect sonication, are shown in Figure 10a,b, respectively. Both sets of samples were synthesized with different molar ratios and utilized different ultrasonic processors. However, both sets varied the M/R ratios from 0.03 to 0.2 for MX3-MX7 and from 0.03 to 0.15 for MX8-MX11. Consequently, the XRD analysis of the magnetic xerogels demonstrated the presence of magnetite, in accordance with the JCPDS card assignments, as described in Figure 1. The diffraction peaks at d311 ($2\theta = 35.68°$) appeared high and sharp for all materials, indicating their magnetic properties [42], the intensity of the iron phase peaks increased with higher M/R ratios in the synthesis. These results are particularly relevant for the analysis of the chemical composition, as presented in Tables 3 and 4.

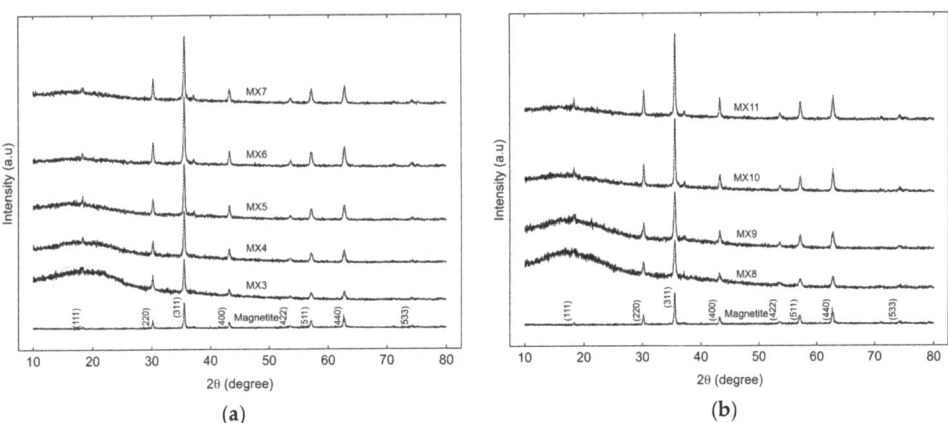

Figure 10. XRD patterns of magnetic xerogels prepared by (**a**) direct (MX3-MX7) and (**b**) indirect (MX8-MX11) ultrasonic with varying M/R ratios of 0.03–0.2.

The FTIR spectra of monolithic resorcinol-formaldehyde xerogels prepared by direct sonication, with varying M/R molar ratios of 0.05, 0.1, 0.15, and 0.2 (referred to as MX4, MX5, MX6, and MX7, respectively) are similar, as shown in Figure 11a. Similarly, Figure 11b presents FTIR spectra of MX8-MX11, prepared by indirect sonication, which exhibit similarities. The resulting FTIR spectrum displays peaks corresponding to different vibrational modes of the molecules in the sample, as discussed in detail in Figure 7. Both groups of materials exhibit an FTIR band at 468 cm^{-1}, attributed to Fe-O stretching vibration [60,61]. Therefore, the use of different sonication methods and power outputs of the ultrasonic processor has no effect on the functional groups and chemical compounds present in the samples of monolithic resorcinol-formaldehyde xerogel, based on the absorption of infrared radiation with wavelength ranges of 1000–400 cm^{-1}.

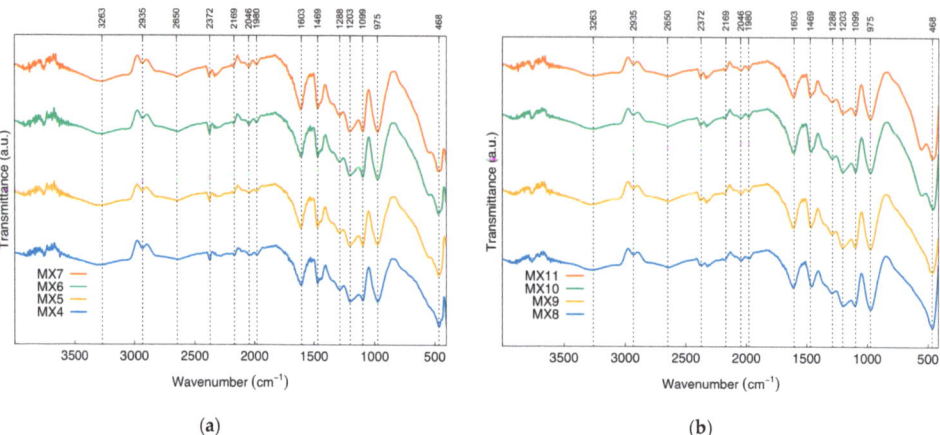

Figure 11. FTIR Spectra of magnetic xerogels prepared by (a) direct (MX4-MX7) and (b) indirect sonication (MX8-MX11).

2.2.2. Performance of Adsorption of Arsenic Using MX4-MX7 and MX8-MX11

Figure 12a presents the removal efficiency of As(V) using MX4-MX7 prepared by direct sonication via a sonicator with 130 watts of power. The removal efficiency of MX4-MX7 was higher than RFX, which was prepared without using magnetite. In particular, MX4 with a lower loading of Fe_3O_4 (M/R = 0.03) gave the highest arsenic removal of 58.78%. Meanwhile, arsenic removals were lower with MX5, then increased and remained constant for MX6 and MX7. This can be explained by the capacity of the sonicator. With a low power output sonication and small diameter tip, it was possible to homogenize the solution well with a low quantity of magnetite. However, with increasing magnetite loading into the RF solution with M/R of 0.05, 0.07, and 0.15, the As removal results were similar. This can be confirmed with SEM/EDX analysis (Figure 9), which showed that magnetite was more homogeneously distributed in MX4 than in the other materials.

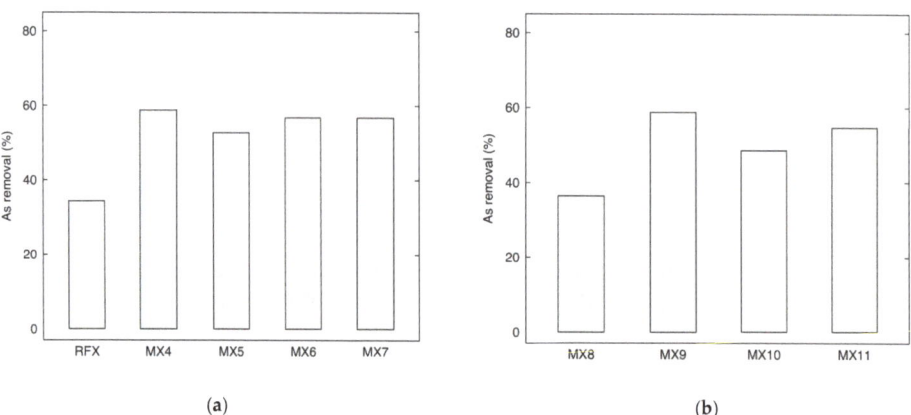

Figure 12. As(V) removal using Fe_3O_4-monolithic resorcinol-formaldehyde xerogels prepared by (a) direct sonication with low power output and (b) indirect sonication with high power output. (Conditions: initial concentration 200 µg/L, pH of 3, dose 2 g/L, 6 h, and temperature 25 °C).

Figure 12b shows the arsenic removal using MX8-MX11 prepared by indirect ultrasonic-assisted synthesis with 700 watts. The removal of MX8 (M/R = 0.03) and MX9 (M/R = 0.05) increased dramatically from 36.49% to 58.78%. With the increasing of M/R to 0.1 and 0.15, their arsenic removal of MX10 and MX11 remained constant, which demonstrated the same behavior as MX4-MX7.

Additionally, the effect of the molar ratio of R/W and M/R on the total solids content of the materials is shown in Tables 3 and 4. The solid content increases with increasing magnetite loading. At the same molar ratios of M/R, the total solids content also increases with increasing R/W. Moreover, low solids contents result in fragile structures, and very high solids contents result in increased densification of the material that lowers porosity. Therefore, the optimum solids content of the xerogel is 20 w/v% [45].

2.3. Fe_3O_4-Monolithic Resorcinol-Formaldehyde Xerogels and Carbon Xerogels by Indirect Sonication

2.3.1. Characterization of Fe_3O_4-Monolithic Resorcinol-Formaldehyde Carbon Xerogels

Some parts of the RF surface of Fe_3O_4-Monolithic resorcinol-formaldehyde xerogels (MXRF) were agglomerated due to the formation of magnetite, as shown in Figure 13a. The presence of Fe in the RF gels was determined to be 14.83 w% by AAS, compared to 24.67% of Fe as quantified by EDX in the solid sample. It can be observed that the morphology and EDAX analysis did not change significantly after the adsorption process (Figure 13b).

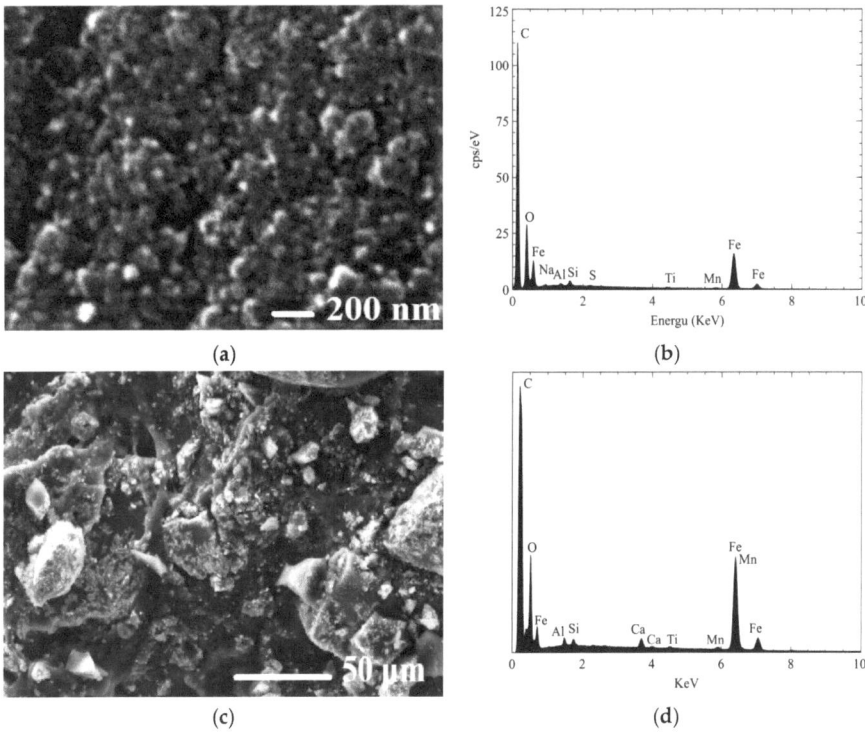

Figure 13. SEM and EDAX images of (**a**,**b**) Fe_3O_4-Monolithic resorcinol-formaldehyde xerogels (MXRF), and (**c**,**d**) Fe_3O_4-Monolithic resorcinol-formaldehyde carbon xerogels (MXRF600).

Figure 14a shows N$_2$ adsorption–desorption isotherms at 77 K, and Figure 4b illustrates the pore size distributions of XRF and MXRF. The analysis results of BET surface area, total pore volume, and average pore size of xerogel adsorbent were 399.19 m^2/g, 0.517 cm^3/g, and 5.228 nm, respectively. When magnetite composites were added to xerogels, the porous properties of MXRF for BET surface area, total pore volume, and average pore diameter were 292 m^2/g, 0.279 cm^3/g, and 3.81 nm, respectively.

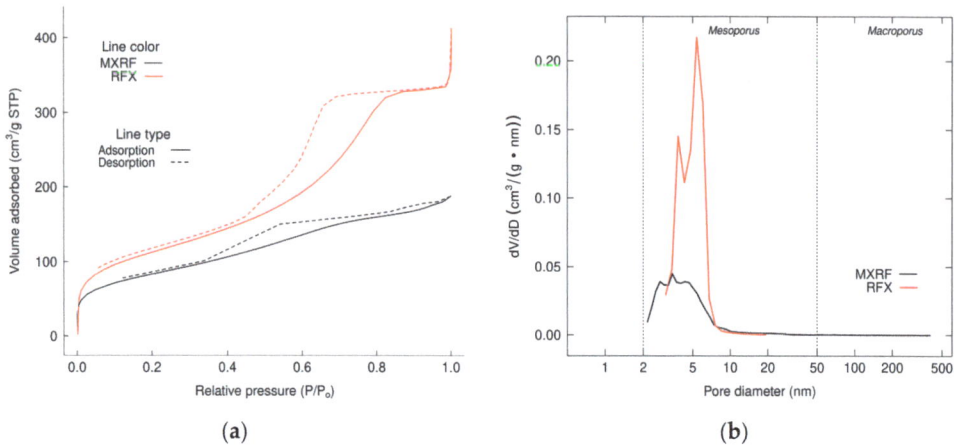

Figure 14. (a) N$_2$ adsorption isotherms and (b) pore size distributions of xerogel (RFX) and Fe$_3$O$_4$-monolithic resorcinol-formaldehyde xerogels (MXRF).

Figure 14a shows that the adsorption isotherms of RFX and MXRF adsorbents at a constant temperature of 77 K with N$_2$ as the adsorptive exhibit a linear relationship between relative pressure and amount adsorbed. RFX and MXRF exhibited type IV adsorption isotherms with H2 and H4 hysteresis loops, respectively. This implies that RFX contained typical mesoporous materials and MXRF contained micro- and mesoporous adsorbents, similar to the results of the pore size distributions.

Figure 14b shows that the main pore diameter sizes of RXF and MXRF are in the range of 2–50 nm, which is defined as mesoporous material. The pore size distribution of MXRF reveals that the average pore diameter was 3.81 nm, which is similar to the results of the narrow centering of PSD of Fe, Co, and Ni doped carbon xerogels [64]. This indicates that the doping with transition metals, such as magnetite, into the xerogels has a similar effect to the composite of magnetite, which affects the reduction of surface areas and total pore volume of the material and makes alterations to their textural properties [65,66]. Similar results were found from SEM analysis, which showed increased agglomeration of particles in RF gels.

As shown in Figure 15, FTIR analysis of MXRF before and after adsorption of As(III) was obtained using attenuated total reflection (ATR) technique. Absorption peaks at 558 cm^{-1} are characteristic peaks of Fe-O-Fe, which are indicative of magnetite, confirming the presence of Fe$_3$O$_4$ on the MXRF adsorbent [67]. The bending vibration of the hydroxyl groups (Fe–OH) confirmed the formation of iron oxide in xerogels [68] and O–H groups on the gel surface. These groups are possible to facilitate the adsorption of arsenic by iron oxides composites in the matrices of RF magnetic xerogels [28].

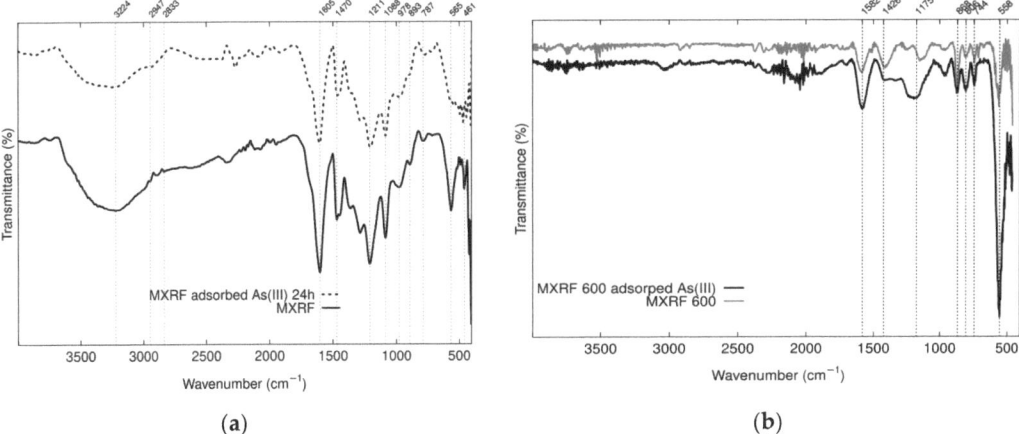

Figure 15. FTIR analysis before and after adsorption of As(III) of (**a**) magnetic xerogels of resorcinol formaldehyde (MXRF) and (**b**) magnetic carbon xerogels of resorcinol formaldehyde (MXRF600).

XRD patterns of MXRF and MXRF600 before and after adsorption (Figure 16) clearly demonstrated that they had high intensity peaks that contained a crystalline phase and corresponded to Fe_3O_4 with the Joint Committee on Powder Diffraction Standards (JCPDS) card No. 19-0629. Therefore, the chemical and structural properties of MXRF and MXRF600 did not change significantly following the carbonization and adsorption process.

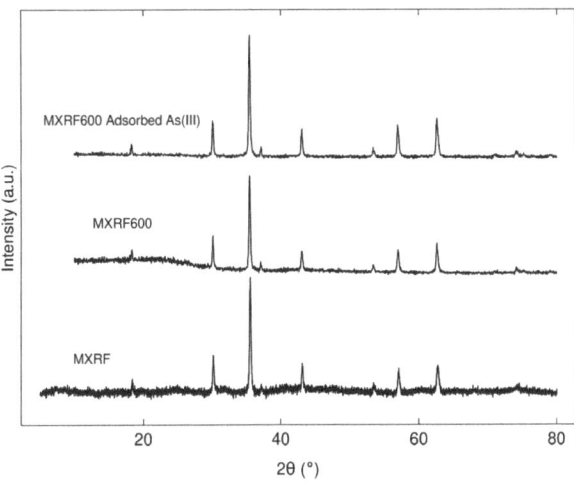

Figure 16. XRD diffractogram of Fe_3O_4-monolithic resorcinol-formaldehyde xerogels (MXRF) and carbon xerogel (MXRF600) before and after adsorption of As(III).

2.3.2. Adsorption of Low and High Concentration of As(III) and As(V) with MXRF and MXRF600

In the adsorption process, contact time is one parameter that is a time-dependent process. Adsorption kinetic studies are important in water treatment. These studies can describe the mechanism of the adsorption process and provide kinetic adsorption constants and valuable information. The experimental data were analyzed with four kinetic models: pseudo first-order, pseudo second-order, Elovich, and Power function.

The effect of contact time on the adsorption process was varied from 10 to 1440 min with different ranges of initial concentration for the low range of As(III) concentrations (25, 50, and 75 µg/L) and high range of concentration for As(III) and As(V) were 514 µg/L and 1034 µg/L, respectively. The adsorption kinetic of As(III) on MXRF is shown in Figure 17. The removal efficiency for As(III) concentration of 75 µg/L increased faster in 10 min and remained constant until 240 min at 97.33%.

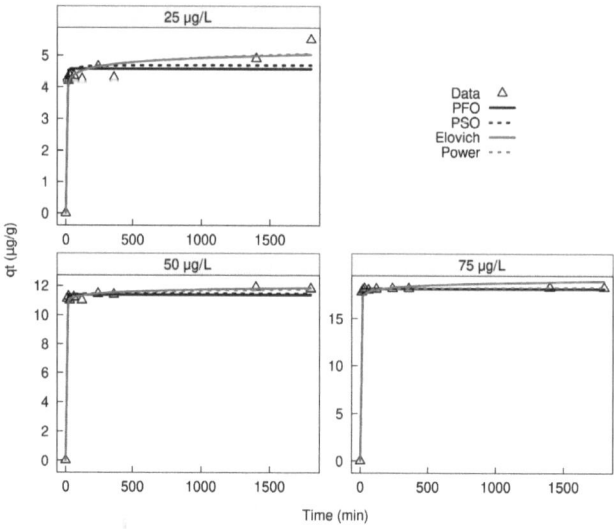

Figure 17. Adsorption kinetics using MXRF with 0.025, 0.05, and 0.075 mg/L of As(III) concentrations, pH of 3, and dosage of 2 g/L.

Kinetic parameters and correlation coefficients for As(III) and As(V) adsorption by using MXRF600 were obtained by nonlinear regression as presented in Table 5, including residual root mean square error (RMSE). The condition of As(III) and As(V) adsorption kinetics were pH of 3, dosage of 2 g/L, and initial concentration of As(III) and As(V) solution of 0.514 mg/L and 1.034 mg/L, respectively. The adsorption kinetic models that presented the best fit in the As(III) and As(V) adsorption process were the Power equation and Elovich chemisorption model.

Table 5. Kinetic parameters and error indices of Pseudo First-Order, Pseudo Second-Order, Elovich, and Power Equation for As(III) and As(V) removal using MXRF600.

Adsorption	Pseudo First-Order				Pseudo Second-Order			
	q_t (µg/g)	k_1	R^2	RMSE	q_t (µg/g)	k_2	R^2	RMSE
As(III)	129.68	0.153	0.374	9.900	134.03	0.002	0.575	8.161
As(V)	230.55	0.147	0.446	17.09	238.00	0.001	0.600	14.51

Adsorption	Elovich Equation				Power Equation			
	α (µg/g min)	β (g/µg)	R^2	RMSE	a	b	R^2	RMSE
As(III)	808,122.16	0.129	0.830	5.166	93.36	0.062	0.842	4.979
As(V)	2,288,305.88	0.075	0.807	10.07	166.34	0.061	0.822	9.681

It can be observed that MXRF600 demonstrated greater adsorption of As(III) and As(V) than MXRF, implying a higher adsorption capacity. The final step of preparing MXRF600 was to produce a carbon xerogel with a carbonization process for removing the rest of the oxygen and hydrogen groups and improving a thermally stable nanostructure [49]. With the use of high temperature under an inert atmosphere, MXRF and MXRF600 demonstrated modifications in their chemical composition and texture properties, which can be identified with the analysis of XRD, FTIR, N_2 physisorption, and SEM/EDAX, as discussed above.

In this study, the experimental data were analyzed with nonlinear equations using the Langmuir and Freundlich isotherm models to describe the adsorption of As(III) and As(V) on MXRF600. The Langmuir isotherm model assumes that a monomolecular layer of adsorbate molecules is formed on the adsorbent surface, with each molecule having the same adsorption energy. The Freundlich isotherm model describes the heterogeneity of the surface and the distribution of adsorption energies.

The conditions for the isotherm adsorption were as follows: adsorbent dose of 2 g/L, initial solution pH of 3.0, and contact time of 24 h. The initial concentrations of the As(III) and As(V) solutions were in the range of 0.05–1.27 mg/L and 0.12–3.0 mg/L, respectively. The Langmuir and Freundlich model parameters and regression coefficients are shown in Table 6. The experimental data for the adsorption of As(III) and As(V) on magnetic carbon xerogel monoliths were fitted to the Langmuir models, and the maximum monolayer adsorption capacity (q_{max}) of As(III) and As(V) were 694.3 and 1720.3 µg/g, respectively, with R^2 values (RSME) of As(III) and As(V) were 0.897 (3.865), and 0.901(9.220), respectively.

Table 6. Isotherm parameters and correlation coefficients for As(III) and As(V) adsorption on MXRF600.

Adsorption	Langmuir				Freundlich			
	q_{max} (µg/g)	K_L (L/µg)	R^2	RMSE	K_F ((µg/g)(L/µg)$^{1/n}$)	n	R^2	RMSE
As(III)	694.3	1.527	0.897	3.865	502.8	1.346	0.894	3.903
As(V)	1720.3	0.641	0.901	9.220	655.7	1.338	0.899	9.309

3. Conclusions

The ultrasonic-assisted synthesis of Fe_3O_4-monolithic resorcinol-formaldehyde xerogels using direct and indirect sonication methods as an easier recovery of adsorbent was shown to reduce the gelation time and improve the textural properties of the final product. The optimal mixing time for magnetite dispersion in an RF aqueous solution was determined to be 5 min using direct sonication and 60 min using indirect sonication, as confirmed by SEM/EDX analysis. This study investigated the effect of different molar ratios of R/C, M/R, R/W, and thermal treatment on RF xerogel. The results show MXRF600 was synthesized by indirect sonication with R/F = 0.5, R/C = 100, R/W = 0.05, and M/R = 0.15 and enhanced adsorption capacity for As(III) and As(V) from groundwater due to the influence of sonication assistance and the carbonization process. However, the optimization of the process parameters for the adsorption of magnetic carbon xerogels should be studied to find out the optimum condition and improve their performance in removing contaminants from the environment. The desorption process, regeneration efficiency, and the lifecycle assessment of magnetic carbon xerogels are suggested for future research.

4. Materials and Methods

4.1. Reagents and Materials

Reagents required to perform the synthesis of Fe_3O_4 nanoparticles were prepared in duplicate, including ferric chloride hexahydrate ($FeCl_3 \cdot 6H_2O$, 98.9%, Fermont), ferrous sulfate heptahydrate ($FeSO_4 \cdot 7H_2O$, 99%, Meyer), and sodium hydroxide (NaOH, 97%, Meyer. Nitrogen gas was purchased from Infra (Morelos, Mexico). Resorcinol ($C_6H_4(OH)_2$, 98%, Chemistry Meyer), sodium carbonate (Na_2CO_3, J.T. Baker, 100%), formaldehyde (HCHO, 37% methanol stabilized Solution, J.T. Baker), acetone (($CH_3)_2CO$, 99.5%, J.T. Baker,), and magnetite Fe_3O_4 (Lanxess, Bayferrox) were used for synthesis of magnetic xerogels. All the solutions used in the synthesis and adsorption experiment were made using ultrapure Type I water from the water purification system (WaterproBT, Labconco, Kansas City, MO, USA).

4.2. Synthesis of Adsorbent Materials

The gels were synthesized by polymerizing resorcinol (R, $C_6H_6O_2$) and formaldehyde (F, CH_2O) in water (W), using sodium carbonate (C, Na_2CO_3) as a catalyst, following the procedure described by [69]. The synthesis utilized molar proportions of R/C = 200, R/F = 0.5, and R/W = 0.06 [70,71]. While keeping other factors constant, the effect of loading of Fe (via direct and indirect sonication), Fe content (M/R = 0.03–0.2), water (R/W = 0.04–0.06), and catalyst (R/C = 50–200) ratios were varied in the realization of the monoliths. Then, they were evaluated for their impact on the physicochemical properties of the resulting materials, as well as their ability to remove As(III) and As(V). Iron oxides (M) used in this study were magnetite obtained from Lanxess, García, Nuevo León, Mexico.

The procedure for synthesizing the gels involved placing half of the deionized water and the mass of R in a 100 mL beaker, which was then vigorously shaken to homogenize the solution. The F solution was added, followed by the addition of C, and the mixture was stirred magnetically until homogeneous. pH of RF solution was controlled between 5.5–6.0 to obtain high surface areas of resulting materials [52]. The resulting solution was then placed in Pyrex® glass tubes, which were sealed with a stopper to prevent evaporation. The temperature of the RF solution during reaction of an ultrasonic processor was controlled to be in the range of 80 to 85 °C. To evaluate the optimal mixing time and the dispersion of magnetite in the RF aqueous solution, three types of ultrasonic devices were applied. First, a digital ultrasonic device (UP400St; Hielscher, Teltow, Germany) with an output of 400 watts, a frequency of 24 kHz, and a 1-inch diameter probe was used in the synthesis of RFX, MC1-MC4, MX1-MX2, and MXRF. The device was equipped with automatic frequency tuning and adjusting an amplitude ranging from 80% to 100%. An ultrasonic processor (VCX 130; Sonics & Materials, Inc., Newton, CT, USA) with a power output of 130 watts, a frequency of 20 kHz, and a $\frac{1}{4}$-inch diameter tip was applied in the synthesis of MX3-MX7. A sonicator (Q700; Qsonica L.L.C, Newtown, CT, USA) with a power rating of 700 watts, a frequency of 20 kHz, and a 1/2-inch diameter probe was used in the synthesis of MX8-MX11. All ultrasonic processors were used for homogenization, dispersal, and deagglomeration of magnetite particles in the RF aqueous solution, using both direct and indirect sonication methods before the gelation process.

4.2.1. Monolithic Resorcinol-Formaldehyde Xerogels Effect of Loading of Magnetite with Direct and Indirect Sonication, and Modification of Catalyst

The study investigated the optimal mixing time and dispersion of magnetite in RF aqueous solution, using both direct and indirect ultrasonication methods prior to the gelation process. Magnetic xerogel monoliths (MCs) were prepared by indirect sonication with molar ratios of R/F = 0.5, R/W = 0.06, and M/R = 0.01, and varying proportions of resorcinol and catalyst. MC1, MC2, MC3, and MC4 were identified based on R/C ratios of 50, 100, 150, and 200, respectively. The homogenization process was carried out using ultrasonic-assisted synthesis, with digital ultrasonic equipment (UP400St; Hielscher, Teltow, Germany), starting at room temperature. After 5 min of sonication, the temperature reached 85 °C. Magnetite (M) was added into the homogeneous RF aqueous solution and subjected to indirect sonication for 60 min to disperse the magnetite particles before the gelation process. Additionally, the variable factors studied in this work include loading of magnetite with direct and indirect ultrasonication. Therefore, MX1 was prepared using the same method as MC4 but with direct sonication to compare their properties and adsorption capacity of arsenic in aqueous solution. Afterward, the materials were placed in the oven at 80 °C for 5 days. In the case of MX2, the gelation and curing process was changed to be left at room temperature for 5 days.

4.2.2. Monolithic Resorcinol-Formaldehyde Xerogels with Direct and Indirect Sonication: Effects of Power Output of Ultrasonic Processor, Varying the Molar Ratios of M/R and R/W

In this study, monoliths of magnetic xerogels (MXs) were prepared by the sol-gel polymerization of resorcinol with formaldehyde, using an alkaline catalyst and direct sonication of magnetite to incorporate them into the xerogels. Different proportions of iron oxides were modified to achieve the maximum adsorption capacity. Initially, batches of magnetic xerogel monoliths (MX3-MX7) were prepared by varying the M/R ratio from 0.03 to 0.2. The molar ratios of R/F = 0.5, R/C = 200, and R/W = 0.04 were maintained for a small portion of the batches. The preparation process involved the use of an ultrasonic processor VCX 130 with a power output of 130 watts and a frequency of 20 kHz.

4.2.3. Monolithic Resorcinol-Formaldehyde Carbon Xerogels by Indirect Sonication

The monolithic resorcinol-formaldehyde xerogels (MXRF) were synthesized in a larger batch using UP400St equipment with the relations of molar ratio of R/F = 0.5, R/C = 100, R/W = 0.05, and M/R = 0.15. Then, the gels were cured in a conventional oven for three days at 80 °C. The gels were taken off the glass tubes and allowed to cool to room temperature. After that, the gels were cut using a diamond disk into pellet forms of 5 mm in diameter. The materials were then exchanged with acetone, sealed in a jar with the lid tightly closed, and wrapped with paraffin film. The jar was placed in a shaking water bath (BS-11; Lab Companion, Daejeon, Republic of Korea) at 150 rpm for two days, with fresh acetone being added daily. Subsequently, the gels were dried for three days in a conventional oven at 80 °C.

MXRF were then pyrolyzed using a tube furnace (STF55346C-1; Lindberg/Blue M, Asheville, NC, USA) with the following conditions: temperature of 600 °C, heating ramp of 3 °C/min, time of 6 h, and nitrogen flow of 100 mL/min. The resulting product was monolithic resorcinol-formaldehyde carbon xerogels, which were labelled as MXRF600.

4.3. Characterization through Analytical Techniques

To assess the physicochemical characteristics of the synthesized materials, the following techniques were employed:

X-ray diffraction (XRD) analysis was used to identify the main constituents and mineralogical phases of the synthesized materials. The analysis was performed using an X-ray diffractometer on MCs, MXs, and MX3-MX11 samples (XPert PW3040; Philips, Almelo, The Netherlands), and on MXRF and MXRF600 (D8 ADVANCE; Bruker, Karisruhe, Germany). Sample preparation involved sieving the sample through a 200-mesh sieve, resulting in an average particle size of 74 μm. A high-temperature chamber attached to the X-ray diffractometer was used to measure diffraction patterns up to 900 °C. Cu(Kα) radiation was applied in a 2θ range from 10° to 80°.

Fourier transform infrared spectroscopy (FTIR) was employed to investigate the surface functional groups of the adsorbents before and after arsenic adsorption, in order to understand the mechanism of ion adsorption. FTIR analysis was conducted using a Shimadzu IRAffinity-1S instrument (Shimadzu Corp., Kyoto, Japan) on dry powder samples. Infrared spectra were measured by connecting to the attenuated total reflection (ATR) contained in the disk of crystal (type IIIa monocrystalline diamond). Before the analysis, the samples were sieved through a standard test sieve No. 142 to obtain a uniform particle size of 106 μm. Subsequently, the powder samples were dried in an oven at 60 °C for 15 h under dry air to avoid interference from water vapor adsorption in the infrared region, which could affect the analysis result. After installing the ATR with infrared spectroscopy, the solid samples were directly added to the crystal plate and pressed for surface analysis. All spectra were recorded between the wavenumbers of 400–4000 cm^{-1}, with 45 scans per sample.

The surface morphology, pore structure, and element analysis of the magnetic xerogels were analyzed using a scanning electron microscope (SEM). MCs, MXs, MXRF, and

MXRF600 were analyzed using a field emission scanning electron microscope (FE-SEM) (7800F Prime; JEOL, Tokyo, Japan) after gold coating. MX3-MX11 were analyzed using a scanning electron microscope (SEM) (JSM IT300; JEOL, Tokyo, Japan). The samples were coated with graphite before the analysis. The acceleration voltages used were between 5 and 20 kV. The textural properties of magnetic xerogels, and magnetic carbon xerogels were characterized by physical adsorption of N_2 at 77 K, using physisorption apparatus (ASAP 2020; Micromeritics, Norcross, GA, USA and NOVA touch 2LX; Quantachrome Instruments, Boynton Beach, FL, USA). The samples were dried at 110 °C for 15 h prior to N_2 physisorption analysis.

Particle Size Distribution (PSD) was determined using Dynamic Light Scattering Analyzers (PMX 500; Microtrac, Meerbuch, Germany), and the data report was generated by FLEX software version 11.1.0.1

The determination of the point of zero charge (pH_{pzc}) and the isoelectric point (IEP) was conducted following the methods described by [72]. The pH solutions were prepared by adjusting deionized water to pH values of 2, 4, 6, 8, 10, and 12 using 0.1 M HCl or 0.1 M NaOH solutions. The pH was measured using a multi-parameter device (Orion Star A211; Thermo Scientific, Beverly, MA, USA). A zeta potential analyzer (PMX 500; Microtrac, Meerbuch, Germany) was employed to measure the zeta potential, with pH variations ranging from 2 to 11 for the determination of the isoelectric point (IEP).

The amount of Fe in the magnetic xerogel monoliths was determined by Inductively Coupled Plasma (ICP) Optical Emission Spectrometer (OES) (Optima 8300; Perkin Elmer, Shelton, CT, USA).

4.4. Batch Adsorption Experiment

Groundwater used in experimental study was obtained from a well approximately 70 m deep located at Jiutepec, Morelos Mexico. Physical and chemical characteristics of groundwater sample used in this study were analyzed. pH (7.6), total dissolved solids (TDS, 172.6 mg/L), turbidity (1.53 NTU), chlorides (Cl^-, 10.1 mg/L), iron (Fe, 0.03 mg/L), fluoride (F^-, 0.25 mg/L), manganese (Mn, 0.001 mg/L), nitrate (NO_3^-, 4.4 mg/L), sulphate (SO_4^{2-}, 37 mg/L), and phosphate (PO_4^{3-}, 0.82 mg/L) were all in the limitation of Mexican stand NOM-127-SSA1-2021 [4]. Since there was no arsenic in the selected water, arsenic was added to the stock solution prepared for adsorption tests on synthetic samples. This water was used to prepare the corresponding arsenic solution to the required concentrations by adding sodium arsenite ($NaAsO_2$, Sigma-Aldrich) and sodium arsenate dibasic heptahydrate ($HAsNa_2O_4 \cdot 7H_2O$, Sigma-Aldrich) for studying As(III) and As(V) adsorption processes, respectively.

The batch adsorption experiment of As(V) using MCs and MXs as adsorbents was conducted to evaluate their adsorption capacities. The effect of solution pH (2–7) on As(V) adsorption was investigated with an initial concentration of 100 µg/L, a dose of 1 g/L, 150 rpm, a contact time of 6 h, and a temperature of 26.2 ± 1 °C.

Batch adsorption of As(V) using MX4-MX7 and MX8-MX11 was studied with direct sonication at low power output, and indirect sonication at high power output, respectively. The following conditions were used: an initial concentration of As(V) of 200 µg/L, pH of 3, a dose of 2 g/L, 150 rpm, a contact time of 6 h, and a temperature of 26.3 ± 1 °C.

XRF600 was carbonized into pellets and used in this form to test kinetics and isotherms. The kinetic study adsorption using MXRF with As(III) concentrations of 0.025, 0.05, and 0.075 mg/L was conducted at a pH of 3, a dosage of 2 g/L, 150 rpm, a temperature of 26.5 ± 1 °C, and contact time ranging from 10 to 1800 min. The adsorption kinetics of As(III) and As(V) using MXRF600 were carried out under a pH of 3, a dosage of 2 g/L, and initial concentration of As(III) and As(V) solution of 0.514 mg/L and 1.034 mg/L, respectively, with a contact time ranging from 10 to 1440 min.

The conditions for the isotherm adsorption using MXRF600 were as follows: an adsorbent dose of 2 g/L, an initial solution pH of 3.0, 150 rpm, a temperature of 26.4 ± 1 °C, and a contact time of 24 h. The initial concentrations of the As(III) and As(V) solutions were in the range of 0.05 to 1.27 mg/L and 0.12 to 3.0 mg/L, respectively.

The importance of kinetic and equilibrium models of adsorption is described in the mechanisms and dynamics of the adsorption system of adsorbents. Adsorption kinetic models that control the adsorption process of arsenic are related to the adsorbate uptake on the adsorbent with chemisorption. Therefore, the Pseudo First-Order (PFO), Pseudo Second-Order (PSO), and Elovich and Power equations were applied to perform the experimental data in this study. The assumptions of the PFO model are: sorption at localized sites, the energy of adsorption is independent of surface coverage, a saturated monolayer of adsorbates, and the concentration of the adsorbate is constant [73]. The assumptions of the PSO model are similar to those of the PFO model. The PSO kinetic equation typically describes metal ion uptake on activated carbons well, as well as the adsorption of dyes, herbicides, oils, and organic compounds from aqueous solutions [73,74]. The Elovich equation is used to describe the kinetics of a heterogeneous diffusion process [74]. It is a semi-empirical equation that is based on the assumption that the rate of diffusion is controlled by the rate of adsorption onto active sites on the heterogeneity of the surface of the adsorbent.

The Langmuir and Freundlich isotherm models are the most commonly used equilibrium models for determining the relative concentrations of the solute adsorbed onto the solid in the solution [75]. The Langmuir isotherm assumes that a solute is adsorbed onto a homogeneous surface with a finite number of similar active sites, forming a monolayer. The Freundlich isotherm is an empirical model that describes multilayer adsorption.

The equations for kinetic and equilibrium models of adsorption used in this study are listed in Table 7.

Table 7. Equation of kinetic and isotherm models of adsorption.

Kinetic Models	Non-Linear Equations	References
Pseudo First-Order	$q_t = q_e(1 - exp(-k_1 t))$	[74,75]
Pseudo Second-Order	$q_t = \frac{q_e^2 k_2 t}{1 + k_2 q_e t}$	[74,75]
Elovich Equation	$q_t = \frac{1}{\beta} ln(1 + \alpha\beta t)$	[76]
Power Equation	$q_t = at^b$	[76]
Isotherm Models	**Non-Linear Equations**	**References**
Langmuir	$q_e = \frac{K_L q_m C_e}{(1 + K_L C_e)}$	[74]
Freundlich	$q_e = K_F C_e^{\frac{1}{n}}$	[74]

q_t and q_e are the amount of adsorbate adsorbed at time t (mg/g) and the equilibrium adsorption capacity (mg/g), respectively. k_1 is the PFO rate constant (min^{-1}), and k_2 is the PSO rate constant (min^{-1}), respectively. t is the contact time (min). α is the initial adsorption rate (mg/g min), β is related to surface coverage (g/mg), and a and b are constants. C_e is the equilibrium concentration of adsorbate in solution (mg/L). q_m is the maximum adsorption capacity (mg/g). K_L is the Langmuir constant that is related to the adsorption energy (L/mg). K_F and n are Freundlich constants that measure the adsorption capacity ((mg/g)(L/mg)1/n) and intensity, respectively.

The arsenic adsorption process was carried out in a batch reactor system. The effect of contact time and initial concentration of arsenic adsorption was investigated on MXRF and MXRF600. Different kinetic and isotherm adsorption models were analyzed using nonlinear regression analysis with the statistical software R v3.5.

4.5. Determination of As(III) and As(V)

The determination of arsenic species was performed using hydride generation atomic absorption spectroscopy (HG-AAS) (Varian; SpectrAA220, Mulgrave, VIC, Australia). To analyze As (III) at trace concentrations, AAS must be combined with the hydride generation (HG) technique with citric-citrate buffer [77].

A total of 10 mL of the samples were adjusted to pH 2. In the case of samples of As(V), 1 mL of HCl and 1 mL of potassium iodide and ascorbic acid were added. This was conducted in order to reduce As(V) to As(III). On the other hand, the arsenic ($NaAsO_2$) calibration curve was prepared. First, a solution was prepared with 1 mL of arsenic standard and 1 mL of HNO_3. This solution was then diluted to a concentration of 1 mg/L. From this, the solutions of 0.001, 0.002, 0.004, 0.006, and 0.0075 mg/L were made in 100 mL flasks. A small amount of deionized water was added to each flask, along with 0.1, 0.2, 0.4, 0.6, and 0.75 mL of stock solution, 6 mL of HNO_3, 10 mL of KI, and 4 mL of HCl. Then, the solutions were calibrated with deionized water. Finally, the solutions were analyzed in a HG-AAS at a wavelength of 193.7 nm.

Author Contributions: Conceptualization, S.E.G.-H. and S.K.; methodology, S.K.; software, A.R.-R.; formal analysis, S.K. and A.M.L.-M.; investigation, S.K. and D.-E.P.-C.; resources, S.E.G.-H. and P.G.-M.; writing—original draft preparation, S.K.; writing—review and editing, S.E.G.-H., P.G.-M. and A.R.-R.; visualization, A.R.-R.; supervision, S.E.G.-H. and P.G.-M. All authors have read and agreed to the published version of the manuscript.

Funding: This research was funded by Mexican Institute of Water Technology, grant number DP2101.1.

Institutional Review Board Statement: Not applicable.

Informed Consent Statement: Not applicable.

Data Availability Statement: Not applicable.

Acknowledgments: Khamkure S. acknowledges the financial support of the National Council of Humanities Science and Technology (CONAHCYT), Mexico, under the Catedras-CONAHCYT program, project number 159. The authors would like to thank: Manuel Sanchez-Zarza (Mexican Institute of Water Technology) and Jazmin A. López Díaz (Autonomous University of Guerrero) for providing technical analysis; José Martín Baas-López (Yucatan Scientific Research Center) for operating equipment; Benjamín Campos Pacheco for providing equipment and helping in the characterization of materials; Eulalio Rodriguez-Jacobo and Arael Torrecilla-Valle for their help and support in the laboratory.

Conflicts of Interest: The authors declare no conflict of interest.

References

1. Agency for Toxic Substances and Disease Registry (ATSDR). Toxicological Profile for Arsenic. *Atlanta* **2007**, 1–559.
2. Environmental Protection Agency (EPA) National Primary Drinking Water Regulations: Long Term 1 Enhanced Surface Water Treatment Rule. *Fed. Regist.* **2002**, *75*, 1812–1844.
3. WHO. *Guidelines for Drinking-Water Quality: Fourth Edition Incorporating the First and Second Addenda*; WHO: Geneva, Switzerland, 2022; ISBN 978-92-4-004506-4.
4. Secretary of Health (SSA). *Official Mexican Standards NOM-127-SSA1-2021, Environmental Health, Water for Human Use and Consumption—Permissible Limits of Quality and Treatments to Which Water Must Be Submitted for Its Drinkability*; Official Daily of the Federation: Mexico City, Mexico, 2021; pp. 1–121.
5. Pal, P.; Sen, M.; Manna, A.; Pal, J.; Pal, P.; Roy, S.; Roy, P. Contamination of groundwater by arsenic: A review of occurrence, causes, impacts, remedies and membrane-based purification. *J. Integr. Environ. Sci.* **2009**, *6*, 295–316. [CrossRef]
6. Singh, R.; Singh, S.; Parihar, P.; Singh, V.P.; Prasad, S.M. Arsenic contamination, consequences and remediation techniques: A review. *Ecotoxicol. Environ. Saf.* **2015**, *112*, 247–270. [CrossRef] [PubMed]
7. Shankar, S.; Shanker, U. Shikha Arsenic contamination of groundwater: A review of sources, prevalence, health risks, and strategies for mitigation. *Sci. World J.* **2014**, *2014*, 304524. [CrossRef]
8. Limón-Pacheco, J.H.; Jiménez-Córdova, M.I.; Cárdenas-González, M.; Sánchez Retana, I.M.; Gonsebatt, M.E.; Del Razo, L.M. Potential co-exposure to arsenic and fluoride and biomonitoring equivalents for Mexican children. *Ann. Glob. Health* **2018**, *84*, 257–273. [CrossRef]
9. González-Horta, C.; Ballinas-Casarrubias, L.; Sánchez-Ramírez, B.; Ishida, M.C.; Barrera-Hernández, A.; Gutiérrez-Torres, D.; Zacarias, O.L.; Jesse Saunders, R.; Drobná, Z.; Mendez, M.A.; et al. A concurrent exposure to arsenic and fluoride from drinking water in Chihuahua, Mexico. *Int. J. Environ. Res. Public Health* **2015**, *12*, 4587–4601. [CrossRef]
10. Rodríguez, R.; Morales-Arredondo, I.; Rodríguez, I. Geological differentiation of groundwater threshold concentrations of arsenic, vanadium and fluorine in El Bajio Guanajuatense, Mexico. *Geofis. Int.* **2016**, *55*, 5–15. [CrossRef]

11. Vega-Millán, C.B.; Dévora-Figueroa, A.G.; Burgess, J.L.; Beamer, P.I.; Furlong, M.; Lantz, R.C.; Meza-Figueroa, D.; O'Rourke, M.K.; García-Rico, L.; Meza-Escalante, E.R.; et al. Inflammation biomarkers associated with arsenic exposure by drinking water and respiratory outcomes in indigenous children from three Yaqui villages in southern Sonora, México. *Environ. Sci. Pollut. Res.* **2021**, *28*, 34355–34366. [CrossRef]
12. Martínez-Acuña, M.I.; Mercado-Reyes, M.; Alegría-Torres, J.A.; Mejía-Saavedra, J.J. Preliminary human health risk assessment of arsenic and fluoride in tap water from Zacatecas, México. *Environ. Monit. Assess.* **2016**, *188*, 476. [CrossRef]
13. Huang, L.; Wu, H.; Van Der Kuijp, T.J. The health effects of exposure to arsenic-contaminated drinking water: A review by global geographical distribution. *Int. J. Environ. Health Res.* **2015**, *25*, 432–452. [CrossRef] [PubMed]
14. Monrad, M.; Ersbøll, A.K.; Sørensen, M.; Baastrup, R.; Hansen, B.; Gammelmark, A.; Tjønneland, A.; Overvad, K.; Raaschou-Nielsen, O. Low-level arsenic in drinking water and risk of incident myocardial infarction: A cohort study. *Environ. Res.* **2017**, *154*, 318–324. [CrossRef] [PubMed]
15. Gamboa-Loira, B.; Cebrián, M.E.; López-Carrillo, L. Arsenic exposure in northern Mexican women. *Salud Publica Mex.* **2020**, *62*, 262–269. [CrossRef] [PubMed]
16. Shih, Y.-H.; Scannell Bryan, M.; Argos, M. Association between prenatal arsenic exposure, birth outcomes, and pregnancy complications: An observational study within the National Children's Study cohort. *Environ. Res.* **2020**, *183*, 109182. [CrossRef]
17. Chen, H.; Zhang, H.; Wang, X.; Wu, Y.; Zhang, Y.; Chen, S.; Zhang, W.; Sun, X.; Zheng, T.; Xia, W.; et al. Prenatal arsenic exposure, arsenic metabolism and neurocognitive development of 2-year-old children in low-arsenic areas. *Environ. Int.* **2023**, *174*, 107918. [CrossRef]
18. Bibi, S.; Kamran, M.A.; Sultana, J.; Farooqi, A. Occurrence and methods to remove arsenic and fluoride contamination in water. *Environ. Chem. Lett.* **2017**, *15*, 125–149. [CrossRef]
19. Ghosh, S.; Debsarkar, A.; Dutta, A. Technology alternatives for decontamination of arsenic-rich groundwater—A critical review. *Environ. Technol. Innov.* **2019**, *13*, 277–303. [CrossRef]
20. Amiri, S.; Vatanpour, V.; He, T. Optimization of Coagulation-Flocculation Process in Efficient Arsenic Removal from Highly Contaminated Groundwater by Response Surface Methodology. *Molecules* **2022**, *27*, 7953. [CrossRef]
21. Rathi, B.S.; Kumar, P.S. A review on sources, identification and treatment strategies for the removal of toxic Arsenic from water system. *J. Hazard. Mater.* **2021**, *418*, 126299. [CrossRef]
22. Abdellaoui, Y.; El Ibrahimi, B.; Abou Oualid, H.; Kassab, Z.; Quintal-Franco, C.; Giácoman-Vallejos, G.; Gamero-Melo, P. Iron-zirconium microwave-assisted modification of small-pore zeolite W and its alginate composites for enhanced aqueous removal of As(V) ions: Experimental and theoretical studies. *Chem. Eng. J.* **2021**, *421*, 129909. [CrossRef]
23. Wang, H.; Qi, X.; Yan, G.; Shi, J. Copper-doped ZIF-8 nanomaterials as an adsorbent for the efficient removal of As(V) from wastewater. *J. Phys. Chem. Solids* **2023**, *179*, 111408. [CrossRef]
24. Lingamdinne, L.P.; Choi, J.S.; Choi, Y.L.; Chang, Y.Y.; Koduru, J.R. Stable and recyclable lanthanum hydroxide–doped graphene oxide biopolymer foam for superior aqueous arsenate removal: Insight mechanisms, batch, and column studies. *Chemosphere* **2023**, *313*, 137615. [CrossRef] [PubMed]
25. Fang, Z.; Li, Y.; Huang, C.; Liu, Q. Amine functionalization of iron-based metal-organic frameworks MIL-101 for removal of arsenic species: Enhanced adsorption and mechanisms. *J. Environ. Chem. Eng.* **2023**, *11*, 110155. [CrossRef]
26. López-Martínez, A.M.; Khamkure, S.; Gamero-Melo, P. Bifunctional Adsorbents Based on Jarosites for Removal of Inorganic Micropollutants from Water. *Separations* **2023**, *10*, 309. [CrossRef]
27. Nguyen, T.H.; Nguyen, T.V.; Vigneswaran, S.; Ha, N.T.H.; Ratnaweera, H. A Review of Theoretical Knowledge and Practical Applications of Iron-Based Adsorbents for Removing Arsenic from Water. *Minerals* **2023**, *13*, 741. [CrossRef]
28. Hao, L.; Liu, M.; Wang, N.; Li, G. A critical review on arsenic removal from water using iron-based adsorbents. *RSC Adv.* **2018**, *8*, 39545–39560. [CrossRef] [PubMed]
29. Mînzatu, V.; Davidescu, C.M.; Negrea, P.; Ciopec, M.; Muntean, C.; Hulka, I.; Paul, C.; Negrea, A.; Duțeanu, N. Synthesis, characterization and adsorptive performances of a composite material based on carbon and iron oxide particles. *Int. J. Mol. Sci.* **2019**, *20*, 1609. [CrossRef] [PubMed]
30. Fierro, V.; Celzard, A.; Fierro, V.; Szczurek, A.; Braghiroli, F.; Parmentier, J.; Pizzi, A.; Suhfxuvruv, S.; Fduerq, R.I.; Dqg, D.; et al. Carbon gels derived from natural resources Geles de carbón de origen natural. 2012, 1–7. Available online: https://digital.csic.es/handle/10261/81796 (accessed on 19 April 2023).
31. Huang, G.; Li, W.; Song, Y. Preparation of SiO_2–ZrO_2 xerogel and its application for the removal of organic dye. *J. Sol-Gel Sci. Technol.* **2018**, *86*, 175–186. [CrossRef]
32. Vivo-Vilches, J.F.; Pérez-Cadenas, A.F.; Maldonado-Hódar, F.J.; Carrasco-Marín, F.; Regufe, M.J.; Ribeiro, A.M.; Ferreira, A.F.P.; Rodrigues, A.E. Resorcinol–formaldehyde carbon xerogel as selective adsorbent of carbon dioxide present on biogas. *Adsorption* **2018**, *24*, 169–177. [CrossRef]
33. Wickenheisser, M.; Herbst, A.; Tannert, R.; Milow, B.; Janiak, C. Hierarchical MOF-xerogel monolith composites from embedding MIL-100(Fe, Cr) and MIL-101(Cr) in resorcinol-formaldehyde xerogels for water adsorption applications. *Microporous Mesoporous Mater.* **2015**, *215*, 143–153. [CrossRef]
34. Kumar, A.; Prasad, S.; Saxena, P.N.; Ansari, N.G.; Patel, D.K. Synthesis of an Alginate-Based Fe_3O_4-MnO_2 Xerogel and Its Application for the Concurrent Elimination of Cr(VI) and Cd(II) from Aqueous Solution. *ACS Omega* **2021**, *6*, 3931–3945. [CrossRef] [PubMed]

35. Pradhan, S.; Hedberg, J.; Blomberg, E.; Wold, S.; Odnevall Wallinder, I. Effect of sonication on particle dispersion, administered dose and metal release of non-functionalized, non-inert metal nanoparticles. *J. Nanoparticle Res.* **2016**, *18*, 285. [CrossRef] [PubMed]
36. Ennas, G.; Gedanken, A.; Mannias, G.; Kumar, V.B.; Scano, A.; Porat, Z.; Pilloni, M. Formation of Iron (III) Trimesate Xerogel by Ultrasonic Irradiation. *Eur. J. Inorg. Chem.* **2022**, *2022*, e202101082. [CrossRef]
37. Maeda, Y.; Hayashi, Y.; Fukushima, Y.; Takizawa, H. Sonochemical effect and pore structure tuning of silica xerogel by ultrasonic irradiation of semi-solid hydrogel. *Ultrason. Sonochem.* **2021**, *73*, 105476. [CrossRef] [PubMed]
38. Dong, Y.; Hauschild, M.Z. Indicators for Environmental Sustainability. *Procedia CIRP* **2017**, *61*, 697–702. [CrossRef]
39. Oyedoh, E.A.; Albadarin, A.B.; Walker, G.M.; Mirzaeian, M.; Ahmad, M.N.M. Preparation of controlled porosity resorcinol formaldehyde xerogels for adsorption applications. *Chem. Eng. Trans.* **2013**, *32*, 1651–1656. [CrossRef]
40. Canal-Rodríguez, M.; Menéndez, J.A.; Arenillas, A. *Carbon Xerogels: The Bespoke Nanoporous Carbons*; Gluib, T.H., Ed.; IntechOpen: Rijeka, Croatia, 2017; Chapter 3; ISBN 978-1-78923-043-7.
41. Tipsawat, P.; Wongpratat, U.; Phumying, S.; Chanlek, N.; Chokprasombat, K.; Maensiri, S. Magnetite (Fe_3O_4) nanoparticles: Synthesis, characterization and electrochemical properties. *Appl. Surf. Sci.* **2018**, *446*, 287–292. [CrossRef]
42. Sulistyaningsih, T.; Santosa, S.J.; Siswanta, D.; Rusdiarso, B. Synthesis and Characterization of Magnetites Obtained from Mechanically and Sonochemically Assissted Co-precipitation and Reverse Co-precipitation Methods. *Int. J. Mater. Mech. Manuf.* **2017**, *5*, 16–19. [CrossRef]
43. Gaikwad, M.M.; Sarode, K.K.; Pathak, A.D.; Sharma, C.S. Ultrahigh rate and high-performance lithium-sulfur batteries with resorcinol-formaldehyde xerogel derived highly porous carbon matrix as sulfur cathode host. *Chem. Eng. J.* **2021**, *425*, 131521. [CrossRef]
44. Pérez-Cadenas, A.F.; Ros, C.H.; Morales-Torres, S.; Pérez-Cadenas, M.; Kooyman, P.J.; Moreno-Castilla, C.; Kapteijn, F. Metal-doped carbon xerogels for the electro-catalytic conversion of CO_2 to hydrocarbons. *Carbon N. Y.* **2013**, *56*, 324–331. [CrossRef]
45. Prostredný, M.; Abduljalil, M.; Mulheran, P.; Fletcher, A. Process Variable Optimization in the Manufacture of Resorcinol–Formaldehyde Gel Materials. *Gels* **2018**, *4*, 36. [CrossRef] [PubMed]
46. Martin, E.; Prostredny, M.; Fletcher, A.; Mulheran, P. Modelling organic gel growth in three dimensions: Textural and fractal properties of resorcinol–formaldehyde gels. *Gels* **2020**, *6*, 23. [CrossRef] [PubMed]
47. Verma, N.K.; Khare, P.; Verma, N. Synthesis of iron-doped resorcinol formaldehyde-based aerogels for the removal of Cr(VI) from water. *Green Process. Synth.* **2015**, *4*, 37–46. [CrossRef]
48. Sharifalhoseini, Z.; Entezari, M.H.; Jalal, R. Direct and indirect sonication affect differently the microstructure and the morphology of ZnO nanoparticles: Optical behavior and its antibacterial activity. *Ultrason. Sonochem.* **2015**, *27*, 466–473. [CrossRef] [PubMed]
49. Calvo, E.G.; Menéndez, J.Á.; Arenillas, A. Designing Nanostructured Carbon Xerogels. In *Nanomaterials*; IntechOpen: Rijeka, Croatia, 2011. [CrossRef]
50. Job, N.; Panariello, F.; Marien, J.; Crine, M.; Pirard, J.P.; Léonard, A. Synthesis optimization of organic xerogels produced from convective air-drying of resorcinol-formaldehyde gels. *J. Non. Cryst. Solids* **2006**, *352*, 24–34. [CrossRef]
51. Mirzaeian, M.; Hall, P.J. The control of porosity at nano scale in resorcinol formaldehyde carbon aerogels. *J. Mater. Sci.* **2009**, *44*, 2705–2713. [CrossRef]
52. Martin, E.; Prostredny, M.; Fletcher, A. Investigating the role of the catalyst within resorcinol– formaldehyde gel synthesis. *Gels* **2021**, *7*, 142. [CrossRef]
53. Al-Muhtaseb, S.A.; Ritter, J.A. Preparation and properties of resorcinol-formaldehyde organic and carbon gels. *Adv. Mater.* **2003**, *15*, 101–114. [CrossRef]
54. Calvo, E.G.; Menéndez, J.A.; Arenillas, A. Influence of alkaline compounds on the porosity of resorcinol-formaldehyde xerogels. *J. Non. Cryst. Solids* **2016**, *452*, 286–290. [CrossRef]
55. Luzny, R.; Ignasiak, M.; Walendziewski, J.; Stolarski, M. Heavy metal ions removal from aqueous solutions using carbon aerogels and xerogels. *Chemik* **2014**, *68*, 544–553.
56. Alonso-Buenaposada, I.D.; Montes-Morán, M.A.; Menéndez, J.A.; Arenillas, A. Synthesis of hydrophobic resorcinol–formaldehyde xerogels by grafting with silanes. *React. Funct. Polym.* **2017**, *120*, 92–97. [CrossRef]
57. Dewes, R.M.; Mendoza, H.R.; Pereira, M.V.L.; Lutz, C.; Gerven, T. Van Experimental and numerical investigation of the effect of ultrasound on the growth kinetics of zeolite A. *Ultrason. Sonochem.* **2022**, *82*, 105909. [CrossRef]
58. Ferri, S.; Wu, Q.; De Grazia, A.; Polydorou, A.; May, J.P.; Stride, E.; Evans, N.D.; Carugo, D. Tailoring the size of ultrasound responsive lipid-shelled nanodroplets by varying production parameters and environmental conditions. *Ultrason. Sonochem.* **2021**, *73*, 105482. [CrossRef] [PubMed]
59. Job, N.; Pirard, R.; Marien, J.; Pirard, J.P. Porous carbon xerogels with texture tailored by pH control during sol-gel process. *Carbon N. Y.* **2004**, *42*, 619–628. [CrossRef]
60. Kim, H.J.; Choi, H.; Sharma, A.K.; Hong, W.G.; Shin, K.; Song, H.; Kim, H.Y.; Hong, Y.J. Recyclable Aqueous Metal Adsorbent: Synthesis and Cu(II) Sorption Characteristics of Ternary Nanocomposites of Fe_3O_4 Nanoparticles@Graphene–poly-N-Phenylglycine Nanofibers. *J. Hazard. Mater.* **2021**, *401*, 123283. [CrossRef]
61. Min, X.; Li, Y.; Ke, Y.; Shi, M.; Chai, L.; Xue, K. Fe-FeS_2 adsorbent prepared with iron powder and pyrite by facile ball milling and its application for arsenic removal. *Water Sci. Technol.* **2017**, *76*, 192–200. [CrossRef] [PubMed]

62. Attia, S.M.; Abdelfatah, M.S.; Mossad, M.M. Conduction mechanism and dielectric properties of pure and composite resorcinol formaldehyde aerogels doped with silver. *J. Phys. Conf. Ser.* **2017**, *869*, 012035. [CrossRef]
63. Canal-Rodríguez, M.; Rey-Raap, N.; Menéndez, J.Á.; Montes-Morán, M.A.; Figueiredo, J.L.; Pereira, M.F.R.; Arenillas, A. Effect of porous structure on doping and the catalytic performance of carbon xerogels towards the oxygen reduction reaction. *Microporous Mesoporous Mater.* **2020**, *293*, 109811. [CrossRef]
64. Liu, Z.; Li, J.; Yang, Y.; Mi, J.H.; Tan, X.L. Synthesis, characterisation and magnetic examination of Fe, Co and Ni doped carbon xerogels. *Mater. Res. Innov.* **2012**, *16*, 362–367. [CrossRef]
65. Alegre, C.; Sebastián, D.; Gálvez, M.E.; Baquedano, E.; Moliner, R.; Aricò, A.S.; Baglio, V.; Lázaro, M.J. N-Doped Carbon Xerogels as Pt Support for the Electro-Reduction of Oxygen. *Materials* **2017**, *10*, 1092. [CrossRef]
66. Ramos-Fernández, G.; Canal-Rodríguez, M.; Arenillas, A.; Menéndez, J.A.; Rodríguez-Pastor, I.; Martin-Gullon, I. Determinant influence of the electrical conductivity versus surface area on the performance of graphene oxide-doped carbon xerogel supercapacitors. *Carbon N. Y.* **2018**, *126*, 456–463. [CrossRef]
67. Sarwar, A.; Wang, J.; Khan, M.S.; Farooq, U.; Riaz, N.; Nazir, A.; Mahmood, Q.; Hashem, A.; Al-Arjani, A.B.F.; Alqarawi, A.A.; et al. Iron oxide (Fe_3O_4)-supported sio2 magnetic nanocomposites for efficient adsorption of fluoride from drinking water: Synthesis, characterization, and adsorption isotherm analysis. *Water* **2021**, *13*, 1514. [CrossRef]
68. Nikić, J.; Tubić, A.; Watson, M.; Maletić, S.; Šolić, M.; Majkić, T.; Agbaba, J. Arsenic removal from water by green synthesized magnetic nanoparticles. *Water* **2019**, *11*, 2520. [CrossRef]
69. Pekala, R.W. Organic aerogels from the polycondensation of resorcinol with formaldehyde. *J. Mater. Sci.* **1989**, *24*, 3221–3227. [CrossRef]
70. Haro, M.; Rasines, G.; MacIas, C.; Ania, C.O. Stability of a carbon gel electrode when used for the electro-assisted removal of ions from brackish water. *Carbon N. Y.* **2011**, *49*, 3723–3730. [CrossRef]
71. Ribeiro, R.S.; Frontistis, Z.; Mantzavinos, D.; Venieri, D.; Antonopoulou, M.; Konstantinou, I.; Silva, A.M.T.; Faria, J.L.; Gomes, H.T. Magnetic carbon xerogels for the catalytic wet peroxide oxidation of sulfamethoxazole in environmentally relevant water matrices. *Appl. Catal. B Environ.* **2016**, *199*, 170–186. [CrossRef]
72. Amaringo, F.A.; Anaguano, A. Determination of the point of zero charge and isoelectric point of two agricultural wastes and their application in the removal of colorants. *Rev. Investig. Agrar. y Ambient.* **2013**, *4*, 27. [CrossRef]
73. Largitte, L.; Pasquier, R. A review of the kinetics adsorption models and their application to the adsorption of lead by an activated carbon. *Chem. Eng. Res. Des.* **2016**, *109*, 495–504. [CrossRef]
74. López-Luna, J.; Ramírez-Montes, L.E.; Martinez-Vargas, S.; Martínez, A.I.; Mijangos-Ricardez, O.F.; González-Chávez, M.d.C.A.; Carrillo-González, R.; Solís-Domínguez, F.A.; Cuevas-Díaz, M.d.C.; Vázquez-Hipólito, V. Linear and nonlinear kinetic and isotherm adsorption models for arsenic removal by manganese ferrite nanoparticles. *SN Appl. Sci.* **2019**, *1*, 950. [CrossRef]
75. Hanbali, M.; Holail, H.; Hammud, H. Remediation of lead by pretreated red algae: Adsorption isotherm, kinetic, column modeling and simulation studies. *Green Chem. Lett. Rev.* **2014**, *7*, 342–358. [CrossRef]
76. Inyinbor, A.A.; Adekola, F.A.; Olatunji, G.A. Kinetics, isotherms and thermodynamic modeling of liquid phase adsorption of Rhodamine B dye onto Raphia hookerie fruit epicarp. *Water Resour. Ind.* **2016**, *15*, 14–27. [CrossRef]
77. Quevedo, O.; Luna, B. Determinación de As (III) y As (V) en aguas naturales por generación de hidruro con detección por espectrometría de absorción atómica. *Rev. CENIC. Ciencias Químicas* **2003**, *34*, 133–147.

Disclaimer/Publisher's Note: The statements, opinions and data contained in all publications are solely those of the individual author(s) and contributor(s) and not of MDPI and/or the editor(s). MDPI and/or the editor(s) disclaim responsibility for any injury to people or property resulting from any ideas, methods, instructions or products referred to in the content.

Article

Effective Removal of Cu²⁺ Ions from Aqueous Media Using Poly(acrylamide-co-itaconic acid) Hydrogels in a Semi-Continuous Process

Jorge Alberto Cortes Ortega [1], Jacobo Hernández-Montelongo [2,3], Rosaura Hernández-Montelongo [3] and Abraham Gabriel Alvarado Mendoza [1,*]

[1] Department of Chemistry, University Center of Exact Sciences and Engineering, University of Guadalajara, Blvd. Marcelino García Barragán #1421, Guadalajara 44430, Mexico; jorge.cortega@academicos.udg.mx

[2] Department of Physical and Mathematical Sciences, Faculty of Engineering, Catholic University of Temuco, Av. Rudecindo Ortega #2959, Temuco 4813302, Chile; jacobo.hernandez@uct.cl

[3] Department of Translational Bioengineering, University Center of Exact Sciences and Engineering, University of Guadalajara, Blvd. Marcelino García Barragán #1421, Guadalajara 44430, Mexico; rosaura.hernandez@academicos.udg.mx

* Correspondence: gabriel.alvarado@academicos.udg.mx

Citation: Cortes Ortega, J.A.; Hernández-Montelongo, J.; Hernández-Montelongo, R.; Alvarado Mendoza, A.G. Effective Removal of Cu²⁺ Ions from Aqueous Media Using Poly(acrylamide-co-itaconic acid) Hydrogels in a Semi-Continuous Process. *Gels* **2023**, *9*, 702. https://doi.org/10.3390/gels9090702

Academic Editors: Daxin Liang, Ting Dong, Yudong Li and Caichao Wan

Received: 14 August 2023
Revised: 25 August 2023
Accepted: 27 August 2023
Published: 30 August 2023

Copyright: © 2023 by the authors. Licensee MDPI, Basel, Switzerland. This article is an open access article distributed under the terms and conditions of the Creative Commons Attribution (CC BY) license (https://creativecommons.org/licenses/by/4.0/).

Abstract: Adsorption is one of the most crucial processes in water treatment today. It offers a low-cost solution that does not require specialized equipment or state-of-the-art technology while efficiently removing dissolved contaminants, including heavy metals. This process allows for the utilization of natural or artificial adsorbents or a combination of both. In this context, polymeric materials play a fundamental role, as they enable the development of adsorbent materials using biopolymers and synthetic polymers. The latter can be used multiple times and can absorb large amounts of water per gram of polymer. This paper focuses on utilizing adsorption through hydrogels composed of poly(acrylamide-co-itaconic acid) for removing Cu²⁺ ions dissolved in aqueous media in a semi-continuous process. The synthesized hydrogels were first immersed in 0.1 M NaOH aqueous solutions, enabling OH⁻ ions to enter the gel matrix and incorporate into the polymer surface. Consequently, the copper ions were recovered as Cu(OH)₂ on the surface of the hydrogel rather than within it, allowing the solid precipitates to be easily separated by decantation. Remarkably, the hydrogels demonstrated an impressive 98% removal efficiency of the ions from the solution in unstirred conditions at 30 °C within 48 h. A subsequent study involved a serial process, demonstrating the hydrogels' reusability for up to eight cycles while maintaining their Cu2+ ion recovery capacity above 80%. Additionally, these hydrogels showcased their capability to remove Cu²⁺ ions even from media with ion concentrations below 100 ppm.

Keywords: semi-continuous process; adsorption by hydrogels; removal Cu²⁺ from wastewater; poly(acrylamide-co-itaconic acid)

1. Introduction

In recent years, the utilization of heavy metals has witnessed a significant increase due to their participation in numerous industrial processes and their incorporation into various products, devices, and equipment developed to enhance people's quality of life, as seen in electronics, for example. However, the excessive use of heavy metals has led to dangerous concentrations of these elements in the soil, air, and water. This represents a serious health problem not only for human beings but also for plants and animals, as heavy metals are non-biodegradable and accumulate within the bodies of living organisms, causing poisoning, gastrointestinal and pulmonary diseases, cancer, and cell abnormalities. Prolonged and severe exposure to these metals can even lead to death [1].

Although some heavy metals serve important biological functions in plants and animals, an increase in their concentration, along with their coordination and oxidation-reduction chemical behavior, leads to serious issues. For instance, in humans, copper is an essential element for various organism functions, including physiological processes, immune system functions, fetal and infant development and growth, brain function, bone strengthening, glucose metabolism, iron and cholesterol regulation, among others [2]. However, the excessive presence of this metal in the human body can cause severe health damage. The initial symptoms that usually subside upon reducing exposure to this metal include nausea, abdominal pain, vomiting, and diarrhea [3]. Prolonged exposure to high copper concentrations has been linked to various conditions such as cancer, dementia, Parkinson's disease, Alzheimer's, childhood cirrhosis, Wilson's disease, kidney disease, cell toxicity, among others [4,5]. The daily intake of copper is determined by one's diet, supplements, and primarily the water ingested. The World Health Organization reported that nearly 104 countries have established an average value of 1.5 mg/L of copper in drinking water. However, copper is widely used in the manufacturing of structural materials, pipes, electronics, heat transfer equipment, the automotive industry, and numerous industrial processes and products, including mining, electroplating, paints, tanneries, and even in the production of fertilizers and pesticides [6]. Consequently, the concentration of this metal has increased in both surface water and groundwater due to these activities.

Due to these concerns, considerable efforts have been made to develop materials and methods capable of removing heavy metals and other pollutants from industrial effluents and water sources. Various methods have been explored, including electrochemical treatments, physicochemical processes such as chemical precipitation and adsorption, as well as more recent advancements such as filtration processes through membranes and photocatalysis [7]. Among these techniques, chemical precipitation stands out as a widely used and economically viable method for heavy metal removal at the industrial level [8,9]. It basically consists of converting a soluble ion into an insoluble compound through a chemical reaction, for example, the formation of metal sulfides, carbonates, and hydroxides. Finally, the insoluble compound is removed from the medium by sedimentation or filtration. In this method, the pH of the medium plays an important role in the recovery of metal ions, and generally, values of pH = 11 are required to increase the amount of metal ions removed [10]. Unfortunately, the chemical substances used and their high concentration necessary in the precipitation process, as well as the sludge obtained that requires certain treatments, can represent a new contamination problem. In that sense, the effectiveness of the chemical precipitation method depends on the type of dissolved ion, its concentration, the precipitating agent, the medium, and the presence of other compounds that can inhibit the reaction [11]. Consequently, this method may not be very effective for low ion concentrations or cases where sedimentation is challenging.

On the other hand, the adsorption process is renowned for its successful application in cases where the concentration of metal ions is low, leading to high-quality treated effluents [12]. This method is not only cost-effective but also highly efficient and easily reproducible and operable. Adsorption relies on mass transfer between the liquid phase and the solid phase (adsorbent), where the negatively charged functional groups on the adsorbent's surface attract positively charged metal ions [13]. A wide variety of adsorbent materials have been developed, including activated carbon, graphene, carbon nanotubes, zeolites, mesoporous silica, clay, biomass, and hydrogels [14,15]. Among these, hydrogels have garnered the most interest.

Hydrogels are three-dimensional networks composed of synthetic or natural polymer chains containing hydrophilic groups. This characteristic enables them to absorb significant amounts of water while maintaining their shape through physical or chemical crosslinking between the chains. These unique properties have sparked significant interest across various fields of application over the last sixty years. For instance, hydrogels have found applications in the biomedical field [16], tissue engineering [17–21], drug transport [22], agriculture [23], and the removal of heavy metals. In the context of heavy metal removal, hydrogels have proven successful in eliminating metal ions such as copper [24,25], nickel [26,27], lead [28], arsenic [29], cadmium [30], chromium [31], among others.

Considering the significance of chemical precipitation and adsorption methods using hydrogels, this work presents a study focused on the effective removal of Cu^{2+} ions from aqueous media using poly(acrylamide-co-itaconic acid) hydrogels in a semi-continuous process. The objective is to determine the optimal ratio of Cu^{2+} ions to hydroxyl groups that enable the removal of the maximum number of ions, resulting in $Cu(OH)_2$ precipitation in the aqueous medium. In this approach, hydrogels serve as transport media for the OH^- ions, preventing metal ions from permeating and becoming trapped within the hydrogel matrix. Consequently, the $Cu(OH)_2$ formed is located on the hydrogel's surface, facilitating the removal and recovery of Cu^{2+} ions from the aqueous solution. The application of this method for removing Cu^{2+} or other metallic ions, performed on the surface of the hydrogel, has not been previously documented in the literature. Finally, the hydrogels were employed in a semi-continuous process, wherein the concentration of metal ions gradually decreased. This approach allowed us to assess the number of times the hydrogels could be reused and determine if their removal capacity remained consistent throughout.

2. Results

2.1. Conversion

The hydrogels obtained after the reaction time were smooth and completely solid, without any residues of the aqueous solution from the reaction mixture. The conversion achieved in the synthesized hydrogels was $97 \pm 2\%$, indicating that the reaction was nearly complete. Figure 1 depicts a pictogram illustrating the possible reaction scheme in the synthesis of the copolymer.

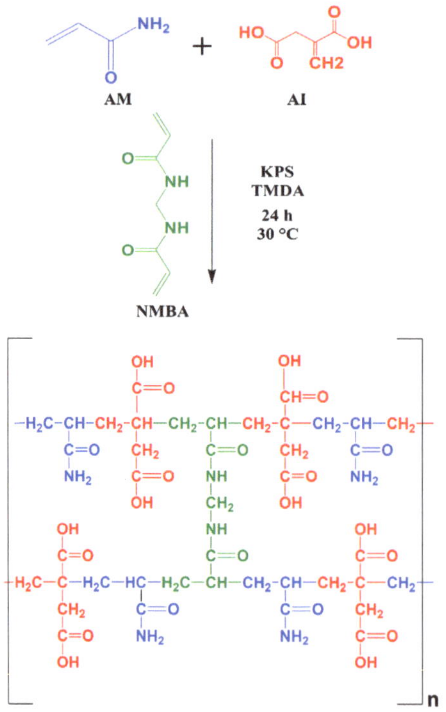

Figure 1. Graphic representation of the network obtained in the copolymerization of AM and AI by means of NMBA. The mass ratio between monomers was not considered.

2.2. Metal Ion Recovery

Table 1 presents the values of the mass of the 0.1 M NaOH solution ($mass_{NaOH}$) and the mass of the mixed $CuCl_2 \cdot 2H_2O$ solution ($mass_{CuCl2}$), along with the percentage of copper ions removed ($R_{Cu^{2+}}$) and the molar ratio between the OH^- and Cu^{2+} groups ($mol_{OH^-}/mol_{Cu^{2+}}$).

Table 1. Removal of copper ions as a function of the mass ratio of NaOH solution/copper solution.

Experiment	1	2	3	4	5	6	7	8	9
$mass_{CuCl2}$ (g)	18	16	14	12	10	8	6	4	2
$mass_{NaOH}$ (g)	2	4	6	8	10	12	14	16	18
$mol_{OH^-}/mol_{Cu^{2+}}$	0.71	1.59	2.74	4.27	6.37	9.52	14.78	25.37	56.36
$R_{Cu^{2+}}$ (%)	30	41	98	99	98	99	95	93	98

It was observed that when the $mol_{OH^-}/mol_{Cu^{2+}}$ ratio reached the value of 2.74, the percentage of copper ions removed was 97.9%. Beyond this ratio, no significant increase in metal removal was observed, even with an increase in the amount of NaOH solution added. In Experiments 7 and 8, the removal percentage decreased. Hence, it was determined that the optimal value for the $mol_{OH^-}/mol_{Cu^{2+}}$ ratio was 2.74. Regarding the adsorption process through the use of hydrogels, maintaining a ratio of 2.74 between the moles of OH^- present within the hydrogel matrix and the concentration of Cu^{2+} ions in the reaction medium resulted in a Cu^{2+} removal efficiency ($Re_{Cu^{2+}}$) of 98.52%. Removal efficiency through the chemical precipitation method was equaled, enabling us to pinpoint the optimal ratio of $mol_{OH^-}/mol_{Cu^{2+}}$ for attaining the utmost elimination of Cu^{2+} ions. Furthermore, this process can be repeated multiple times using the same hydrogel, as the precipitate formed on the surface of the hydrogel can be easily removed by gently shaking the medium and separating it by decantation. The hydrogel can be regenerated by reintroducing it into a NaOH solution, making it suitable for reuse and thereby reducing operating costs.

To evaluate the effect of the concentration of the NaOH solutions in which the xerogels are immersed and swollen, they were immersed in solutions with concentrations of 0.1, 0.2, 0.3, and 0.4 M. It was found that the swelling capacity (W (%)) of the hydrogels decreased as the concentration of the NaOH solution increased, ranging from 64% to 50%. This decrease is attributed to the higher presence of Na^+ and OH^- ions in the medium. Consequently, the interaction between water molecules and the hydrophilic chains of the polymer diminishes, resulting in a reduction in its swelling.

Subsequently, when the hydrogels swollen with the NaOH solution were immersed in the $CuCl_2 \cdot 2H_2O$ solution with 1000 ppm Cu^{2+} at a ratio of Cu^{2+} ion solution mass to xerogel mass of 200/1, it was observed that the mass of copper removal per gram of xerogel ($mg_{Cu^{2+}}/g_{xerogel}$) slightly increased with the increasing concentration in NaOH solutions, from 0.1 M (183 $mg_{Cu^{2+}}/g_{xerogel}$) to 0.4 M (211 $mg_{Cu^{2+}}/g_{xerogel}$). Additionally, the $R_{Cu^{2+}}$ (%) increased from 98% to 99.7%. Table 2 summarizes the values of the swelling capacity, $mg_{Cu^{2+}}/g_{xerogel}$ removed, and the $R_{Cu^{2+}}$ as a function of the concentration of the NaOH solution.

Table 2. Removal of copper ions as a function of the concentration of NaOH solutions, with a lye/xerogel ratio of 125/1 and a copper/xerogel solution of 200/1.

[NaOH] M	W (%)	$mg_{Cu^{2+}}/g_{xerogel}$	$R_{Cu^{2+}}$ (%)
0.1	64	183	98.52
0.2	55	194	98.44
0.3	53	205	99.10
0.4	50	211	99.68

The R_{Cu}^{2+} values were very similar in all cases. This is because the metal removal occurs through the reaction of the OH⁻ ions present inside the gel, which migrate toward the gel's surface and react with the metal ions. The efficiency of the removal was not dependent on the concentration of the NaOH solutions but on the amount of OH⁻ ions within the hydrogels. Therefore, as the absorption capacity of the hydrogels decreased due to the increase in NaOH concentration, the OH⁻ ions inside the hydrogel also decreased, resulting in similar efficiency in copper recovery in all cases. Thus, a 0.1 M concentration of NaOH in the solution was the most suitable for the recovery of copper ions present in the solution, similar to the precipitation method. This proves that hydrogel functions akin to a "sponge," proficient in capturing, transporting, and releasing OH⁻ ions. In that sense, the AI played an important role due to its ability to confer carboxylic groups to the hydrogel. Such groups facilitate pronounced swelling in basic environments, thereby enhancing the influx of OH⁻ ions when diluted solutions of NaOH were employed [32].

To verify this, the process was carried out by immersing the xerogels in bidistilled water under the same conditions mentioned above and in the same proportion. Subsequently, they were placed in 1000 ppm Cu^{2+} solutions at a ratio of Cu^{2+} solution mass to xerogel mass equal to 200/1. In this case, the ions were not recovered on the surface but within the gel matrix. The cations penetrated the gel matrix, causing it to saturate, and as a result, the network closed, leading to the collapse of the hydrogel. Additionally, as seen in the residual water, Cu^{2+} ions were still present (Figure 2).

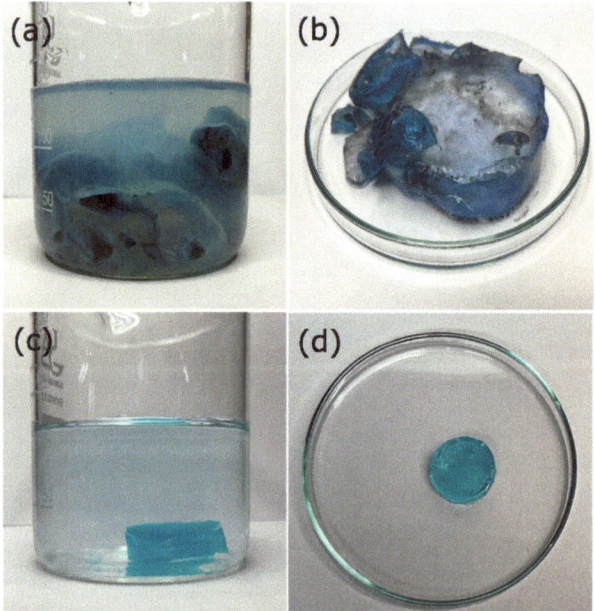

Figure 2. The xerogels immersed in NaOH solutions removed Cu^{2+} in the form of $Cu(OH)_2$, which forms on its surface (**a**). Only a small part of these ions managed to enter the matrix; thus, the hydrogel remained swollen (**b**). In contrast, in the case of water-swollen xerogels, (**c**) Cu^{2+} ions entered the matrix, causing the hydrogel to collapse (**d**).

Moreover, when xerogels were directly introduced into a 1000 ppm Cu^{2+} solution, without a previous immersion in 0.1 M NaOH solution, at a mass ratio of 200/1 maintained at 30 °C without agitation and left for a duration of 48 h, the R_{Cu}^{2+} value equated to 14.67% or 29.29 mg Cu^{2+}/g of xerogel. This is only 16% of the obtained R_{Cu}^{2+} compared to the previous immersion in NaOH: 98.52% or 183 mg Cu^{2+}/g of xerogel, which confirms the key role of OH^- ions in the process.

In another experiment, it was demonstrated that if the total volume of the metal ion solution is increased while keeping the $mol_{OH^-}/mol_{Cu^{2+}}$ ratio constant at the value of 2.74, the $R_{Cu^{2+}}$ value remains around 97% with no significant changes until a concentration of 1000 ppm (Figure 3). Below that concentration, the $R_{Cu^{2+}}$ value decreased to about 75% for concentrations of 30 ppm (inset Figure 3). Another study was conducted to determine the detection limit of Cu^{2+} ions by the hydrogel, and it was found that this process can cause the precipitation of metal ions in solutions containing up to 10 ppm Cu^{2+} (Figure 4c). The value of $R_{Cu^{2+}}$ was not displayed for concentrations lower than 30 ppm due to reading and precision limitations in UV-vis analysis. On the other hand, it can be observed that the standard deviation was much higher for concentrations lower than 100 ppm due to reading limitations and the precision obtained with the technique used in the analysis. Figure 4 shows photographs of the experiment, where the formation of $Cu(OH)_2$ on the surface of the hydrogel can be observed in solutions with concentrations of 30, 20, and 10 ppm.

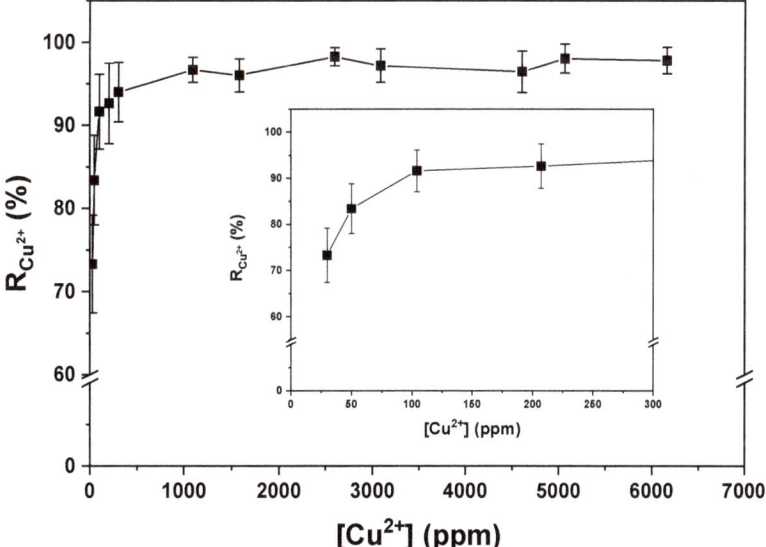

Figure 3. The amount of Cu^{2+} removed through the use of hydrogels while maintaining a ratio of $mol_{NaOH}/mol_{Cu^{2+}}$ equal to 2.74, as the total volume of the Cu^{2+} ion solution is increased.

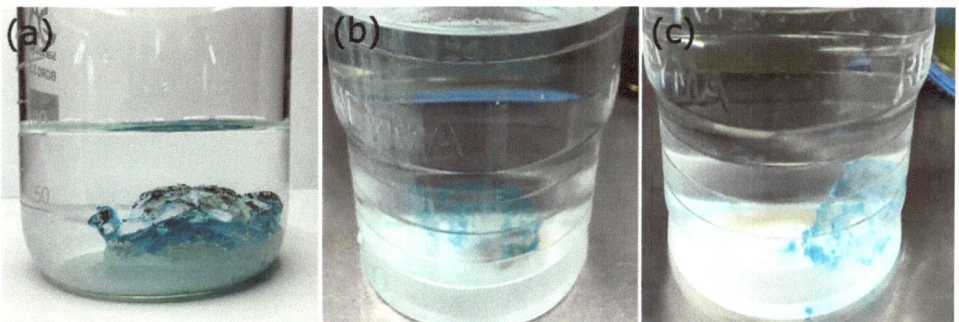

Figure 4. Recovery of copper ions in aqueous solutions of cupric chloride at different concentrations: 30 (**a**), 20 (**b**), and 10 ppm (**c**), while maintaining a ratio of $mol_{NaOH}/mol_{Cu^{2+}}$ at the value of 2.74.

The effect of the initial concentration of dissolved Cu^{2+} ions on the $R_{Cu^{2+}}$ and the value of mg Cu^{+2}/g of xerogel are shown in Figure 5. It was observed that as the concentration of initial cations in the medium increased, the value of $R_{Cu^{2+}}$ remained, on average, at 98.61% for concentrations ranging from 200 to 1750 ppm. However, for concentrations greater than 1750 ppm, the $R_{Cu^{2+}}$ value gradually decreased, falling below 80% for initial concentrations higher than 2000 ppm. This decline is attributed to the insufficient amount of OH^- ions present inside the hydrogel to effectively react with the high concentration of Cu^{2+} ions. In the range of concentrations where the $R_{Cu^{2+}}$ was greater than 95%, the $mol_{OH^-}/mol_{Cu^{2+}}$ ratio was found to be higher than 2.74. Conversely, for values where the $R_{Cu^{2+}}$ decreases below 80%, the $mol_{OH^-}/mol_{Cu^{2+}}$ ratio was less than 2.74. Therefore, the two most important parameters for effective copper ion removal were the initial cation concentration and the amount of OH^- ions present within the hydrogel.

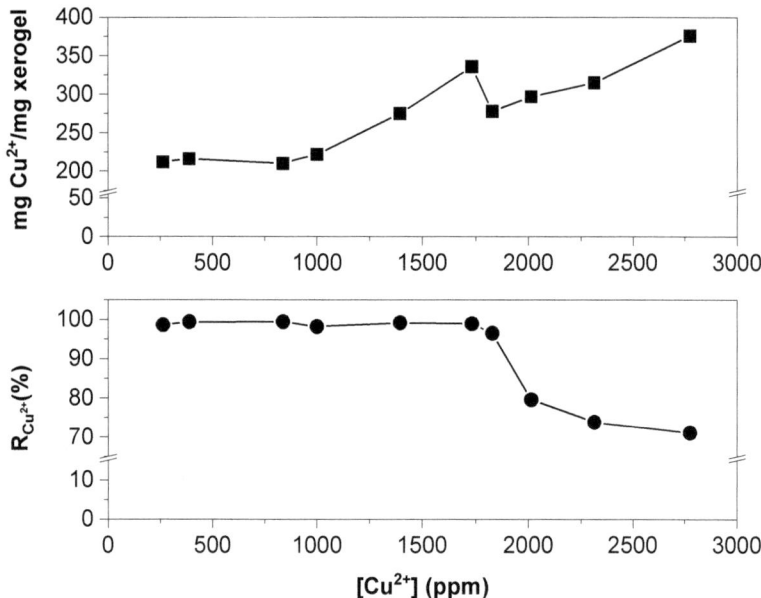

Figure 5. Effect of the initial concentration of Cu^{2+} ions on the $R_{Cu^{2+}}$ and mg of Cu^{2+}/g of xerogel.

Regarding the maximum mg Cu^{+2}/g of xerogel ratio obtained within acceptable $R_{Cu^{2+}}$ values, it was 336 mg Cu^{+2}/g of xerogel. Although in higher initial concentrations, the maximum ratio of mg Cu^{+2}/g of xerogel was also 336 mg of mg Cu^{+2}/g of xerogel, the $R_{Cu^{2+}}$ was no longer acceptable because it reached only 71%.

Finally, in the semi-continuous study, it was found that the same hydrogel sample can be used up to eight times (Figure 6). The process was categorized as a semi-continuous process because the hydrogel required a retention time at each stage to facilitate the elimination of Cu^{2+} ions (which constitutes a batch process). Following this, the hydrogel was regenerated and proceeded to the subsequent stage, wherein the concentration of Cu^{2+} ions was lower compared to the previous stage (representing a continuous process).

On average, the $R_{Cu^{2+}}$ value was 93.40% in the first six stages, subsequently decreasing to 87.30% and finally to 83% in the last stage (Figure 7). This decrease in removal capacity is attributed to the fact that in each stage where the hydrogel was used, a small fraction of the metal ions remain trapped in its matrix (Figure 6c,f), which accumulated with each reuse. As a result, the amount of OH^- ions inside the hydrogel decreased, leading to a reduction in its ability to remove the cations. Furthermore, the lowest $R_{Cu^{2+}}$ value was found in solutions with the lowest amount of initial Cu^{2+} ions. As previously demonstrated, the two main factors for maintaining high $R_{Cu^{2+}}$ values were the concentration of OH^- ions in the hydrogel and the initial concentration of Cu^{2+} ions in the solution.

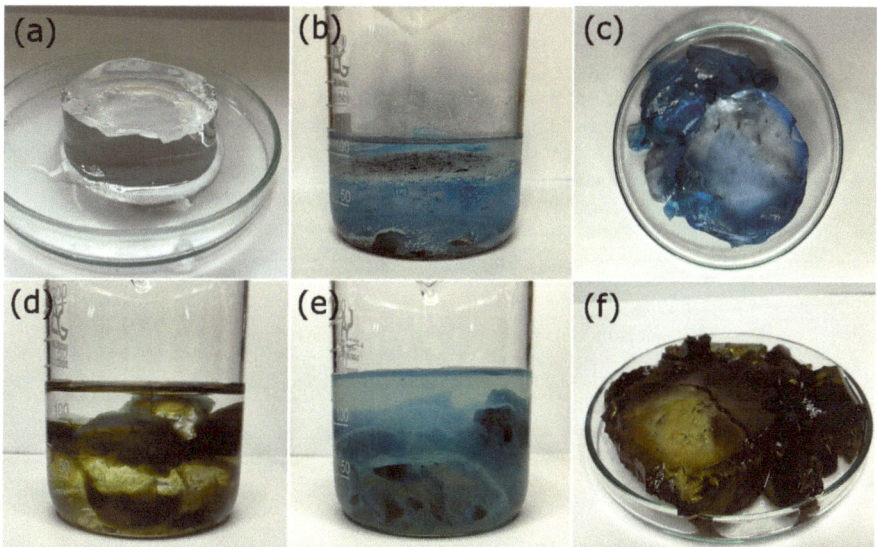

Figure 6. Semi-continuous metal ion removal process: obtained xerogel swollen after immersion in a 0.1 M NaOH solution (**a**), hydrogel immersed in the Cu^{2+} ion solution (**b**), hydrogel after the first stage (**c**), hydrogel regeneration in the NaOH solution (**d**), hydrogel immersed in ion solution Cu^{2+} (**e**), and hydrogel after being used eight times (**f**).

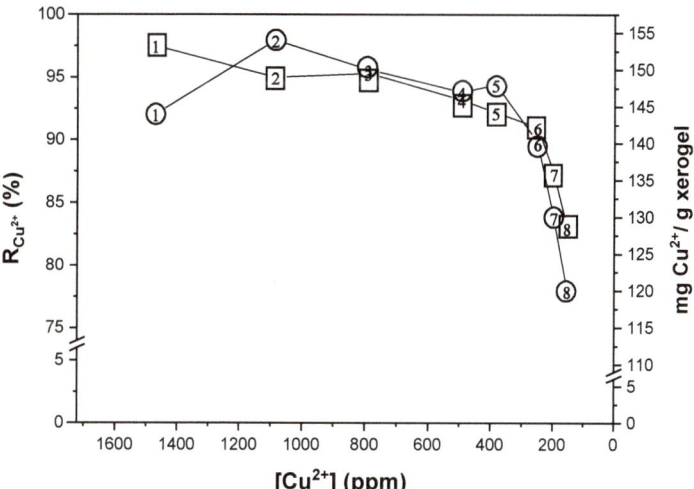

Figure 7. Values of $R_{Cu^{2+}}$ (□) and $mg_{Cu^{2+}}/g_{xerogel}$ (○) as a function of the initial concentration of Cu^{2+} ions in each stage.

The achieved percentage removal of copper ions using this method was 98.52% at a temperature of 30 °C with a residence time of 48 h, no agitation, and with a mass ratio of Cu^{2+} solution to xerogel of 200/1. This corresponds to the maximum removal capacity (Qmax) of 183 mg Cu^{2+}/g xerogel, recovered on the surface of the hydrogel in the form of Cu(OH)2, which is easily removed by slightly shaking the container or simply rinsing with the minimum amount of distilled water to the hydrogel. This avoids the use of acidic solutions for the recovery of the metal ions, a practice employed in processes in which the metal ions remain within the gel matrix. Table 3 provides a compilation of various adsorption studies along with their respective Qmax values. While this study does not boast the highest Qmax value, it does mark the pioneering utilization of the hydrogel surface for recovery. Furthermore, it shows a method for the easy recovery of metal ions in a process similar to industrial water treatment, where the concentration of the ions decreases in the process, deviating from the conventional practice of recovering throughout the entire matrix.

Table 3. Adsorption capacities of different adsorbent materials for Cu^{2+} from other studies.

Adsorbents Materials	Q_{max} (mg/g)	Reference
Graphene oxide-polyethylene glycol and polyvinyl alcohol (GO-PEG-PVA) triple network hydrogel	917	[33]
Hybrid hydrogel of acrylic acid monomer/wheat bran/montmorillonite	17.64	[34]
Hydrogels comprised of polysaccharide salecan injerted with poly(3-sulfopropyl methacrylate potassium salt).	107.2	[35]
Aerogels comprised of carboxylated cellulose and $MnFe_2O_4$.	73.70	[36]
Hydrogels comprised of Loess of clay/Itaconic acid/2-Hydroxyethyl methacrylate/N-vinyl-2-pyrrolidone	594.43	[37]
Poly(acrylic acid-co-itaconic acid)/NaOH hydrogel	85	[38]
Polyvinyl alcohol/alginate/iron oxide nanoparticles (PAI) hydrogels	60	[39]
Carboxymethylcellulose sodium/polyvinyl alcohol (PVA)/Cellulose nanocrystals hydrogels	108.8	[40]
Corn starch/acrylic acid/itaconic acid ion exchange hydrogel	699.31	[41]

3. Conclusions

The present work demonstrated that hydrogels composed of poly(acrylamide-co-itaconic acid) were capable, efficient, and economical for the removal of Cu^{2+} ions from aqueous solutions. Once the xerogels were swollen in 0.1 M NaOH aqueous solutions, they acted as carriers for OH^- ions, which reacted with dispersed Cu^{2+} ions in the solution to form $Cu(OH)_2$. This $Cu(OH)_2$ adhered to the hydrogel surface, preventing its dispersion in the solution. The supernatant-diluted solution was removed by decantation, and the solid formed was recovered by rinsing the hydrogel. The hydrogel was regenerated by submerging it again in 0.1 M NaOH solutions and was used up to eight times while maintaining its removal capacity above 80%. It was demonstrated that when the $mol_{OH^-}/mol_{Cu}{}^{2+}$ ratio was equal to 2.74, the $R_{Cu}{}^{2+}$ was 98%. As long as this ratio was maintained, the total volume of the solution in which the hydrogel was immersed did not affect its cation removal capacity. Based on the experiments carried out, the factors that greatly influenced the removal capacity of metal ions were the amount of OH^- ions inside the hydrogels and the initial concentration of Cu^{2+} ions in the medium. Additionally, the maximum $mg_{Cu}{}^{2+}/g_{xerogel}$ ratio obtained was 336 when the initial concentration of Cu^{2+} was 1750 ppm, with an average value of W = 64.24%. Finally, it was demonstrated that this process can be used in very diluted solutions as 30, 20, and 10 ppm of Cu^{2+}.

4. Materials and Methods

4.1. Materials

The monomers acrylamide (AM), itaconic acid (AI), and the $CuCl_2 \cdot 2H_2O$ salt, all with a purity of 99%, were obtained from Aldrich (St. Louis, MO, USA). The initiator used in the polymerizations was potassium persulfate ($K_2S_2O_8$) (KPS), also with a purity of 99%, sourced from Aldrich (St. Louis, MO, USA), along with the crosslinking agent N,N′-methylenebisacrylamide (NMBA). To carry out the polymerization reactions at 30 °C, N,N,N′N′ groups (mol-tetramethyl-ethylenediamine (TMDA) from Tokyo Kasei (Shanghai, China) served as an accelerator. Finally, sodium hydroxide (NaOH) with 99%

purity from Aldrich (St. Louis, MO, USA) and bidistilled water from Productos Selectropura (Guadalajara, Mexico) (pH = 6.36) were used as the reaction medium. All reagents were used as received.

4.2. Hydrogel Synthesis Reactions

The synthesis reactions were conducted in glass vials under temperature control using a LAUDA Eco Silver (LAUDA DR. R. WOBSER GMBH & CO. KG, Germany) brand overboard thermostat set at 30 °C, with a reaction time of 24 h. The composition of the reaction mixture in all cases was 90% water by mass and 10% monomers by mass, with a mass ratio of 80/20 AM/AI (0.1125/0.0154 molar ratio). For every total mass of monomers, 1% KPS, 2% TMDA, and 1% NMBA were added by mass (percentage molar ratio corresponds to 0.289 KPS, 0.5074 NMBA, and 1.3448 TMDA with respect to the total amount of monomers). Afterward, the hydrogels were removed from the vials and cut into 0.5 cm-thick discs, identified with three sections: upper, middle, and lower. Subsequently, the hydrogels were immersed in bidistilled water to wash and eliminate all traces of the reaction. The water was replaced every 6 h for 3 days and then every 24 h for a further 5 days. Previous research has demonstrated that this process is sufficient for cleaning the materials [42].

4.3. Conversion Determination

To measure the conversion, the hydrogels were removed from the vials and cut into 0.5 cm-thick discs. These discs were then placed in Teflon Petri dishes and subjected to a convection oven at 50 °C until a constant mass was achieved, resulting in xerogels (completely dry hydrogel). Subsequently, the xerogels were immersed in bidistilled water to clean the hydrogels following the previously described procedure. Finally, the samples were placed back in the drying oven until they reached a constant mass again. The yield of the reaction was determined by gravimetry using the following equation:

$$X\% = \frac{M_{x,0} - M_{x,t}}{M_{x,0}} \times 100 \quad (1)$$

where $M_{x,0}$ is the mass of the xerogel before being washed, and $M_{x,t}$ is the mass of the xerogel after undergoing the washing process.

4.4. Batch Study of Removal of Cu^{2+} Ions in Aqueous Solution

First, in order to determine the optimum ratio of moles of OH^- used to moles of Cu^{2+} present ($mol_{OH^-}/mol_{Cu^{2+}}$) for enhanced metal ion precipitation, solutions of $CuCl_2 \cdot 2H_2O$ with 1000 ppm of Cu^{2+} were prepared and mixed with aqueous solutions of 0.1 M NaOH in varying proportions. This allowed us to identify the $mol_{OH^-}/mol_{Cu^{2+}}$ ratio ranging from 0.71 to 56.36. The residence time for the experiments was 48 h, conducted at a constant temperature of 30 °C. These experiments were performed without gels.

Once the most suitable ratio was determined, the obtained xerogels in Section 4.2 were weighed using an OHAUS (OHAUS CORPORATION, Parsippany, NJ. USA) brand balance with a precision of 0.0001 g. Subsequently, they were immersed in 0.1 M NaOH aqueous solutions for 24 h at 30 °C without stirring, with a mass ratio of NaOH solution/xerogel (NaOH/xerogel) set at 125/1. The amount of NaOH solution absorbed by the hydrogels was calculated using the following equation:

$$W = \frac{m_t - m_0}{m_0} \times 100 \quad (2)$$

where m_t is the mass of the hydrogel swollen at time t, and m_0 is the mass of the xerogel.

After 48 h, the hydrogels were removed from the NaOH solution and placed in $CuCl_2 \cdot 2H_2O$ solutions containing 1000 ppm Cu^{2+}, maintaining the optimal $mol_{OH^-}/mol_{Cu^{2+}}$ ratio identified in the chemical precipitation process. The subsequent study focused on evaluating the effect of the concentration of NaOH solutions used to swell the hydrogels, the total volume of copper solution while maintaining a constant $mol_{OH^-}/mol_{Cu^{2+}}$ ratio, and the initial concentration of copper ions on the hydrogels' ability to remove metal ions. The overall process is illustrated in Figure 8.

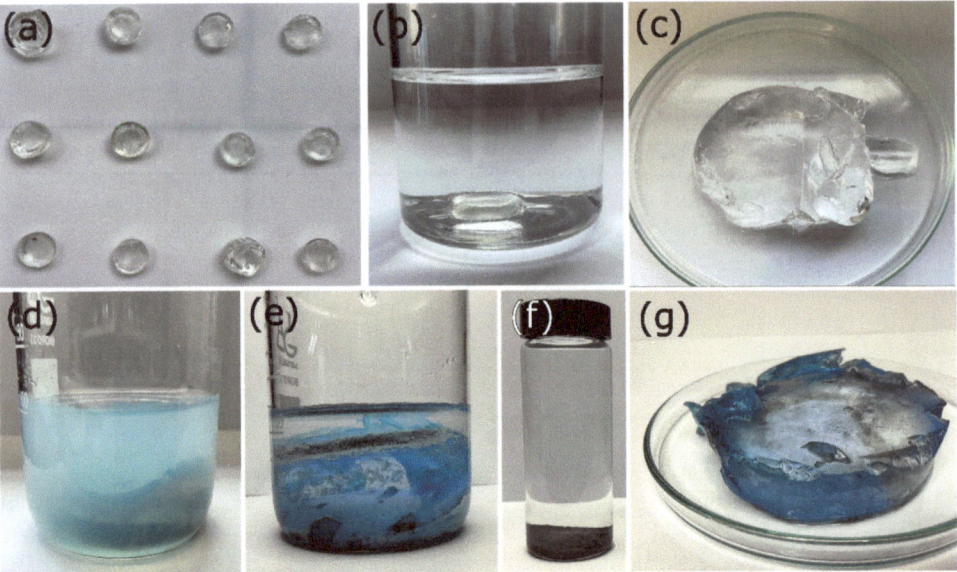

Figure 8. Obtained xerogels (**a**). A xerogel is immersed in a 0.1 M NaOH solution with a mass ratio of 125/1 for 48 h (**b**), resulting in a swollen hydrogel (**c**). The hydrogel is then immersed in a copper ion solution of 1000 ppm for 48 h (**d**), facilitating the migration of copper ions from the solution to the hydrogel (**e**). The generated $Cu(OH)_2$ is separated by decantation (**f**), and the hydrogel is subsequently regenerated for reuse (**g**).

4.5. Semi-Continuous Study of Removal of Cu^{2+} Ions in Aqueous Solution

To determine the reusability of hydrogels for metal ion recovery, the same hydrogel sample was immersed in different $CuCl_2 \cdot 2H_2O$ solutions, each with a decreasing concentration of Cu^{2+} ions from one container to another (from one stage to another). The process involved removing $Cu(OH)_2$ from the hydrogel in the first container, followed by washing with distilled water. Subsequently, the hydrogel was placed back in a 0.1 M NaOH solution at a 125/1 NaOH to xerogel solution ratio. Afterward, this hydrogel was placed in a new aqueous solution of $CuCl_2 \cdot 2H_2O$ with a mass ratio of Cu^{2+} solution to xerogel of 200/1. The concentration of the metal solution was progressively reduced to simulate a semi-continuous metal ion removal process using the synthesized hydrogels.

In both the batch study and the semi-continuous study, the amount of Cu^{2+} ions removed was quantified using UV-visible spectroscopy (UV-vis). To achieve this, the supernatant of each solution was decanted, and the first 15 mL were utilized for residual copper measurement. A previously established calibration curve on a UV-visible spectrophotometer, UNICO model UV2150 (United Products & Instruments Inc., Dayton, NJ, USA), at the wavelength of 800 nm, aided in determining the concentration of Cu^{2+} ions. The percentage of Cu^{2+} ions removed ($R_{Cu^{+2}}$) from the medium was calculated using the following equation:

$$R_{Cu^{+2}}(\%) = \frac{[Cu_0^{2+}] - [Cu_f^{2+}]}{[Cu_0^{2+}]} \times 100 \qquad (3)$$

where $[Cu_0^{2+}]$ represents the initial concentration of Cu^{2+} ions, and $[Cu_f^{2+}]$ is the residual concentration. Throughout all experiments, five samples were used, and the average values were reported.

Author Contributions: Conceptualization, A.G.A.M., J.A.C.O. and R.H.-M.; investigation, J.A.C.O. and R.H.-M.; data curation, A.G.A.M. and J.A.C.O.; writing—original draft preparation, A.G.A.M. and J.H.-M.; writing—review and editing, A.G.A.M. and J.H.-M.; visualization, A.G.A.M. and J.H.-M.; Funding Acquisition, R.H.-M. All authors have read and agreed to the published version of the manuscript.

Funding: This research received no external funding.

Institutional Review Board Statement: Not applicable.

Informed Consent Statement: Not applicable.

Data Availability Statement: The data presented in this study are available on request from the corresponding author.

Conflicts of Interest: The authors declare no conflict of interest.

References

1. Jaishankar, M.; Tseten, T.; Anbalagan, N.; Mathew, B.B.; Beeregowda, K.N. Toxicity, mechanism and health effects of some heavy metals. *Interdiscip. Toxicol.* **2014**, *7*, 60–72. [CrossRef]
2. Hordyjewska, A.; Popiołek, L.; Kocot, J. The many "faces" of copper in medicine and treatment. *Biometals* **2014**, *27*, 611–621. [CrossRef]
3. Araya, M.; Chen, B.; Klevay, L.M.; Strain, J.J.; Johnson, L.; Robson, P.; Shi, W.; Nielsen, F.; Zhu, H.; Olivares, M.; et al. Confirmation of an acute no-observed-adverse-effect and low-observed-adverse-effect level for copper in bottled drinking water in a multi-site international study. *Regul. Toxicol. Pharmacol.* **2003**, *38*, 389–399. [CrossRef]
4. Committee on Copper in Drinking Water and National Research Council. *Copper in Drinking Water*; National Academies Press (US): Washington, DC, USA, 2000.
5. Gaetke, L.M.; Chow-Johnson, H.S.; Chow, C.K. Copper: Toxicological relevance and mechanisms. *Arch. Toxicol.* **2014**, *88*, 1929–1938. [CrossRef]
6. Lipowsky, H.; Arpaci, E. *Copper in the Automotive Industry*, 1st ed.; Wiley-VCH: Weinheim, Germany, 2007; pp. 3–9.
7. Azimi, A.; Azari, A.; Rezakazemi, M.; Ansarpour, M. Removal of Heavy Metals from Industrial Wastewaters: A Review. *ChemBioEng Rev.* **2017**, *4*, 37–59. [CrossRef]
8. Ku, Y.; Jung, I.L. Photocatalytic reduction of Cr(VI) in aqueous solutions by UV irradiation with the presence of titanium dioxide. *Water Res.* **2001**, *35*, 135–142. [CrossRef]
9. Basha, C.A.; Bhadrinarayana, N.S.; Anantharaman, N.; Begum, K.M.M.S. Heavy metal removal from copper smelting effluent using electrochemical cylindrical flow reactor. *J. Hazard. Mater.* **2008**, *152*, 71–78. [CrossRef]
10. Fu, F.; Wang, Q. Removal of heavy metal ions from wastewaters: A review. *J. Environ. Manag.* **2011**, *92*, 407–418. [CrossRef]
11. Pohl, A. Removal of Heavy Metal Ions from Water and Wastewaters by Sulfur-Containing Precipitation Agents. *Water Air Soil Pollut.* **2020**, *231*, 503. [CrossRef]
12. Bilal, M.; Shah, J.A.; Ashfaq, T.; Gardazi, S.M.H.; Tahir, A.A.; Pervez, A.; Haroon, H.; Mahmood, Q. Waste biomass adsorbents for copper removal from industrial wastewater—A review. *J. Hazard. Mater.* **2013**, *263*, 322–333. [CrossRef]
13. Sayadi, M.H.; Salmani, N.; Heidari, A.; Rezaei, M.R. Bio-synthesis of palladium nanoparticle using Spirulina platensis alga extract and its application as adsorbent. *Surf. Interfaces.* **2018**, *10*, 136–143. [CrossRef]
14. Burakov, A.E.; Galunin, E.V.; Burakova, I.V.; Kucherova, A.E.; Agarwal, S.; Tkachev, A.G.; Gupta, V.K. Adsorption of heavy metals on conventional and nanostructured materials for wastewater treatment purposes: A review. *Ecotoxicol. Environ. Saf.* **2018**, *148*, 702–712. [CrossRef]
15. Renu; Agarwal, M.; Singh, K. Heavy metal removal from wastewater using various adsorbents: A review. *J. Water Reuse Desalin.* **2017**, *7*, 387–419. [CrossRef]
16. Hoffman, A.S. Hydrogels for biomedical applications. *Adv. Drug Deliv. Rev.* **2012**, *64*, 18–23. [CrossRef]
17. Shapiro, J.M.; Oyen, M.L. Hydrogel composite materials for tissue engineering scaffolds. *JOM* **2013**, *65*, 505–516. [CrossRef]
18. Daniele, M.A.; Adams, A.A.; Naciri, J.; North, S.H.; Ligler, F.S. Interpenetrating networks based on gelatin methacrylamide and PEG formed using concurrent thiol click chemistries for hydrogel tissue engineering scaffolds. *Biomaterials* **2014**, *35*, 1845–1856. [CrossRef]
19. Spicer, C.D. Hydrogel scaffolds for tissue engineering: The importance of polymer choice. *Polym. Chem.* **2020**, *11*, 184–219. [CrossRef]
20. Liu, M.; Zeng, X.; Ma, C.; Yi, H.; Ali, Z.; Mou, X.; Li, S.; Deng, Y.; He, N. Injectable hydrogels for cartilage and bone tissue engineering. *Bone Res.* **2017**, *5*, 17014. [CrossRef]

21. Chen, G.; Tang, W.; Wang, X.; Zhao, X.; Chen, C.; Zhu, Z. Applications of Hydrogels with Special Physical Properties in Biomedicine. *Polymers* **2019**, *11*, 1420–1437. [CrossRef]
22. Vigata, M.; Meinert, C.; Hutmacher, D.W.; Bock, N. Hydrogels as Drug Delivery Systems: A Review of Current Characterization and Evaluation Techniques. *Pharmaceutics* **2020**, *12*, 1188. [CrossRef]
23. Ghobashy, M.M. The application of natural polymer-based hydrogels for agriculture. In *Hydrogels Based on Natural Polymers*; Chen, Y., Ed.; Elsevier: Amsterdam, The Netherlands, 2020; pp. 329–356.
24. Wang, W.B.; Huang, D.J.; Kang, Y.R.; Wang, A.Q. One-step in situ fabrication of a granular semi-IPN hydrogel based on chitosan and gelatin for fast and efficient adsorption of Cu^{2+} ion. *Colloids Surf. B Biointerfaces* **2013**, *106*, 51–59. [CrossRef] [PubMed]
25. Zhu, Y.; Zheng, Y.; Wang, F.; Wang, A. Monolithic supermacroporous hydrogel prepared from high internal phase emulsions (HIPEs) for fast removal of Cu^{2+} and Pb^{2+}. *Chem. Eng. J.* **2016**, *284*, 422–430. [CrossRef]
26. Abdelwahab, H.E.; Hassan, S.Y.; Mostafa, M.A.; El Sadek, M.M. Synthesis and characterization of glutamic-chitosan hydrogel for copper and nickel removal from wastewater. *Molecules* **2016**, *21*, 684–698. [CrossRef]
27. Firdaus, V.; Idris, M.S.F.; Yusoff, S.F.M. Adsorption of Nickel Ion in Aqueous Using Rubber-Based Hydrogel. *J. Polym. Environ.* **2019**, *27*, 1770–1780. [CrossRef]
28. Sahraei, R.; Ghaemy, M. Synthesis of modified gum tragacanth/graphene oxide composite hydrogel for heavy metal ions removal and preparation of silver nanocomposite for antibacterial activity. *Carbohydr. Polym.* **2017**, *157*, 823–833. [CrossRef]
29. Ramos, M.L.P.; González, J.A.; Albornoz, S.G.; Pérez, C.J.; Villanueva, M.E.; Giorgieri, S.A.; Copello, G.J. Chitin hydrogel reinforced with TiO_2 nanoparticles as an arsenic sorbent. *Chem. Eng. J.* **2016**, *285*, 581–587. [CrossRef]
30. Zhou, G.; Luo, J.; Liu, C.; Chu, L.; Ma, J.; Tang, Y.; Zeng, Z.; Luo, S. A highly efficient polyampholyte hydrogel sorbent based fixed-bed process for heavy metal removal in actual industrial effluent. *Water Res.* **2016**, *89*, 151–160. [CrossRef]
31. Wu, B.; Yan, D.Y.S.; Khan, M.; Zhang, Z.; Lo, I.M.C. Application of Magnetic Hydrogel for Anionic Pollutants Removal from Wastewater with Adsorbent Regeneration and Reuse. *J. Hazard. Toxic Radioact. Waste.* **2017**, *21*, 04016008. [CrossRef]
32. Darban, Z.; Shahabuddin, S.; Gaur, R.; Ahmad, I.; Sridewi, N. Hydrogel-Based Adsorbent Material for the Effective Removal of Heavy Metals from Wastewater: A Comprehensive Review. *Gels* **2022**, *8*, 263. [CrossRef]
33. Serag, E.; El Nemr, A.; El-Maghraby, A. Synthesis of highly effective novel graphene oxide-polyethylene glycol-polyvinyl alcohol nanocomposite hydrogel for copper removal. *J. Water Environ. Nanotechnol.* **2017**, *2*, 223–234. [CrossRef]
34. Vesali-Naseh, M.; Barati, A.; Vesali Naseh, M.R. Efficient copper removal from wastewater through montmorillonite-supported hydrogel adsorbent. *Water Environ. Res.* **2019**, *91*, 332–339. [CrossRef] [PubMed]
35. Qi, X.; Liu, R.; Chen, M.; Li, Z.; Qin, T.; Qian, Y.; Zhao, S.; Liu, M.; Zeng, Q.; Shen, J. Removal of copper ions from water using polysaccharide-constructed hydrogels. *Carbohydr. Polym.* **2019**, *209*, 101–110. [CrossRef] [PubMed]
36. Wang, X.; Jiang, S.; Cui, S.; Tang, Y.; Pei, Z.; Duan, H. Magnetic-controlled aerogels from carboxylated cellulose and $MnFe_2O_4$ as a novel adsorbent for removal of Cu (II). *Cellulose* **2019**, *26*, 5051–5063.
37. Shen, Y.; Wang, Q.; Wang, Y.; He, Y.F.; Song, P.; Wang, R.M. Itaconic copolymer modified loess for high-efficiently removing copper ions from wastewater. *J. Dispers. Sci. Technol.* **2019**, *40*, 794–801. [CrossRef]
38. Olvera-Sosa, M.; Guerra-Contreras, A.; Gómez-Durán, C.F.; González-García, R.; Palestino, G. Tuning the pH-responsiveness capability of poly (acrylic acid-co-itaconic acid)/NaOH hydrogel: Design, swelling, and rust removal evaluation. *J. Appl. Polym. Sci.* **2020**, *137*, 48403–48416. [CrossRef]
39. Nie, L.; Chang, P.; Liang, S.; Hu, K.; Hua, D.; Liu, S.; Sun, J.; Sun, M.; Wang, T.; Okoro, O.V.; et al. Polyphenol rich green tea waste hydrogel for removal of copper and chromium ions from aqueous solution. *Clean. Eng. Technol.* **2021**, *4*, 100167.
40. Wang, H.; Fang, S.; Zuo, M.; Li, Z.; Yu, X.; Tang, X.; Sun, Y.; Yang, S.; Zeng, X.; Lin, L. Removal of copper ions by cellulose nanocrystal-based hydrogel and reduced adsorbents for its catalytic properties. *Cellulose* **2022**, *29*, 4525–4537. [CrossRef]
41. Lin, Z.; Li, F.; Liu, X.; Su, J. Preparation of corn starch/acrylic acid/itaconic acid ion exchange hidrogel and its adsorption properties for copper and lead ions in wastewater. *Colloids Surf. A Physicochem.* **2023**, *671*, 131668–131679. [CrossRef]
42. Hernández, J.A.; Zárate-Navarro, M.A.; Alvarado, A.G. Study and comparison of several methods to remove Ni(II) ions in aqueous solutions using poly(acrylamide-co-itaconic acid) hydrogels. *J. Polym. Res.* **2020**, *27*, 238–245. [CrossRef]

Disclaimer/Publisher's Note: The statements, opinions and data contained in all publications are solely those of the individual author(s) and contributor(s) and not of MDPI and/or the editor(s). MDPI and/or the editor(s) disclaim responsibility for any injury to people or property resulting from any ideas, methods, instructions or products referred to in the content.

MDPI AG
Grosspeteranlage 5
4052 Basel
Switzerland
Tel.: +41 61 683 77 34

Gels Editorial Office
E-mail: gels@mdpi.com
www.mdpi.com/journal/gels

Disclaimer/Publisher's Note: The title and front matter of this reprint are at the discretion of the Guest Editors. The publisher is not responsible for their content or any associated concerns. The statements, opinions and data contained in all individual articles are solely those of the individual Editors and contributors and not of MDPI. MDPI disclaims responsibility for any injury to people or property resulting from any ideas, methods, instructions or products referred to in the content.

www.ingramcontent.com/pod-product-compliance
Lightning Source LLC
LaVergne TN
LVHW072328090526
838202LV00019B/2370